About the author

Peter Marlow repaired his first TV sets, procured from jumble sales, at an early age. In his teens he constructed a 3 inch TV set from a design by the great F. J. Camm in *Practical Television Circuits*. At college for his third year project he built a black and white to colour television converter which used a rotating tri-colour disk – it worked but he still has the scars to prove it! He has been a frequent contributor to both *Electronics World* and *Television* magazine since 1986, with articles ranging from a set-top teletext decoder project to an Internet guide for TV and video engineers. In 1992 he set up *SoftCopy* to market the idea of putting trade catalogues and other databases on to floppy disk, CD-ROM and the web, in order to save vast quantities of paper. *SoftCopy* currently sells an index on disk for *Television* magazine and *Electronics World with Wireless World*. Peter lives in Gloucestershire and is married with two children, who understand how to programme the video recorder much better that he does. You can contact him by email at *peter.marlow@softcopy.co.uk*.

VCR Fault-Finding Guide

VCR Fault-Finding Guide

Selected VCR fault reports, tips and know-how from *Television* magazine's popular VCR Clinic column.

Edited by

Peter Marlow

Newnes
OXFORD AUCKLAND BOSTON JOHANNESBURG MELBOURNE NEW DELHI

Newnes
An imprint of Butterworth-Heinemann
Linacre House, Jordan Hill, Oxford OX2 8DP
225 Wildwood Avenue, Woburn, MA 01801-2041
A division of Reed Educational and Professional Publishing Ltd

A member of the Reed Elsevier plc group

First published 2000

British Library Cataloguing in Publication Data
A catalogue record for this book in available from the British Library

ISBN 0 7506 4634 9

Library of Congress Cataloguing in Publication Data
A catalogue record for this book is available from the Library of Congress

Composition by Genesis Typesetting, Laser Quay, Rochester, Kent
Printed and bound in Great Britain by Biddles Ltd, www.biddles.co.uk

FOR EVERY TITLE THAT WE PUBLISH, BUTTERWORTH-HEINEMANN
WILL PAY FOR BTCV TO PLANT AND CARE FOR A TREE.

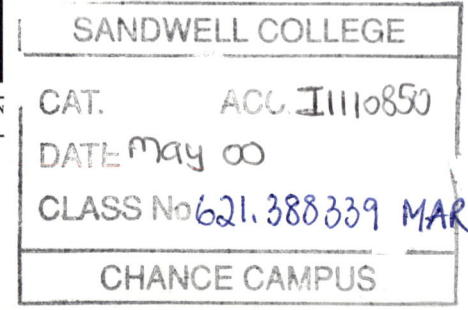

Contents

List of contributors

Hugh Allison
Ronald Aranha
G. N. Ashcroft
Chris Avis
Richard J. Avis
E. M. Beddow
Steve Beeching
Nick Beer
David Belmont
Harvey Benson
Philip Blundell
Ronnie Boag
Simon Bodget
David Botto
Ian Bowden
Derek Bracknell
Edward D. Branch
Michael Brett
Donald Bullock
Roger Burchett
Nigel Burton
Steve Cannon
David A. Chaplin
Paul J. Charlton
Joe Cieszynski
J. P. Cleak
Martin Cleaver
John Coombes
David Corcoran
V. W. Cox
Ray Crockit
J. R. Cutts
Savio Da Costa
Alfred Damp
Basil Davidson
Brian Davidson

D. H. Davies
Mervyn Deeley
Peter M. Delaney
Glynn Dickinson
Michael Dranfield
Dave Dulson
Mick Dutton
E. J. Edwards
John Edwards
Keith Evans
Noel E. Evans
Geoff Fardon
Adrian P. Farnborough
S. A. Featherstone
K. E. Fellingham
Andrew J. Finn
Richard Flowerday
Andy Gallagher
Malcolm George
Owen Green
Les Grogan
Pete Gurney
Steve Hague
Paul Hardy
Michael Harris
Chris Hawkins
John Hepworth
Geoff Herbert
Christopher Holland
John Hopkins
Ben Hosseinally
Shane Humphrey
Phil H. Ireland
Steven Johnstone
Edward Joyce
Keith T. Keeton

Khalied Kwimry
Terry Lamoon
Mike Leach
Steven Leatherbarrow
J. LeJeune
Jim Littler
William G. Lockitt
Robert J. Longhurst
Barry Loughran
Dave Mackrill
Hafidh Mahmood
Phil Marrison
Robert Marshall
Michael Maurice
Bob McClenning
Colin McCormick
Bob Meade
R. S. Narwan
Richard Newman
Christopher Nunn
J. Olijnyk
Mike Orr
Denis R. Parsons
Robert Philpot
John Pitt-Francis
Chris Plaice
Ray Porter
J. K. Potts
John C. Priest
Mike Pritchard

Jim Rainey
Mike Rathbone
Graham Rees
Brian Renforth
Graham Richards
B. Ross
Ed Rowland
K. Rutherford
Estelle Sandford
Alan Shaw
E. Shirt
Alan Smith
Gerald Smith
Justin Smith
Adrian Spriddell
Steve Stamford
Brian Storm
Fauz Ahmed Sumar
Andrew Tebbutt
Graham Thomson
Alan Travers
Eugene Trundle
Della Verita
Chris Watton
T. J. Welford
Gerald White
Roger F. White
George Whiteside
N. J. Williams
Andy Worrall

Introduction

Television is a monthly magazine concerned with the technical aspects of domestic TV/video and associated equipment, particularly servicing. It includes monthly fault-finding reports on TV sets, VCRs, camcorders and satellite receivers, and articles on the technology used in and servicing particular chassis. Readers are kept up to date with technical developments, reviews on test equipment and new technology. There is also a long-distance/satellite TV section.

The notes contained in this Guide are based on material originally published over the last ten years in the VCR Clinic pages that are a regular monthly feature in *Television*. They have been re-edited and collated for ease of reference. We have concentrated on the common models and decks – over 1900 faults are listed from more than 360 models from 35 manufacturers.

We would like to thank the many engineers who have contributed to the *Television* VCR Clinic over the years – their names appear in the List of contributors. Grateful thanks also to the editor of *Television* magazine, John Reddihough, for his help and advice.

Indexes for *Television* magazine covering the last 12 years are available in hard or soft copy from SoftCopy Limited, 1 Vineries Close, Cheltenham GL53 0NU, UK. The soft copy version comes on CD-ROM which also contains the text of 12 000 fault reports, covering TVs, VCRs, satellite TV receivers, camcorders, monitors and CD players. See SoftCopy's advert in *Television* magazine or the web site *http://www.softcopy.co.uk* for more information.

Disclaimer

The fault reports listed in this book come from real-life cases as observed in the workshop. There is no suggestion implied that certain models are more fault prone than others or that particular faults are likely. Where we have listed a large number of reports for a VCR it is because of the popular and widespread use of the VCR rather than any shortcomings in its design. In some cases we have recommended modifications, some of which come from the manufacturers of the VCRs. However, there is no implication intended that these modifications are necessarily authorized or condoned. You should not try to modify or repair a VCR under warranty as this could invalidate the warranty. The equipment should be taken to an authorized dealer.

In some fault reports certain component distributors are named as being able to supply specific spare parts. This is not meant to be an exhaustive list – their names are included for guidance only and are believed to be correct at the time of writing. Of course circumstances do change and the reader is advised to consult the latest TV & VCR Spares guide as published every spring in *Television* magazine.

There can be high voltages on the chassis of video cassette recorders, even when disconnected from the mains – a repair should only be attempted if you are competent. Many components in video cassette recorders are designated safety components: they are designed to fail in the event of an overload, e.g. resistors in power supply rectifier circuits and feeds, or to fail safely. Replacements must be parts obtained from or approved by the manufacturer. Failure to use approved components could have legal implications in the event of a subsequent fault occurring.

Neither the editor nor the publisher can accept responsibility for any loss, damage, death, injury or litigation which arises directly or indirectly from the use of information presented in this book. All information has been given in good faith and is believed to be correct at the time of writing.

VCR Cross-reference

Some VCR models listed in this guide have similarities with other models. In these cases, fault reports may relevant although there may be some differences in component references and values, and internal layout.

Model:	Similar:
Alba VCR6000X	Sentra VX8400
Amstrad VCR6000	Hinari VXL8, Proline 9000
Amstrad VCR6100	Tashiko VVF934
B & O VHS80	Hitachi VT17
Baird 8940	Ferguson 3V30, JVC HR7350
Bush VCR185	Alba VCR6700, Goodmans GVR3400
Ferguson 3V23	JVC HR7700
Ferguson 3V24	JVC HR2200
Ferguson 3V29	JVC HR7200, Baird 8930
Ferguson 3V30	JVC HR7300, Baird 8940
Ferguson 3V31	JVC HR7650
Ferguson 3V32	JVC HR7655, Baird 8942
Ferguson 3V35	JVC HRD120, Telefunken 1930i
Ferguson 3V36	JVC HRD225
Ferguson 3V38	JVC HRD110
Ferguson 3V39	JVC HRD110
Ferguson 3V42	JVC HRD455
Ferguson 3V43	JVC HRD725
Ferguson 3V44	JVC HRD140, Toshiba V65
Ferguson 3V45	JVC HRD150
Ferguson 3V48	JVC HRD565
Ferguson 3V53	JVC HRD755
Ferguson 3V55	JVC HRD120
Ferguson 3V57	JVC HRD755
Ferguson 3V58	JVC HRD370
Ferguson 3V59	JVC HRD180
Ferguson 3V65	JVC HRD170
Ferguson FV11R	JVC HRD170
Ferguson FV12L	JVC HRD230

Ferguson FV26D	JVC HRD700
Ferguson FV42L	JVC HRD520
General VGX520	Panasonic NV430
Hinari VXL8	Amstrad VCR6000
Hitachi VT11	GEC V4100
Hitachi VT17	B & O VHS80
Hitachi VT33	GEC V4004H
ITT VR3906	JVC HRD140
JVC HR7200	Ferguson 3V29
JVC HR7300	Ferguson 3V30
JVC HR7700	Ferguson 3V23
JVC HRD110	Ferguson 3V38, 3V39
JVC HRD120	Ferguson 3V35, 3V55, Tatung VRH8400
JVC HRD140	Ferguson 3V44, ITT VR3906, Toshiba V65
JVC HRD150	Ferguson 3V45
JVC HRD170	Ferguson 3V65, Ferguson FV11R
JVC HRD180	Ferguson 3V59
JVC HRD225	Ferguson 3V36
JVC HRD230	Ferguson FV12L
JVC HRD455	Ferguson 3V42
JVC HRD520	Ferguson FV42L
JVC HRD530	Ferguson FV14T
JVC HRD580	Ferguson FV43H
JVC HRD700	Ferguson FV26D
JVC HRD725	Ferguson 3V43
Logik VR950	Samsung VI611
Logik VR955	Samsung VI710
Matsui VX3000	Saisho VR4300
Matsui VX755	Saisho VR3600
Matsui VX800	Saisho VR1000
Matsui VX820	Saisho VR1200, Hinari VXL35
Matsui VX880	Saisho VR1600, Hinari VXL4
Mitsubishi HS304	Salora SV8500, SV8300
Mitsubishi HS337	Salora SV8600
Nokia VR3615	Daewoo V200
Osaki VCR33	GoldStar GHV1232I
Panasonic NV370	Philips VR6520
Panasonic NV430	General VGX520
Philips VR6180	Pye DV186
Philips VR6185	Granada VHSHP7
Philips VR6460	Pye 64VR60
Philips VR6462	Finlux VR1010, B & O VHS63, Pye DV464
Philips VR6585	Granada VHSGP7
Philips VR6760	Pioneer VR707

Philips VR6870	Pioneer VR727
Pioneer VR727	Philips VR6870
Saisho VR1000	Matsui VX800
Saisho VR1200	Matsui VX800A, VX820, Hinari VXL35
Saisho VR1600	Matsui VX880, Hinari VXL4
Saisho VR2500	Matsui VX990
Saisho VR705	Lloyd LV400
Saisho VR805S	Hinari VXL2
Salora SV6500	Sanyo VHR1100
Salora SV6600	Sanyo VHR1300
Salora SV8500	Mitsubishi HS304
Salora SV8600	Mitsubishi HS337
Salora SV9300	Mitsubishi HS330
Samsung SI7220	Goodmans VCR2500
Samsung VI611	Logik VR950, ITT VR3907
Samsung VI710	Logik VR955, Sentra VX8100HQ
Sanyo VHR1100	Salora SV6500
Sanyo VHR1300	Salora SV6600
Tatung TVR6111	Amstrad VCR9410

Akai

AKAI VS1
AKAI VS22
AKAI VS23
AKAI VS25
AKAI VS35
AKAI VS4
AKAI VS425
AKAI VS427
AKAI VS485
AKAI VS5
AKAI VS55
AKAI VS765
AKAI VS967
AKAI VSF200
AKAI VSF33
AKAI VSG64

Akai VS1

No results: The 'always' 12 V and 5 V rails were found to be present and a check at pin 9 of the STK5325 chip IC1 revealed that the correct momentary low command came from the microcomputer chip when the power-on switch was operated. This instructs IC1 to switch on provided all is well. A check on its output rails proved that the supplies appeared momentarily then switched off. Obviously the MB88401 microcomputer chip was detecting a problem. Checks here revealed that the 'B down' signal at pin 5 was permanently low instead of at 5 V. As a check the NTSC switching pin 6 was shorted to pin 5. This enabled the power supply, and we were on our way. The fault was traced to TR9 being open-circuit as someone in the past had shorted out the sensor lamp. A function check on the deck then revealed a further problem. Although the VCR seemed to acknowledge the fast forward command, no motors turned in this mode. All the other functions worked correctly. The switching signal was traced up to the BA6109 reel drive chip IC6 and was found to be normal.

There was no output from the chip in the fast forward mode, however. We interchanged it with the loading chip as this is of the same type and there was then no tape loading, proving that the chip was faulty. A replacement put matters right.

Machine would lace but not run: The head drum rotated but the capstan motor didn't start. It's driven by the BA6209 chip IC7, which contains a logic circuit, preamplifier and the motor drive stage. Checks on the driver stage, after finding that the logic inputs were OK, showed that the servo input was high at 24 V instead of 1.4 V. This input comes via an operational amplifier in IC11 which seemed to be OK when checked. We concluded that there was a short-circuit in IC7 and a replacement provided the cure.

No record: Playback and E-E were OK. On test all modes worked fine except when record was selected. The machine's record indicator then blinked on and off a few times after which the cassette was ejected. The anti-record safety leaf switch mounted on the carriage turned out to be the culprit. Its plastic mounting hooks had broken and were hanging loose. The part number is 11906SW from Chas Hyde.

On-screen display problem: A fault that can only be described as the Space Invaders syndrome – in both the E-E and playback modes the monitor screen displays a rectangular box which contains rows of letter As (180 of them in a 20×9 block to be exact) and flashes on/off at the clock rate. Outside the box there is just a blank raster. The character generator chip is IC2, type MB88303M. It's mounted on the front clock/timer board. Voltage checks at its pins produced correct readings as per the manual except at pins 12 and 13, which were at 5 V instead of 0.1 V. Replacing IC2 cured the problem. We've had this fault on several occasions.

Cassette stuck half way: The power LED came on as soon as the machine was plugged in, and the cassette lamp was lit, but there was no lift power. After poking about in the motor control system and worrying about the big i.c.s in the system control we spotted it – one of the wires had fallen off the cassette lift motor. Don't you feel a fool? – but relieved.

No vision in E-E and playback modes: Loss of the playback signal was due to loss of the 12 V PB supply because fuse SF5 had blown. Absence of the E-E signal was the result of loss of tuning memory as one of the tags had corroded off the NiCad battery – these machines give a blank raster when not tuned in, not snow.

Poor sync, tape damage: If the problem is poor sync with a single line on the tape about a quarter of an inch (6 mm to the youngsters) from the bottom edge, look closely at the first tape guide. You'll find that the centre pin has pulled out of the plastic subdeck moulding and that the spring used is reminiscent of that used on the good old pogo stick. The waste bin is the only place for it (the spring). Replace it with something a lot lighter – old retractable pen springs work wonders. Glue the shaft back in place, pressing it well home, then reassemble using the new spring. I've done lots of these repairs – when you've seen the spring for yourself you'll see why the fault occurs.

Akai VS22

Power supply problem: We've had our fair share of these nice machines in recently because of no results. The cause is invariably failure of one of the transistors on the power supply panel. We strongly recommend fitting only exact replacements – otherwise problems can arise. The numerous small electrolytics on this panel regularly fail, producing various symptoms such as dead with localized transistor overheating or display OK but with no power up. We've had C6 causing interference/patterning and C20 (100 μF, 16 V) causing no power-up. Our policy is to remove and test all the electrolytics on this board. It saves time in the long run as you avoid bounces. After doing this recently with two of these machines and successfully repairing the power supplies we found that they would both load tapes normally but neither the drum nor the capstan motor operated. There was no rewind or fast forward drive either. In both cases the BU2735AS chip IC503 was the cause.

More power supply problems: The design of the power supply section of this machine is not of the best. As they age, we are getting lots of these VCRs in for repair with symptoms that range from ripple, hum and interference on the picture to intermittent and 'weary' deck operation or complete loss of functions. Akai can supply a reasonably priced replacement PSU board, part number 99002209, but we find it less trouble to replace all the electrolytics on the board. There are lots of them, but they are small, inexpensive ones. No machines have bounced after this treatment.

Transistors overheating in power supply: If the problem with one of these machines is that TR10 and TR11 in the power supply are either overheating or short-circuit, replace the following electrolytics: C7 (100 μF, 10 V); C6 (220 μF, 10 V); and C21 (47 μF, 16 V). We use replacements rated at 25 V or above. It seems that excessive ripple upsets things and results in TR10 and TR11 conducting when they shouldn't.

Clock out of sync and hum bars on the picture in the E-E mode: The power supply in these machines can be a little trying to say the least. This one wasn't too bad, however. I would normally expect to find some leaky transistors and overheated print, but the cause was simply C6 (220 μF, 10 V).

Bad hum bar on E-E pictures: We found that C4 (47 μF, 25 V) on the power supply PCB was leaky.

Tape load OK, no drum or capstan: The tape loaded correctly but there was no drum or capstan rotation in either direction because the BU2735AS chip IC503 was faulty. We've had failure of this chip on a number of occasions.

Akai VS23

Mechanical fault: This machine came in with the complaint 'not working'. On removing the top cover we found that the loading arms were in the fully loaded position but the cassette house was in the eject position. There were comments in the workshop about how the cassette could have been removed, and that the problem looked like being a difficult one. We removed the loading block, reset the timing and then left the machine to play. Later that day another VS23 came into the workshop in the same state. It responded to the same treatment, working after the loading block had been removed and the mode timing reset. Taking no chances, as both machines were still within the guarantee period, we ordered and fitted new mode switches. Neither machine has been seen since.

Crunched tape: Inspection showed that the mechanism was at odds: the FL cradle was in the fully ejected position while the tape guides were in the loading complete positions. Yet they are all driven by the same motor! We rephased the deck mechanics, tested the machine and returned it. With hindsight this was a foolish act . . . A week later the machine was back with another chewed tape. The mechanics were in the same contradictory state as before. After much testing and head-scratching we found the cause of the trouble. On the front left-hand side of the deck there's a vertical steel plate which carries the plastic cogs and pinions that drive the FL cradle. The plate had bent to the left, the result being that the FL pinion could – maybe once in 50 loading operations – jump out of phase.

Hum on sound and unstable vision: The cause was ripple on the 5 V line from the chopper power supply. Replacing C7 (100 μF, 10 V), C6 (220 μF, 10 V) and C60 (100 μF, 25 V) which were all low in value cured the trouble. Beware of a misleading indication, however. A finger placed across pins 2 and 3 of the NJM2352 chopper chip IC1 may seem to provide a cure and lead you to look elsewhere, as we did. All this does is to increase the chopper frequency, making the electrolytics more effective. Repeated failure of TR12 (2SD1292) in the voltage doubler circuit that produces the −35 V supply for the display digitron is caused by shorted turns in choke L8. When you replace TR12, always replace L8 at the same time.

Severe vertical jitter on playback: The E-E and playback pictures suffered from severe vertical jitter, with line tearing and a ragged hum bar across the centre of the screen. A check on the voltages at the power supply output plug P1 showed that pin 4 was at 17 V instead of 9 V. The 2SA1286 transistor TR7 turned out to be leaky all round. When this had been replaced the picture was steady but a slight hum bar was still visible. Replacing C4 (47 μF, 25 V) and C6 (220 μF, 10 V) cured this final problem.

If TR12 is short-circuit collector-to-emitter: If FR1 is also open-circuit with power supply V1084B502A, the cause is likely to be shorted turns in L8. Replace L8, TR12 and FR1.

Wavy horizontal bars across picture: This machine has a rather complex power supply, with a mains transformer, chopper circuits and voltage doublers. One of the more obscure faults that arises in this area is partial failure of C6 (220 μF, 10 V). The symptoms are wavy horizontal bars (like r.f. interference) across the picture and on-screen captions and intermittent colour in the E-E and playback modes. It's worth noting that if the audio/preamplifier PCB behind the drum isn't earthed the syscon shuts the deck down within a few seconds in all modes. Beware of this!

Mechanism out of sync: The loading arms were in the fully laced-up position but the carriage (loading block as Akai calls it) was in the eject position. We've had this fault before, so it was simply a matter of removing the carriage, resetting the timing and fitting a new mode switch.

Tape speed too fast in play: This machine had had previous unsuccessful treatment elsewhere. The tape speed in the play mode was too fast, the symptoms being no line lock and muted sound. Our first check was to see whether capstan FG pulses were present at pin 21 of the BU2735AS chip IC503. They weren't. On tracing the path of the pulses from the capstan motor to IC503 we found that they were present at pin 2 of IC502 but not at pin 3 of IC507. As the resistor in between (R625) was OK, it seemed that IC507 was faulty at its input. Closer examination showed that its pins had all been freshly soldered. All was then clear – it had been fitted the wrong way round! Putting this right produced the correct playback speed – and a sigh of relief!

No fluorescent clock display and no functions: The auto-play function worked when a pre-recorded tape was inserted. Playback picture quality was good, but there was no sound. The tape had to be removed from the machine manually, as the stop button had no effect. We found that pin 5 of the clock microcontroller chip IC901 was at 0 V instead of 5 V. This is the power-down detect line, which senses loss of mains power by monitoring the always 12 V supply. This supply was low at 4.2 V and was the cause of the problem. TR7 (2SA1286) and TR15 (2SB1010) on the power supply PCB were both found to be leaky. When they were replaced, the 12 V supply was restored along with the clock display and playback sound. We wonder how many machines can load up and play a tape with an always 12 V supply virtually missing?!

Akai VS25

Intermittent 'won't play': This machine came from another dealer with a ticket that simply said 'won't play'. When we tested it there wasn't a no-play fault initially but we did notice a slight drum servo twitch. The cause of this couldn't be pinned down as the symptom quickly disappeared. Later, on soak test, the machine intermittently switched to standby, leaving the tape threaded up. Extensive power supply checks failed to reveal anything amiss here, so we left a meter connected to pin 61 (function off) of the main microcontroller chip IC506. This proved that the chip was intermittently issuing the off command. On test next day another symptom appeared; the head switching point wandered up and down the picture. This was the last straw, so we referred the problem to Akai Technical. A very nice man suggested that as the machine went into standby only when the tape was playing it would be a good idea to check the continuity of the drum PG pulse feed – the machine will go to standby if these pulses are missing. A scope left connected to pin 7 of the BU2735AS digital servo chip IC503 showed that the drum PG pulses were OK here, but the story was very different when the scope was connected to pin 9. The mark–space ratio of the 25 Hz head switching square wave varied intermittently then, a bit later, the waveform started to disappear completely from time to time, the result being that the machine switched to standby. A new BU2735AS chip cured all the faults. Phew!

Horizontal dark lines across screen: There were horizontal dark lines across the entire screen, more noticeable during playback and in the E-E mode with the aerial disconnected, i.e. the picture muted. C6 (220 μF, 10 V) and C7 (100 μF, 10 V) on the power supply board were responsible for this. Over 300 mV of spiky noise was measured across C6. Both capacitors measured OK when checked with a capacitance meter, but when they were compared with a known good capacitor using the scope's component tester function a marked difference was displayed. We're very impressed with this facility on the scope, and find it much more accurate and trustworthy than the majority of separate component testers.

Thin lines on playback: The playback picture was marred by thin lines that looked like the result of r.f. interference. We'd had a similar problem before, caused by defective capacitors in the power supply, but the symptom remained as bad after replacing all the electrolytics. To cut rather a long story short, the cause of the trouble was the MSM9565–3 chip IC202.

Akai VS35

Refusal to accept cassette: Checks were made on the mechanism timing, which was found to be correct. We then looked for 'cassette detect' switches in the cassette housing but couldn't find any. In these machines the cassette is detected by the end-sensing phototransistors. The one on the supply side was open-circuit.

White dashes on playback: The playback picture was marred by numerous white dashes which were similar to the spots produced by e.h.t. arcing in a TV set. The cause of this interference was traced to the fact that the head preamplifier screening-plate's chassis earthing screw was loose. Retightening it solved the problem.

Machine stops at random in record: This machine would stop at random in the record mode. The tape would remain threaded and the pinch wheel would be engaged on the capstan shaft. To try to simulate the fault we disconnected the take-up reel pulses while the machine was working correctly. The machine stopped, but the pinch wheel was disengaged and the tape was released from the drum. So absence of the reel pulses was not the cause of the problem. We next tried stopping the capstan motor physically, whereupon the normal stop mode was entered. So the capstan department was not the cause of the fault. When the drum was stopped physically the symptoms matched those of the fault condition. So we had a drum servo fault. We replaced the BU2735S servo chip IC503 but after recording for 2 hours the fault reappeared. Voltage and waveform checks were of no use because of the intermittent nature of the fault. So we replaced the motor circuit board on the lower drum assembly. Numerous test recordings proved that the fault had been cured.

Akai VS4

'Rewinds then dead': On the bench we found that it was totally dead, with no functions whatsoever and no channel display number. Our first check was on the power supply. The 12 V and 14 V outputs from the STK5325 regulator IC1 were missing because the control line (pin 9) was at 4.5 V instead of 0 V. We've had this before when the cassette deck lamp has been open-circuit. A quick check confirmed that it was, but a replacement left things as before. We decided to short-circuit pin 9 of IC1, and when this was done we had 2 V and 3 V respectively on the 12 V and 14 V lines. As there didn't seem to be excessive current a new STK5325 was fitted. This produced the 12 V and 14 V supplies – but only with pin 9 held at 0 V. Also the reel motor went into continuous rewind. With the short-circuit removed from pin 9 of IC1 the machine stayed dead. There was no response when the function-on button was pressed, so a check around the key scanning chip IC3 seemed to be a logical course of action. A scope showed that there were no key scanning pulses on pins 12–15. Replacing the MB88401 micro-computer chip restored the scanning pulses, but I still had no 12 V and 14 V rails. Back to system control chip IC5, another MB88401, that provides the turn on control signal for the regulator. Scope and meter checks here were inconclusive, so the chip was replaced. When the function button was pressed the function LED and the channel display lit, but the reel motor still went into rewind for a few seconds after which the machine shut down again. Time was wasted on a fruitless search through the motor and system control circuitry before we found that the cause of the final fault was a slipping loading belt. Somehow it managed to get the system control confused. Once it was replaced everything worked correctly. We assume that this was the original fault and that a second fault, the defective regulator, had killed the lamp and the two microcomputer chips. Just our luck!

No screen clock, no tuning: The only characters displayed on the screen could best be described as a child's drawing of a house. Both faults were cured by replacement of the MB88303 character generator chip IC2.

Problem with on-screen display: These machines have an on-screen display. A fault that can only be described as the Space Invaders syndrome – a screen full of As that flicker at clock rate. The culprit is the character generator chip IC2 on the operation PCB (front panel). It's an MB88303. We've had this fault on several occasions.

Distorted video: There was distorted video in the playback and E-E modes with this machine. The symptoms suggested a fault on the video panel, where most of the signal processing is carried out. On screen the 'picture' lacked contrast, with no sync. A scope check showed that the sync pulses were badly crushed. We found that TR31 was short-circuit all ways. The 2SA1115 fitted in this position was replaced with a BC212L.

Akai VS425

Intermittent remote control operation: We've often found when working on the bench with these machines that the remote control operation is intermittent or doesn't work at all. This is not a fault: if the handset is too close to the machine there's no response. At normal operating distances you'll find that the outfit works perfectly. The head drum motors in these machines can give trouble ranging from erratic speed and phase to a runaway condition in which the drum continues to whiz at high speed even when the rest of the deck has shut down in the stop mode.

Wouldn't accept tapes: When we'd removed the top of the machine we found that a small label from a cassette had stuck itself to the head and lower drum, thus locking the drum solid. But after removing the label and cleaning the head it still wouldn't turn. We carried out some further checks then removed the motor drive PCB from under the drum motor. A hole had been blown through the control chip. The PCB is available complete, and when a new one had been fitted the drum ran at the normal speed. My troubles weren't over, however, as the head itself had been damaged. All was well when this had been replaced. We always advise our customers not to put labels on the tops of cassettes. They can easily come off and cause expensive damage, as this customer found out to his considerable cost.

Dim display: This machine was brought in because the display was dim. As usual, the cause was the 56 μF capacitors C446 and C447. Once these had been replaced, however, the machine was dead. Great! The power supply connector 201 had correct voltages, but the power-down detector was operating because the 15 Ω 23 V feed resistor R221 was open-circuit. It supplies TR205.

Akai VS427

Dead: Checks in the power supply revealed that the safety fusible resistor FR1 was open-circuit while the associated 1N4007 rectifier diode was short-circuit. Replacing these components, using a new resistor obtained from Akai, completed the repair. This power supply circuit looks a lot more friendly than that in the VS22 series.

No playback sound: The sound circuits were being switched off because the main microcontroller chip was sending out the wrong signal. Must be the chip we thought, but to be on the safe side we tried a new front operations panel. This cured the fault. The μPD75216A-268 timer microcontroller chip IC1 was the culprit, a replacement restoring the sound.

Hum bar would appear on eject: This machine came in because it required the usual pinch roller replacement and a general clean up. We also noticed a secondary fault. A hum bar would appear briefly when a cassette was ejected and sometimes during rewind. The symptom was worse when pressure was applied to the capstan motor, with the hum bar permanent. Various smoothing capacitors in the power supply were tried, but the cause of the fault was eventually traced to D2 (1N4002) which was leaky – it's also in the power supply.

Akai VS485

E-E sound disappeared after a few minutes: After an hour the E-E video disappeared as well, leaving a blank raster. The culprit was the UPD75216A-OA6 timer chip. Pin 37, the tuner mute line, was at an indeterminate state (1.2 V) instead of 0 V when a signal was detected.

The 1.2 V was enough to turn on the digital sound and video muting transistors when they had become warm. A new timer chip cured the fault.

Hum bar and picture disturbance: The picture disturbance only appeared only when the loading motor ran, i.e. when ejecting or inserting a cassette. Apart from that and the hum bar the machine performed normally. Scope checks showed that there was significant ripple on the IDL 10 V supply. There was a smaller amount of ripple on the IDL 5 V line. As the ripple was at 50 Hz it seemed that there was an open-circuit rectifier diode somewhere in the supply, but investigation showed that D9 was short-circuit while the fusible resistor FR6 (0.1 Ω) was open-circuit. Aren't these resistors expensive?!

Tape stuck inside: When this machine was switched on we found that it was in limbo, with a tape stuck inside. When the tape was released and the machine was once more powered up it worked perfectly. But it's wise with these machines to check the loading gear on the cassette carriage. We found that the half-moon gear was in poor condition while the metal running gear that operates it was very sloppy. A replacement gear cured the first problem, but the other fault required a simple modification that comes from Akai. It consists of a small U-shaped piece of plastic which you glue into position just beneath the metal gear retaining clip. This has the effect of tightening up the gears and preventing damage to the plastic ones. Once it was dry we replaced the carriage and the machine worked perfectly.

No clock display: There was no clock display because the d.c.–d.c. converter was faulty. If you get inside this and find that TR408, TR409 etc. are OK an economy repair can be achieved by replacing the electrolytic capacitors – all eight of them! They are C432, C434, C446, C447, C448, C449, C450 and C451. The problem is that they dry up because of the heat.

Loop of tape left at eject: This machine would very intermittently leave a loop of tape hanging out when the cassette was ejected. The cause was the capstan motor, which wasn't turning freely. A replacement motor cured the fault.

Akai VS5

Would not play, automatic shutdown: This machine worked properly most of the time. Occasionally, however, it would thread the tape then go 'beep-beep-beep' like a VCR possessed when asked to play or record.

This would be followed by unthreading, with the breakdown caption lighting up. To escape from this impasse the owner had to switch off and on again, then once more key in play or record. The automatic shutdown was due to the loading switch not being triggered and tape threading wasn't always fully completed. The cause was the loading belt slipping. Fitting a new one cured the problem.

A.f.c hopping: This machine was in good condition despite its age. But on switching from channel to channel, a popping on sound, bars over the picture and loss of colour indicated that the a.f.c. was hopping about. When the sweep tuning was tried we found that it wouldn't rest on located programmes but swept on. The cause of the fault was tracked down to the AN6362 chip IC8, whose output is evidence to the tuning microcomputer chip that a legitimate signal has been found.

No colour in playback or record: There was no colour in either the playback or the record mode. As the machine was getting on in years repair might not have been economic if we'd had to spend time on fault-finding. We'd come across the fault in the past, however, and as we note such things in the relevant manual we were able to restore the colour by going straight to C60 (0.01 μF) and replacing it. The capacitor had become leaky.

Akai VS55

Horizontal patterning in all modes and tearing on the characters of the on-screen display: The cause was C10 (22 μF, 50 V) in the power supply – it was open-circuit.

No record colour: Checks were carried out around the LA7330 chroma processing chip IC400. The video input, filter input and output, sync signals and d.c. levels appeared to be OK but there was no record output signal at pin 9. We then found that the frequency of the 4.433 MHz voltage-controlled oscillator was erratic – about 1.2 kHz low on average. Because of failure of the oscillator to lock, the colour-killer was active. Careful monitoring of the oscillator's control voltage (nominally 2.2 V) at pin 15 showed that it was about 100 mV low. After some further checks we found that steering diode D401 had a reverse leakage of 50 kΩ. A replacement restored normal operation.

Noise bars on playback/E-E picture: Both the playback and E-E pictures had noise bars across them. In addition the machine would sometimes fail to play, record, eject or fast wind in either direction.

We suspected the power supply for the noise bars but also felt that the capstan motor might be faulty as it sometimes refused to start to rotate. The various power supply outputs seemed to be OK, although the 6 V output had a small amount of ripple on it (the voltage was correct). Following advice on similar Akai models given in previous issues of *Television* magazine, we checked all the small electrolytics in the power supply. Capacitance meter tests showed that a number of them were low in value. Fitting replacements failed to cure the problems, however. What did eventually clear the faults was replacement of C15 (220 μF 16 V) which sits on the output side of the 6 V supply. This cleared the ripple. Incidentally C15 tested OK for value and leakage with the capacitance meter, so we've no idea why it failed to work when in circuit.

Reluctant to come out of standby: If one of these machines is reluctant to come out of standby, or it does but the capstan seems to operate sluggishly, check the capacitors in the power supply, especially C15 (220 μF, 25 V). These electrolytics cause many problems as they dry out.

Dead: As there was ripple on all the major supply lines I replaced C6, C10, C15 and C17 – they all measured low value when checked with a capacitance bridge. At power-up the machine still refused to do anything and there was no clock display. There was no response to front panel or remote control commands, and the machine wouldn't accept a cassette. Both the oscillators associated with IC900 were running, but there was no serial clock or data and there were no pulses on the switch strobe lines. As we were working on it the deck started to shuffle and finally initialized. Everything except the clock then worked. If the power was disconnected from the machine and it was left for half an hour, the above process would be repeated. We suspected IC900. When we went to remove the fluorescent display to get at the chip one of the end legs fell off. Someone had obviously had a go, and had broken the display while trying to get at IC900 – which hadn't been replaced. Whoever it was had even attempted to solder the broken ends back on to the bits of leg that were still visible at the edge of the display! We replaced both items, then confidently powered up. But there were still no functions. After half an hour everything started to work. The cause of the trouble was eventually traced to transistor TR104 on the main board – it was dry-jointed. We had to remove the board to solder the joints, which are hidden beneath a couple of cables. TR104 provides the BU (back-up) 5 V supply for the operations panel.

Akai VS765

Intermittent quiet ticking noise in play and record modes: We've had the same problem with three of these machines now. In each case the cause has been the pinch roller assembly, which has an unusual forked support bracket. The replacement types now supplied are probably modified ones – we've had no trouble with them. The part number is BL387501J.

Tuning drift: After a very involved, costly and time-consuming job elsewhere in this machine we were dismayed to find that the u.h.f. tuning drifted and dithered when the machine warmed up. The cure was to replace all the capacitors (C11–14) in the low-pass RC filter that produces the tuning voltage.

No-go, power section OK: A very puzzling no-go situation, with the main power supply section working correctly, can be caused by loss of the BU (back-up) 5 V supply to the microcomputer chip on the front control panel. The usual cause is R221 (15 Ω) on the main PCB going high-resistance or open-circuit.

No functions, no clock: This machine wouldn't respond to any function requests. There was also no clock display. A start was made in the power supply, to see whether there was anything obvious amiss, but the voltages at plug WP201 were all correct. Attention was next turned to the microcontroller chip IC403 and the front control panel. We found that the 5 V supply to the chip and the panel were OK, but the 5 V reset voltage to the clock microcontroller chip was very slow in coming up while there was no reset voltage at all at the microcontroller chip on the main board. The reset pulse is derived from the BU 5 V supply, which was very low. It comes from TR205 on the main panel. This transistor should receive a 23 V input, but the voltage had fallen to only 5–6 V. The cause was the 15 Ω safety resistor R221, which had gone high in value. A replacement produced normal operation.

Akai VS967

Fluorescent display panel wouldn't light up: This machine worked perfectly in every respect apart from this. Very low heater drive was the cause of the problem, and replacement of C446/7 in the voltage-doubler circuit on the main PCB provided a complete cure. The faulty capacitors measured OK when removed from the machine. .. This and several contemporary models can suffer from premature display panel wear unless the circuit in the area of C446/7 is modified – details are available from Akai in modification sheet AV10015.

Fluorescent display failure: We have replaced many failed fluorescent displays in this series of VCRs, and have carried out Akai's modifications to the display power supply circuit, around L4W. After carrying out these repairs this machine's display was almost invisible. A check on the display filament voltage produced a reading of only 1.3 V d.c.: the modification usually reduces this voltage from 5 V to about 3.5 V. So what had gone wrong? We eventually discovered that with this particular coil (L404) the 'centre tap' was not at the centre of the winding but at an approximately 2:1 ratio. Correct operation was obtained when the cut connection to C447 was restored and the connection to the opposite end of L404 was cut and linked to the middle pin. Was this just a freak case caused by an incorrectly wound coil? Perhaps, but it might be worth noting for future reference.

Capstan motor problem: If the capstan motor works in the fast forward and rewind modes but won't turn in the play mode check the fusible resistor FR4 in the power supply. It will almost certainly be open-circuit.

Nicam sound very low level: When the sound was switched to Nicam it would almost completely disappear. Replacing the SAA7320GP chip IC4 restored the Nicam sound.

Akai VSF200

Dead: If one of these machines comes in with the no-go symptom you may well find that fusible resistor FR1 is open-circuit. If so, check zener diodes D13 and D16 in the motor supply stabilizer circuit for leakage or being short-circuit. D13 is a 16 V type and D16 a 10 V device.

Dead, 'L' showing in display: The customer thought that this machine was faulty. In fact it was in the child-lock mode. Nothing then works, with just 'L' showing in the display. Press the handset's play button for 10 seconds to return to normal operation.

Tape creasing at the top: Replace the pinch roller with the modified type. If the old type with a U-shaped insert and the spring tension on the side is fitted, replace it with the new type. Make sure that you read the modification sheet carefully and set the machine up correctly for an even FM video envelope.

Akai VSF33

Hum bar on E-E picture in playback: When play was selected there was a hum bar on the E-E picture while the playback picture consisted of noise and a hum bar. All the supply lines were correct except for the 'limit 12 V' rail which was low at 8.5 V. A check at the junction of the limit 12 V reservoir capacitor C7 and resistor FR4 showed that a large ripple was present, but replacing C7 made no difference. This point is also connected to the collector of TR7, the power-on transistor for the 'idle 5 V' supply which is derived from TR6. We found that the supply to TR6 was missing because FR2 was open-circuit. The 'idle 5 V' supply was drawing current via TR6 and TR7, overloading the 'limit 12 V' supply.

Dead machine: If the power supply is working check TR408 (2SD1292) on the main PCB. If this is short-circuit check the 4.7 Ω safety resistor FB498 and replace the two 56 μF, 16 V electrolytics C446/7. Failure to replace these two capacitors will result in a very dim clock display with TR408 overheating and leading a short life. It's also a good idea to remove that blob of brown glue near C446/7 – the one with the blue wire running through it – as it absorbs moisture. This results in corrosion of the PCB beneath.

Low or no clock display: This was caused by the fact that the power supply wasn't oscillating. Akai has a modification sheet and kit, AV10015, part number EX744015JOAP, to correct this.

Dead machine and then a poor display: A nice easy fault for a change: the 16 V zener diode D8 in the power supply was short circuit. The Fluorescent display left a lot to be desired, however. It's a common problem with VCRs in this series. The cause of the problem is that the display is overdriven. The cure is as follows. Fit 120 μF, 10 V capacitors in positions C446 and C447: they have to be high-temperature types rated at 105°C. Change D416 and D417 to type RB-100AT. If L404 has a centre tap, cut the print between C447 and L404 and connect a jumper wire between the negative side of C447 and L404's centre tap. If L404 is not centre tapped, cut the print between C447 and L404 and add a 2.2 Ω resistor across the cut. Modification kits are available from Akai. Sometimes this modification improves the operation of an existing display, but for best results a new display should be fitted.

Power would go off when any operation selected: This would happen via either the machine's own controls or the remote control unit, leaving just the clock working. When the machine was switched on again it would stay on until a tape was inserted or the channel was changed. It would then power down again. The cause of the fault lay in the 23.5 V regulator circuit, where R221 (120 Ω) had gone high in value.

Akai VSG64

Mechanism out of alignment: This machine bounced in and out of the workshop like a yo-yo. Each time the mechanism was out of alignment. In the end we decided that it was easiest to replace the mechanism block complete. Akai is the only manufacturer, as far as we know, that will supply a mechanism, which is complete apart from the drum and audio/control head assembly. After fitting this and setting it up we returned the machine to the customer and haven't heard from him since. The part number for the mechanism is BBV1172A020A. It can be used with all models that use this deck.

Jammed mechanism: This machine had been to the workshop on several occasions for this fault. As we'd already carried out the usual modifications we decided to plump for a complete mechanism block (see above). We fitted this and set up the head alignment – and haven't seen the machine since.

Most tapes would not play properly: The off-tape pictures rolled. The customer's husband, who is a computer engineer, mentioned that if he twiddled the left-hand guide this would sometimes temporarily stop the rolling. Our rule is that if the guides are tight, why should they have moved from the manufacturer's settings? In fact the cause of the fault was poor head-to-tape contact. The user confessed that the machine led a hard life, being constantly in use. A new upper drum, together with slight realignment of the entry guide, restored normal working.

Akura

AKURA VX140
AKURA VX150

Akura VX140

Failure to eject tape, no functions: This was cured by replacing the BA209N chip IC601 and the 2.2 Ω resistor R601.

Kept going to standby: This machine kept going to standby and ejected any tape that was inserted. To be more precise, the motor that drives the lift ran continuously then the machine went to standby. We stripped down the lift assembly and refitted a spring on the switch-arm assembly (item 244). This triggers the lift eject stop switch 'C-OUT'. The machine was then up and running.

No on-screen display: If the PCB was tapped, OSD letters would jumble or flash. The problem was cured by resoldering ICC101 and LC01, the voltage feed coil to the OSD section.

Akura VX150

Failure to accept a tape: Everything was OK around the microcontroller chip, but there was no drive from pin 12 of IC702. R762 in the feed to pin 11 was open-circuit.

Loading motor chip burn-out: We've seen a number of these machines under different names all with the same fault – the BA6209N loading motor chip burnt out. Sometimes the PCB is badly scorched. The cause of the fault is the loading motor going short-circuit intermittently. Sometimes the bearing seizes up. Replace the motor with a different type, part number MOTOR4305, from SEME. But note that you have to reverse the leads, as it's wired in the opposite polarity. Replace the BA6209N chip with the uprated BA6209, which has a small heatsink tab.

Appeared that heads had failed: This VCR gave the impression that the heads had failed. As a replacement upper drum was in stock we fitted it, but there was no improvement to the picture. Further checks, around the head amplifier, revealed the cause of the fault: the snap-fit connector that joins the foil wire from the lower drum was loose. Effectively, only one half of the drum was connected. All that was required to restore the machine to full working order was to refit the connector.

No E-E or playback video: Scope checks showed that the video waveform at pin 28 of IC201 was missing. It reappeared when pin 14 of ICC01 (type LC7475) was desoldered. A new LC7475 chip cured the fault.

Alba

Alba VCR6000X

Would accept cassette, no other functions: Eject would not even work. The cause of the trouble was the DMB5208VT chip IC701 on the front panel.

Machine dead then suffered from tuning drift: After replacing the fuse with the correct semi-delay type we tested the unit and found that the tuning was very erratic running at high speed through and about the correct point. The tuning voltage was varying of course, the cause of the trouble being leakage between print tracks. We cured this by cutting out the VT line and hardwiring it instead. On test we discovered that after a few hours' use the tuning on all channels would shift by a tiny amount. Use of freezer proved that C134 (0.1 μF), which decouples the tuning line, was the cause of the trouble. As C133 and C135 are identical components that perform the same task at various stages of the d.c. line these were also replaced.

Tuning drift: As mentioned above these machines very often suffer from tuning drift. Decoupling capacitors C133/4/5 for the VT line are prone to being leaky. In addition hardwiring the VT line to cure leakage will indeed provide a cure. But the reason for this tuning drift isn't leakage between the print tracks: it's caused by leakage on the component legs themselves! – around C134. The problem is caused by the quantity of glue that's put around the components in this area of the PCB during manufacture (top upper left-hand side with the bcard hinged up). I suspect that this glue absorbs moisture and then slowly becomes conductive. Thus rather than hardwiring it's easier and quicker to remove this glue and replace C134 (0.1 μF). The μ574 33 V regulator on this board can also be the reason for tuning drift.

Amstrad

AMSTRAD DD8900 TWIN DECK
AMSTRAD TVR1
AMSTRAD TVR3
AMSTRAD UF20
AMSTRAD UF40
AMSTRAD VCR4500
AMSTRAD VCR4600
AMSTRAD VCR6000
AMSTRAD VCR6000/6100
AMSTRAD VCR6100
AMSTRAD VCR7000
AMSTRAD VCR9000

Amstrad DD8900 twin deck

Top deck would stop after 2 seconds: The cause of this problem was the reel sensor, which is not available separately. It comes with the plate sensor, part number 250827.

No fluorescent display: R29 (15 Ω, fusible) in the power supply was open-circuit. You are advised to replace D29 (BA157) as well, or the repair will probably bounce. It's wise to mount these two components off the board to increase the air circulation around them.

Intermittent recording: We found that the cause of the problem was dry joints at IC53 (78M05) on the lower video PCB. Another of these devices, on the upper PCB, was OK.

Wrong control functions, flickery display: Various display segments flickered and the control keys were either very slow to act or produced the wrong function. The stop – eject key initiated forward wind, for example. A look at the circuit diagram showed that the display drivers operate with a – 27 V supply. The display outputs are also connected to the key lines. So a supply problem was suspected. A check showed that the voltage was low at only −15 V. Replacing C27 (47 μF, 50 V, 105°C) cured both problems.

No tape slack removal: Also the wind operations were erratic. As an alternative to a clutch arrangement, these decks have a limited drag lubricant applied to the reel arms during manufacture. With this machine the lubricant appeared to have become runny. As a result, it didn't apply sufficient torque to make or retain secure reel drive cog engagement. To resolve the problem we cleaned off the old lubricant and applied a coating of 'Kilopoise' which is available from Farnell.

Amstrad TVR1

No forward functions: Rewind was operational. The cause was a leaky tape-end phototransistor.

Intermittent mechanical operation: A fault we've had on several occasions with these TV/video combinations is intermittent operation of the various VCR mechanical functions – fast forward, rewind etc. The usual cause of this trouble is malfunction of the mode switch. A cure can usually be effected by dismantling the switch assembly, cleaning and retensioning the contacts, and applying a small amount of Servisol before reassembling. It's essential to use the correct size of Phillips screwdriver when doing all this as the small retaining screws are usually quite tight and their heads are easily damaged. If cleaning fails to the cure problem, as is sometimes the case, the switch must be replaced.

Two lines present during playback: Two well-defined lines, approximately half and two-thirds of the way down the screen, were present during playback. They looked very much like head-switching point signals. Initial panic was followed by recollection that we'd come across this one before: C2 (10 μF, 16 V) on the lower drum PCB assembly was open-circuit. This fault could well occur with the similar drums fitted to other TVR models.

Played tapes but wouldn't record: The record button had no effect because the switch was leaky between pins. A replacement from a scrap panel cured the fault. The customer had been using the timer override instant record button for ages to delay repair, but had finally got fed up with having to press the button every half hour to continue recording!

Power supply breakdown: Most engineers will be familiar with the usual power supply breakdown in the TV section, requiring renewal of the fuse, the four bridge rectifier diodes, the surge-limiter resistor and the STK7348 regulator chip. This unit was no different except when the items just listed had been replaced it remained lifeless. Individual checks on each component in the power supply revealed that C1509 (1 μF, 63 V) was short-circuit. This capacitor is well known for going open-circuit or changing value: I've never had one go short-circuit before. Unfortunately I had to replace the regulator chip again as well as the capacitor to restore normal operation.

Amstrad TVR3

TV controls do not operate: Here's an interesting one we've had with several of these new combi-units (TV plus VCR). The remote control handset operates the VCR functions but not the TV ones. The fault lies within the VCR section, associated with the remote control receiver. Ribbon cable CL8 to the front of the tuned circuit can should have six leads but a five-wire cable is fitted, leaving a vacant hole at either end of the ribbon. Fitting a short length of wire cures the problem. So much for quality control . . .

Plays films in green: The strange report with this VCR/TV set combination read 'plays films in green although the TV pictures are all right'. Sure enough, this was so. After some contemplation we decided to replace HIC101 (1812421), which restored normal colour in all modes. The CPC part code is AM152030.

TV section dead: Loss of the −27 V supply from the VCR section prevented the TV section from working. We replaced zener diode D603 and the timer chip IC803, which was dragging down the −27 V supply, fitting a 14DN487 in place of the 14DN332A, also the unnumbered 1N4148 protection diode that feeds it. After testing the unit for two days it was returned to the customer. Two weeks later it was back again, with apparently the same fault. Couldn't be, we thought. But it was. All the same components had failed for the same reasons and were replaced FOC, this time together with the real culprit – the fluorescent display.

Amstrad UF20

No fast forward or rewind: This has been the fault with several of these machines, whose design leaves a lot to be desired. If you insert a dummy cassette the reels turn but the torque is low because the clutch is engaged. The culprit is the M lever (part number 2549601), which is under the cam gear. The clutch here loses its grip. Replacement restores normal operation.

No E-E or test signals, or playback: This is one of those centre-load machines. Whoever thought of the idea doesn't like repair technicians. The problems were no E-E or test signals and no playback. As the supply to the r.f. modulator was present it was a fair bet that the modulator had failed. It was replaced with some difficulty, but the fault remained. We then noticed that there's a power-on 12V line to this module. When checked it was found to be low. Tracing the source back brought us to Q01 which was leaky – it's in the power supply. A replacement restored normal results. There's a lot of heat stress in this area of the UF20, so the fault could become a common one.

Would power down when tape loaded: The display would flash erratically. Checks on the supplies showed that the voltage on the 12V rail was low. The cause of the trouble was the 12V regulator IC01 which was breaking down. A replacement restored normal operation.

No wind or rewind: We were told by Amstrad to replace the M lever, which did the trick. It's available from CPC under part number AM255034 and is mounted under the cam and the loading motor – replacement is quite easy, however.

Refused to accept a cassette: A faulty cassette push switch (CPC part number AM255151) was the cause. It's not easy to replace this – the whole machine, including the PCB, has to come out.

Buzz on sound: There was buzz on the E-E sound and tuner signal recordings: recordings via the SCART socket were fine. In this and later Amstrad models the tuner, i.f. section and r.f. converter are housed in one can, part number 254873. A replacement cured the fault.

Amstrad UF40

Dead machine: Fortunately the power supply hadn't blown up – R1018 was dry-jointed. When this had been attended to we had no E-E or record sound. The r.f. block, which consists of an r.f. converter, a tuner and i.f. strip all in one can, had to be replaced.

Would eject tapes at the wrong time: This machine would eject tapes, but not when it was in the service mode. By blanking off the sensors (not easy) we found that when a cassette was inserted, the tape would be wound forwards slowly then stop, followed by ejection. Checks showed that the voltage at the take-up sensor was permanently low. The cause was a $3\,k\Omega$ leak to chassis at pin 46 of the microcontroller chip IC6001. Replacing IC6001 restored normal operation.

Power supply failure due to short: The spring associated with the cassette flap opener had come out of the carriage and landed in the power supply section of the PCB, where it had done major damage that included blowing the optocoupler in half! All the semiconductor devices on the primary side of the power supply were short-circuit, with the fusible resistors open-circuit. When these items had been replaced the power supply worked but we found that the main microcontroller chip IC6001 had also failed. Replacing this got the machine going again.

Amstrad VCR4500

Tape rides down the capstan: Several of these machines have come in recently with the same fault: after playing for a few minutes the tape rides down the capstan, giving loss of sound and a picture with tracking lines. It creases the tape a treat! The pinch rollers appear to be OK but the take-up torque is excessive. So far replacement of the large clutch drum has done the trick. Don't forget to mark the position of the mode control switch before removing the large plate!

Crinkling of lower tape edge: Crinkling of the lower edge of the tape as it re-enters the cassette has become a problem with these machines. The official modification involves reducing the take-up torque by replacing the clutch gear and fitting a smaller clutch spring. In one or two instances, however, some tapes continue to be damaged. We've found that a more certain modification can be carried out by fitting, in addition to the previously mentioned items, the pinch wheel tape support as used in the VCR4600. The only problem here is that the nylon nut doesn't have a part number and is thus presumably not available. A steel nut can be used provided the thread is varnished after setting up. Replacing the pinch roller arm doesn't appear to be necessary since in all the machines we've come across the arm has been drilled ready to accommodate this part. The relevant part numbers are as follows: tape support 151482; tape shaft

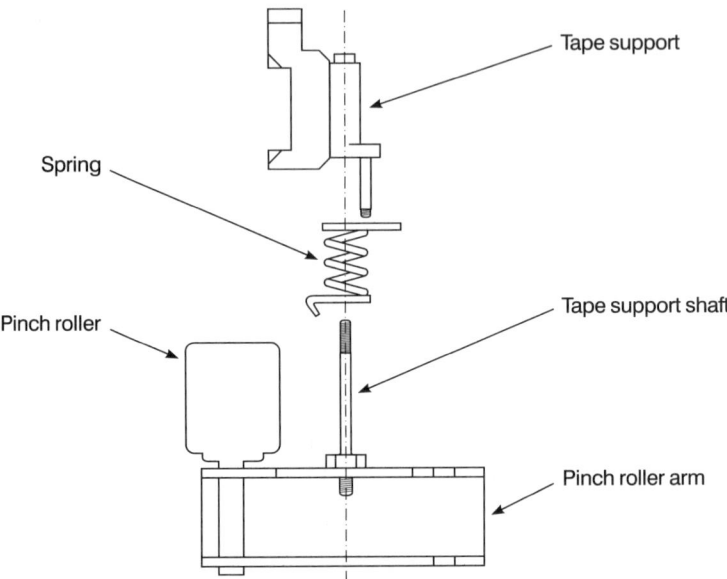

Figure 1 *The pinch wheel tape support assembly used in the Amstrad Model VCR4600*

151483; tape spring 152484; pinch roller arm if required 150769. Figure 1 shows these items.

No clock display and no functions: The function and pause LEDs were permanently lit. Voltage checks around the power supply revealed that the A/T 12 V rail at pin 1 (red lead) of plug CL4 was low, the reading being 2.8 V. Further investigation brought us to R661 (4.7 Ω) which was open-circuit. Fitting a replacement cleared up the trouble.

No clock display, 'weird' operation: There was indeed no clock display, although the channel indicators worked. The machine accepted a cassette but when fast forward was selected it entered the stop mode after a few seconds. When rewind was selected the rewind indicator came on and a clunk was heard from the mechanism, but there was no operation and again it went to stop. We decided not to bother about checking the play mode as without any rewind function we would probably have ended up with a chewed tape. Eject worked OK. A syscon or power supply fault was suspected. So checks were made on the supply lines. This showed that the 12 V supply to the timer board, at pin 5 of plug CN16, was missing. Hence no clock display. This feed is tapped from pin 1 of plug CN15 via a 27 Ω resistor, which was open-circuit. As no short to chassis could be measured, the resistor, circuit reference R662, was replaced. Normal functioning was then restored. A long soak test revealed that there was tape chewing. So in went the pinch wheel modification kit and the waste bin received another tape. In future, no matter what the problem, we are going to change the pinch wheel kit before returning the machine to its owner.

Wouldn't accept cassette: The function LED was permanently alight. Voltage checks in the power supply showed that the all-time 12 V line was low at just over 2 V. As application of 12 V d.c. from an external source restored all functions, checks were carried out around the 12 V regulator Q651. Although resistance readings failed to reveal anything amiss, when Q651 and D655 were replaced normal operation was restored.

Amstrad VCR4600

Went into forward search in play: This machine was brought to us brand new in a box. Its owner had travelled 350 miles from London where he'd bought it at a very discounted price. It was too much trouble for him to take it back under guarantee, so we got the job. The problem was that when play was selected the machine would go straight into forward search. All other functions worked correctly. A circuit diagram was obtained eventually. We then had to find a magnifying glass to sort out the very small print layout and wiring diagram. This was on the outside back page and was already tatty when we received it. We noticed that there are capstan forward, reverse and fast commands from the microcomputer chip. These appeared to be correct when the relevant keys were pressed. When play was selected the voltage at pin 7 of the BA718 operational amplifier IC302 was lower than when search was selected. The output at pin 8 didn't alter, however, so we suspected the i.c. This was duly ordered and after several weeks arrived. Fitting it cured the fault, and the customer was given a bill which meant that his trip to London turned out to be expensive. This was the first time we've seen the inside of one of these machines. We were struck by how well they are laid out and manufactured. Picture reproduction is also excellent.

Recorded and played back at a very slow speed: Whether the standard-play or long-play mode was selected, the slow speed never varied. So the capstan motor drive voltage at pin 2 of the output driver chip IC303 was checked. It was low at about 1.2–1.5 V. The regulated 18 V rail was correct and the next voltage check we made was at pin 8 of the dual operational-amplifier chip IC302. The voltage here was low but the input voltage at pin 4 was high due to QR302 being nearly cut off by the pulse-width output from the servo chip IC301. This should have made the capstan motor run faster. The cause of the trouble was the first operational-amplifier in IC302 (type BA718). With 4 V at its input there should have been 3 V at its output, but the output voltage at pin 2 was only 1.2 V. We subsequently had two more machines with the same fault.

Nothing could be tuned in when using the thumbwheels: As all the tuner supplies were correct a replacement was fitted. This enabled all stations to be tuned in but the station selector buttons didn't change channel. The indicator LEDs changed but the picture remained the same. The AN5015K RAM chip on the front panel proved to be at fault.

Failure of left-hand sensor: A very common fault that's recently started to occur is failure of the left-hand sensor. Complaints range from no play or fast forward to intermittent stopping in the record mode.

No servo action: This was fixed by replacing the BA718 chip IC302. It's a dual operational-amplifier that drives the main servo chip. This seems to be a stock fault.

Two thin lines on screen: The lines were approximately a third and two-thirds of the way down the screen. It affected both record and playback. The fault persisted when a known good recording was tried, but improved as the machine warmed up. In addition varying the back tension changed the look of the fault. After a long search we discovered that a small 10 μF, 16 V electrolytic on the drum motor assembly was the cause. But how did the effect get into the signal circuits? Who knows with an Amstrad . . .

No playback or record colour: The usual cause of no playback or record colour, possibly intermittent, is the chroma module HC1–201. First check for dry-joints. If the soldering is all right fit a replacement.

Capstan speed varied: The result was wow on the playback sound. Replacing IC302 (BA718) provided a complete cure.

No E-E picture just blank raster: The Proline 5000XR looks like the Amstrad VCR4600. The one that came to us had the same fault we've experienced with many VCR4600s, no E-E picture, just a blank raster, but the sound OK. In the Amstrad VCR4600 the 1000 μF, 6.3 V video coupling capacitor C817 at the i.f. block's output pin is the cause of the trouble. This capacitor has a different reference number, C710, in the Proline model. On test we found that it wasn't short-circuit, but we replaced it anyway, using a 25 V type for good measure. This made no difference. The cause of the trouble was actually inside the i.f. block: there was a dry-joint at the earthing pad to can, connected to a small, blue surface-mounted component near TR3 which is presumably the video output buffer transistor. Resoldering the joint put matters right.

Machine dead, fuse blown: This machine was dead with the 2 A fuse F603 open-circuit. We checked the rectifiers in the main power supply and as they all read OK a new fuse was fitted. It blew only a few seconds after switching the machine on again. The cause of the fault turned out to be C836 (3.3 μF, 35 V) which is in one of the voltage regulator circuits on the main servo/system control panel.

Crinkled pictures: And crinkled they were – almost like a Thom 3000 chassis when the 140 μF electrolytic had failed (those were the days!). We first checked the tape path for anything that might make the tape shudder. But everything seemed to be running nicely and evenly and looked clean. The lower drum wasn't damaged and didn't look at all worn. As we had one available, however, we decided to change it. To our pleasure this cured the fault – except for one thing. The head switching point now appeared at the top of the picture. The lower drum had been taken from a new full deck assembly that we'd recently purchased for about £25. It must have been for a different model, however, as the magnet assembly at the bottom had the pulse magnets in a different position with respect to the drum's position of rotation. Taking off this assembly and fitting the one from the old drum allowed the switching point to be set up perfectly.

Severe patterning on TV channels: There was severe patterning on the TV channels, present only when the VCR was switched off. We suspected the booster module but a replacement made no difference and when the original one was tested out of the machine with a bench supply it proved to be OK. As the only supply that's present when the machine is switched off is the always 12 V line we checked the source of this. C508 (100 μF, 16 V) on the main panel was low in value.

Power down on function selection: When a function was selected the mechanics would start but the machine would almost instantaneously power down. Scope checks on the supply lines showed that there was a large negative-going pulse on the 5 V rail. As a result, the supply to the microcontroller chip was being regularly removed. Thus functions wouldn't latch on. Our first suspect, the 5 V regulator, proved to be OK. The culprit was eventually found to be the bridge rectifier.

Corrugated picture: The fault occurred with both its own recordings and pre-recorded tapes. There was obviously something amiss around the head drum. We checked the guides and the tape path for cleanliness. All was well here, so suspicion fell on the lower drum/ motor assembly. As we had one in stock it was quickly fitted. This cured the ragged picture, but the head switching point was about 2 inches from the top of the screen and couldn't be adjusted. Inspection of the motor then revealed that it had a different magnet assembly. So off with the one from the old motor – only a couple of screws. When this was fitted to the new unit and everything was set up, the machine worked as it should.

When play was selected VCR went straight into forward picture search: Replacing IC3D2 (BA718) cured the problem.

When a deck function was selected the mechanism started to operate then shut down: This happened whichever function was selected. A clue was that the clock would also reset. Whenever a function was selected the 5 V supply went low. Replacing the bridge rectifier and the fuse to the 5 V supply cured the fault.

Would accept cassette, nothing else (Aiwa G700GPS): This model uses the same deck as the Amstrad VCR4600. The one we had would accept a cassette but nothing else worked because the belt between the capstan motor and the intermediate idler had fallen off. A new belt kit cured that, but while the machine was on test it began to crinkle the tape. So in went the modified clutch/pinch roller kit. When the machine was put back on test there was a reasonable playback picture for about half an hour after which the colour suddenly flickered on and off a few times then disappeared. We didn't panic, honest! Memory took over; the colour was restored when we'd resoldered all the pins of the chroma module, HC1201. Unfortunately the customer had been quoted for only a belt. Oh well, what's new!?

Erase fault: We don't get many machines with erase faults. A quick waveform and voltage check revealed that R405 (22 Ω) was open-circuit. Note that it's a safety component.

Capstan speed slow in fast forward and rewind: Play was even worse. We found that the voltage across the capstan motor was around 2.5 V, the current being only 50 mA. This ruled out the motor. Our next checks were at IC504, where the 18 V supply was OK at pins 7 and 8 but the voltage at the control pin 4 was low. The voltage status here is determined by the BA718 chip IC302, which turned out to be the cause of the fault.

Sound problems and video signal variations: There seemed to be two problems with this machine, but they proved to have the same cause. If any deck mode was selected while the machine was in the E-E mode, the sound would be either muted or its level would vary momentarily. Deck mode changes also produced video signal level variations. The obvious thing to do seemed to be to check the supply voltages. When we did this we found that the AL 12 V supply was at 18 V and varying. This supply is produced by Q802, along with the 5 V regulator IC801 and the 8.2 V zener diode D810. The culprit was Q802.

Sound warble: The symptom was more apparent in the LP mode. A previous engineer had replaced the capstan motor, the belts, the pinch roller and the capstan drive chip. A check showed that the capstan control waveform at TP22 was incorrect, with a couple of extra negative-going pulses present. These drifted and, when coincident with the control pulse, reduced its effective amplitude. The capstan control loop then failed. We eventually found that the extra pulses were being produced by the BA718 dual opamp chip IC302.

Amstrad VCR6000

Intermittent poor picture and sound: The call-out note said 'intermittent poor picture and the sound grunts'. On playing back a tape in the SP mode it was obvious that the machine was switching randomly between SP and LP. So the tape path was examined and the audio/control head was cleaned and its alignment checked. This didn't clear the fault but we next found that putting the machine into the forward search mode then reverting to play would clear it. After doing this the machine would play normally for several minutes or until stop was pressed. On next pressing play the fault would again be present. We checked the A/C head connections and followed the path back to plug and socket CN-E on the servo board. As this area of the board seemed to be sensitive to pressure some time was spent looking for dry-joints, checking the coupling electrolytics C417 and C419 and bypassing CN-E by wiring a lead direct to the board. As none of this produced a solution the machine was taken back to the workshop. With the machine on the bench we used a scope to check the CTL pulses from the A/C head right through to pin 25 of IC402. Everything was OK. The waveforms at TP401 and TP402 were next checked. RF SW (f) at TP401 was OK but the CTL pulses (g) at TP402, although of correct amplitude and in the correct phase relationship with RF SW (f), were accompanied by an awful lot of noise on the base line. Checks on the peripheral components in this area (CTL-Amp) of the chip failed to reveal anything amiss so a new 14DN363 servo chip (IC402) was obtained and fitted. Result, a nice clean waveform at TP402 and fault-free performance. The apparent dry-joint or board sensitivity to pressure had been a red herring, the cause of this being hand capacitance effects. In fact it was possible to trigger the machine into and out of the fault condition by applying a fingertip or the end of a screwdriver to the can of C419. The resulting massive square wave that showed momentarily at TP402 booted the LP/SP switching into the opposite mode in the same way that selecting forward search would also clear the fault for short periods.

Crinkling of bottom edge of tape: As usual with calls from remote and exposed places, the symptoms hadn't been very clearly explained over the phone. A quick glance at the owner's tapes showed a familiar sight, however – crinkling of the bottom edge. But this wasn't a 4500 or a 4600. It seems that the fault had been present from new. Recordings made in the LP mode were unwatchable as the machine kept switching cyclically into the SP mode. Easing off the back tension showed that the tape was being pulled down by the pinch roller, which was some way off vertical.

Random switching between SP and LP mode: We had two of these machines in recently, both with the same complaint but with different faults. Both machines would switch between the SP and LP modes at random. The cause of the problem on the first machine was easy to see: the tape was being pulled down across the audio/sync head because the pinch roller bracket was bent. A new bracket and pinch roller put matters right. Tracing the cause of the problem with the second machine was more difficult. The tape path was perfect, and the heads were clean. A scope check at test point HP1 (TP402), however, showed that the sync pulses were of very low and varying amplitude, with a lot of noise present. As checks on components in this area failed to show that anything was amiss we decided to replace the 14DN363 servo chip IC402. This cured the problem and cleaned up the waveform.

Poor playback pictures: It turned out that the heads were faulty, but the symptoms were misleading. Playback of a test tape with colour bars produced a display that was clear but with violently juddering verticals, as though there was a shuddering drum motor or a bent drum motor shaft. While a recording made by the machine could be played back on another good machine at an acceptable level. Quite some time was spent before we got round to trying a new drum: why don't we invest in a head-checking machine?

Could not programme tuner: The machine would tune in only BBC-1! Replacing Q601 (2SA1038) in the 32 V series regulator section of the power supply, followed by memory reprogramming, got things back to normal.

Remote control problem: Remote control operation of these machines has always been a problem: the batteries last for only a few weeks, making RC operation very expensive. When the battery voltage drops to 5.7 V the handset's programming section becomes inoperative: if you try to transfer information to the recorder the display clears and nothing is sent. A scope check on the battery connections with new batteries fitted produced a d.c. reading of 6.4 V. The voltage bounced when one of the keys was pressed, the lowest point being 6 V, with the old batteries fitted the same checks produced readings of 5.9 V and 4.9 V respectively. Although the handset still transmitted instructions such as play with the old batteries fitted, the programming information held in the chip was lost when the voltage fell below 5.1 V. The simple solution was to connect a 1000 μF capacitor across the battery terminals, soldered to the print in a position where it didn't foul the case moulding. We then tried the handset for 20 consecutive programmes, held the play key down for 1 minute and finally programmed again, using batteries that would previously have been discarded. As a matter of curiosity, to see how low the battery voltage could go without the unit failing to transmit a programme, we tried it with only three of the new batteries, the other position being linked across. The handset still worked.

Randomly switched from LP to SP (Goodmans TX1101 etc.): The deck used in this machine is the same as that in the Amstrad VCR6000 and many other, often obscure, models. There are many similarities in the electrical/electronic systems as well. The following fault, which has become common, applies to them all. The usual symptom is switching from the LP to the SP mode without being asked. Ripple on the 'mcou +5 V' supply is the usual cause – it's at the right-most power supply plug pin. The component that's responsible is either the bridge rectifier or C509 (220 μF, 25 V). We replace them both, also the pinch roller, then set up the control head. There have been no comebacks after doing this. It's also common to find that the audio/control head tilt adjustment was incorrect from new (as it was with the Amstrad machines), to the extent that there is insufficient tension at the top edge of the tape. This produces rippling at the top edge and thus poor tape/head contact. C509 can deteriorate further. As a result the 5 V supply is so poor that the microcontroller chip fails to reset. The deck continues to carry out its initial shuffle, however.

Goes to standby when a tape function is selected: The loading belt is suspect if the machine goes to standby when a tape function is selected. Often, however, the control cam's brush assembly is the cause. This fits into the cam. When fitting a replacement, clean the static part of the mode switch and smear a small amount of silicone grease on it.

Amstrad VCR6000/6100

No colour in LP mode: We've had some no chroma sillies with these machines. As in previous models an HIC101 chip is used and this is often faulty. On many occasions recently the problem has been no colour in the LP mode – the sound has also been distorted. Both faults have been due to missing items. A missing link causes the lack of LP chroma – it's just by the HIC101 chip. A lot of spaces where resistors or capacitors should have been were found on the audio panel. It's easy to spot these as next to each space there's an LP symbol. Unstable playback with the colour dropping in and out caused some confusion. Hum on the 5 V rail to the 14DN300 was responsible due to a diode in the discrete l.t. bridge going open-circuit when warm. Intermittent or permanent shutdown with the cassette symbol flashing can be caused by several things. The carriage itself is often faulty. Usually the cassette-in switch fails to make, the result being ejection and shutdown. If a VCR6100 accepts a cassette and half loads, then unloads, ejects and shuts down, the usual cause is an intermittent half-load switch. This is situated beneath the audio/control head and is sometimes broken, or the associated gears are stripped. It's not very easy to change. By far the most common reason for the dreaded cassette symbol flashing is grease on the mode switch contacts. Stripping and cleaning is all that's required, but do it with care, checking the deck timing at the same time. If, afterwards, the machine refuses to play/rewind/fast forward (no tape movement) you've fitted the brake actuation arm incorrectly. It should run on the outside of the main gear.

No forward wind or rewind as the reel brakes were on: They would come off if the brake plate, reference 261, or the brake actuator,

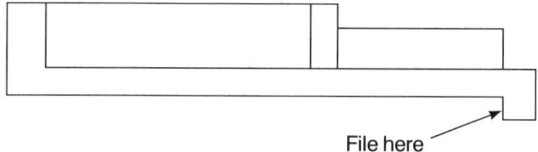

Figure 2

reference 262, was operated manually before selecting a wind mode. But when a wind mode was stopped then selected again the reel brakes once more remained on. The cause of the problem was that the brake actuator plate moved but the brake plate didn't. Careful observation showed that the lever trigger, reference 260, didn't return to its correct resting place, thus inhibiting movement of the brake plate. Filing at the point shown in Figure 2 cured the fault.

Amstrad VCR6100

No playback colour: The lady customer who brought in this machine said there was no playback colour. She also said 'my husband thinks it's just a wire off' – that kind of comment can put a tenner on the bill! On test there was just a slight green haze on the playback picture, but the machine's recordings were OK when played back on a good machine. After trying all the usual things that cause this fault, i.e. the 4.43 MHz crystal, various filters and the HIC101 chroma module, we retired the scope and engaged the meter. A check on the 9 V supply at fuse F602 on the power supply produced a reading of 6.5 V. This 9 V supply goes to several i.c. regulators on the video/chroma board. Disconnecting various lines brought the voltage up slightly, but the fault wasn't being caused by shorts or leakage. What had happened was that some smart person had fitted a 200 mA instead of a 500 mA fuse in the F602 position. The internal resistance of a 200 mA fuse is higher of course, hence the voltage drop. A new 500 mA fuse restored playback in living colour. So next time one of these machines comes in with no playback colour, remember that it's not a wire off but probably just a fuse! For those interested in the charge, it was seven times more than that for the lady's comment!

Machine chewed tapes: This machine had been in storage and now chewed tapes. We found that the larger of the two half-loading belts had decomposed, causing erratic drive – particularly when unloading. A new pair of belts, fitted in the rather clumsy way made necessary by the ridiculous design of this part of the mechanism, cured the problem.

Tape loop left after fast forward: As a result there would be chewed tape if eject was then selected. The cause of the problem was inefficient braking of the supply spool – little wonder as the brake pad had been removed! When a replacement had been fitted there was permanent supply spool braking in fast forward and rewind, as the supply brake arm wasn't being released. In these modes it should be released by the take-up arm, which is moved by the brake actuating lever, being released by the lever on its own in the play and record modes. The cause of the problem was that the take-up arm was the wrong shape. A new arm of the type used in the VCR6000 put matters right. One wonders whether the wrong arm had been fitted during production or whether it was simply incorrectly moulded. Also who had removed the pad instead of dealing with the problem properly!

Wow on sound: We replaced the take-up clutch and pinch wheel but the problem persisted. So we ordered an HIC401 hybrid servo chip from CPC. On inspecting the faulty chip we saw that C3 ($33\,\mu\mathrm{F}$, $10\,\mathrm{V}$) was leaking and thus of low capacitance value. It's not visible until the chip is removed from the PCB: next time a cheaper repair will be possible. On another occasion, the 'wow on sound' symptom was caused by a 'phantom repairman' who had fitted the flywheel belt so that it ran on the wrong part of the motor pulley.

Tape loop on eject: This tape loop caught on the lift and broke. Tape reclaim works in two ways in these decks. From the play to the stop mode the tape is drawn back into the cassette by reverse rotation of the supply reel turntable, driven by the rewind drive gear assembly. The drive for the final reclaim is at the take-up reel, operated by the half-loading wind gear. If this is either loose or sticks on the pillar the tape won't be drawn back in: it turns only about twice, and unless it engages instantly a small amount of tape is left hanging out of the cassette. The item to replace is number 613 in the service manual.

Amstrad VCR7000

Intermittent reel stopping: If you come across one of these machines that suffers from intermittent reel stopping and the problem isn't caused by the reel motor or idler, check relay RY2001 for dirty contacts.

No colour but monochrome recordings OK: No fault could be found in the colour circuits, apart from the fact that the 40FH a.f.c. potentiometer ran out of range. I then found that the 9.3 V rail was at

7.11 V – yes, we know, we should have checked the supply lines first. After restoring it to 9.3 V the rest of the settings were all found to be wrong. Much time had to be spent setting the record a.g.c., E-E level, white clip, dark clip, carrier, deviation and luminance playback level controls. We've had this sort of problem before, haven't we? It seems to us that owners of what we'll refer to as low-cost recorders tend to take them to dubious low-cost repairers – or a relation. They come back with an apology that the cause of the problem can't be found, although a charge is made for trying. The machine then comes to someone like us to sort out, only to find that the phantom twiddler has been at work. The next hurdle is that the owner can't see why he has to pay something like a third of the purchase price for the repairs and subsequent setting up. If the machine had come to me in the first place the cost of the repair would have been more in the £30 region. We might add that after investing in a lot of equipment the charge we have to make is £25 an hour and it matters little to me whether the machine is an £800 JVC or a £250 Taiwan special. It still costs the same to repair – and in a lot of cases the spares for the more expensive models are cheaper. Two examples: JVC spares are cheaper than anything from Mastercare; a set of VEH218 heads for a Panasonic NV370 cost £34.50 while the heads for the Philips clone cost £49.96. Work it out for yourselves.

No take up, poor fast wind etc.: We had to replace the reel idler in this machine. No surprise here of course, but we were surprised to find that the symptoms were still present after the replacement. The drive circuitry is very simple and the fault was quickly traced to Q15 (2SD1348) which was short-circuit base-to-collector. For those who may not know this, the idler is the same as the one used in the Sharp VC9300/VC381 range of VCRs.

Cyclical noise bars on the display: The symptom could be mistaken for absence of the control track pulses but was caused by failure of the 2.2 µF, 50 V electrolytic capacitor C503 in the power supply. It smooths the regulated 12 V supply to the servo chip IC2007. A scope check showed that there was a 1 V square wave on this line. A new capacitor restored normal operation.

No functions, tape inside: This Orion machine had a partly laced-up tape in it. No functions worked. The cause of the trouble was an open-circuit N20 circuit protector on the power board. It doesn't seem to have a circuit reference number. The only other problem was a loose (slipping) loading belt.

Amstrad VCR9000

Something heavy dropped on machine: The main board had flexed sufficiently for Q307 to come into contact with the capstan motor. Apparently the machine had been super-sensitive for some time, even to people walking across the room. No wonder – Q307 had been pushed up from the board, breaking all three solder pads. When we received the machine for repair it would usually thread up and then unthread immediately. We didn't have a manual at the time but it seems that this transistor is in the servo's reference pulse circuit. It would appear to be a stock fault in the making, so bear it in mind when tackling intermittent faults on this machine.

No E-E playback or sound: A check at pin 6 of IC701 showed that its 8 V supply was missing, although there was voltage at the other end of R735. This resistor was OK, the cause of the problem being that C711 (100 μF, 10 V) was short-circuit. We fitted a replacement rated at 16 V.

Tape stuck in, wouldn't eject: When the machine was switched on, the stop symbol and channel indicator lit, the tape laced up about half way then the machine powered down and the cassette symbol flashed. If you pushed the power button the machine would unlace then relace fully, but with no drum rotation. The machine then shut down with only the cassette symbol flashing. Checks in the subpower supply, on the main PCB, revealed that the power-on 5 V was low at only 1 V. The cause was transistor Q1505, which had a leaky base-emitter junction. Although a bit hefty, a BD131 proved to be a worthy substitute.

'Won't play tapes:' This is a common enough complaint from our customers. On this occasion the cause wasn't a mechanism fault. When we inserted our trusty blank cassette we saw that the drum wasn't rotating. Checks on the various voltage rails showed that there was no 5 V supply. The 78M05 regulator chip IC1 on the main panel had failed.

B & O

B & O VHS80

Chewing tapes: Easy we thought – the idler is faulty. However, we found that the capstan motor didn't move in fast forward and rewind, nor did it move to give take-up in play or record. But it did thread. There was also a loud scratching noise during threading. A check on the circuit revealed that we'd had this problem before, but the resistor was OK this time. So time for some proper fault-finding. There was no output from the capstan drive switching chip IC1151 though it was being switched correctly. This was because there was no motor supply input. There should be a regulated 4.2 V input which is obtained from the 16 V rail via Q1158. This transistor's collector voltage was high as it had no base bias because R1153, a tiny 1 kΩ resistor, was open-circuit. Replacing this resistor restored motor drive. In the fast modes the reel drive was sluggish. A new idler upped the torque, but the deck didn't lace completely and because of this there was no take-up in the play and record modes. The cause was much simpler this time: the loading belts were very worn – as were the pinch roller and the other belts. The mechanism is very similar to that in the Hitachi VT11.

Record button broken: The mechanism and most of the electronics in this machine are the same as the Hitachi VT11. This one had come to us about 18 months previously when it required the usual belts and idler. It had also needed a new front flap escutcheon as the record button had disappeared inside. The controls are robust enough, so we thought it was a one-off failure. But here it was back again with the complaints that the button had once again broken and that it intermittently failed to lace. The customer admitted that the broken buttons were due to hamfistedness, and having been in the trade he was able to give us a very accurate description of the lacing problem. This was just as well, as the fault wouldn't show up for us initially. As the machine had obviously seen a lot of use we suspected that the loading belts were worn. When the fault did finally show up replacing them was the obvious course of action. One thing to remember is to get the similarly sized belts the right way round. We once spent ages looking for this intermittent fault only to find that another engineer had fitted the belts the wrong way round. After we'd replaced the belts the machine performed faultlessly on soak test for several days. Then it started to play up again. The motor or mode switch were the next suspects. Voltage checks in the syscon circuit ruled out the mode switch, so a motor was ordered. Fitting this cured the fault – now all that remains is to replace that control flap again.

Intermittent capstan motor drive: This machine of Hitachi origin suffered from intermittent capstan motor drive. As it's not uncommon we expected to find a dead spot in the capstan motor. The cause of the fault turned out to be diode D511 on the top panel, however. It responded to the hairdryer and freezer every time.

Baird

Baird 8940

'**One channel down**': So said the job card. These guessing games are part of workshop life. The TV channels were all OK. So we turned our attention to the audio channels – this machine has stereo capability on longitudinal tracks. The front-mounted audio monitor switch gave us good sound when pushed to the left, reasonable sound at its centre setting and no sound at all when pushed to the right-most position. During playback there was no output from pin 3 of the audio DIN socket unless the monitor switch was pushed to the left: the E-E sound here was OK, however. Plainly the LH (channel 1) playback channel was in trouble. With two identical circuits, fault-finding in one of them is easy. IC2 on the audio/video panel looks after the LH sound, and we found a big oscillation at some supersonic frequency at pin 1. There was a bigger dose of the same at pin 4. Playback equalization is applied here and C9 (22 μF, 16 V), which is part of the feedback network, turned out to be open-circuit.

Snapped tapes: The four tapes snapped belonged to the customer (will some people never learn?!). The cause was a badly fitted cassette lamp. It was so far down its holder that the sensor received no light when the leader appeared. Thus the tape got ripped out of its clip on the supply spool.

Intermittent loss of playback picture: This was caused by the track of the noise-cancelling potentiometer R208 (470 Ω) being open-circuit.

No capstan motor operation: C62 and Q8 were both short-circuit. The component reference numbers for this and above four fault reports were taken from the JVC HR7350 manual.

No erase or recorded sound: This symptom prompted a gleeful leap on to the erase head connector, only to find that it had already been bypassed and removed. The cause of the problem was that the bias oscillator was receiving no supply voltage in the record mode because switching transistor Q10 was not being turned on. Its base bias resistor had risen in value from 5.6 kΩ to 53 kΩ. On a previous occasion Q10 had gone short-circuit and caused similar symptoms.

No rewind or fast forward operation: Checks showed that CP2 on the mechacon panel was open-circuit. So we changed IC12 (10VT05) for good measure. After this the machine functioned perfectly.

No E-E or playback sound: After checking for any obvious switch position sillies I traced the audio output from the i.f. strip to the AN6394 chip IC2. The signal was present here but got no further because this chip's supply at pin 14 was missing. It's derived from Q11 (2SC2673). There was no 11 V supply at its emitter because of an open-circuit junction.

Bush

Bush VCR185

No E-E signals: Checks showed that the u.h.f. tuner received a 0–30 V ramp voltage while search tuning but, as voltage UB was missing, no signals were selected. R132 (10 kΩ) in the band-switching system was open-circuit.

No playback FM: We'd even cleaned the heads first! Careful checks on the record and playback 12 V supplies to the head amplifier showed that in both modes the rec. 12 V line was stuck at 2.5 V. Although it read OK, we replaced the rec. supply switching transistor Q505. This cleared the fault. We used a BC546.

Accepted tape then ejected: The VCR was actually a Goodmans GVR3400. It would accept a tape but then immediately ejected it. After a detailed inspection we discovered that the eject button was resting on the switch on the panel behind it. We presume that as a result of constant use the plastic hinge had lost its springiness – we've seen this fault in quite a few of these machines (and their clones) that have had odd symptoms (intermittently going off etc.) because either the eject or the power button is stuck.

Ferguson

FERGUSON 3V23	FERGUSON 3V58
FERGUSON 3V24	FERGUSON 3V59
FERGUSON 3V29, 30	FERGUSON 3V65
FERGUSON 3V29	FERGUSON FV10B
FERGUSON 3V30	FERGUSON FV11
FERGUSON 3V31	FERGUSON FV11R
FERGUSON 3V32	FERGUSON FV12L
FERGUSON 3V35, 36	FERGUSON FV20
FERGUSON 3V35	FERGUSON FV21
FERGUSON 3V36	FERGUSON FV26D
FERGUSON 3V38	FERGUSON FV30
FERGUSON 3V39	FERGUSON FV31
FERGUSON 3V42	FERGUSON FV31R
FERGUSON 3V43	FERGUSON FV32
FERGUSON 3V44, 45	FERGUSON FV32L
FERGUSON 3V44	FERGUSON FV41
FERGUSON 3V45	FERGUSON FV42L
FERGUSON 3V48	FERGUSON FV61LV
FERGUSON 3V53, 55, 57	FERGUSON FV62
FERGUSON 3V53	FERGUSON FV68TX
FERGUSON 3V54	FERGUSON FV71
FERGUSON 3V55	FERGUSON FV71LV
FERGUSON 3V57	

Ferguson 3V23

'Sound is slow in the E-E mode': The engineer who carried this machine into the workshop exclaimed 'you won't believe me but I've heard it with my own ears!'. There was a general muttering of 'heard what?' around the workshop. 'The sound is slow in the E-E mode!' Amidst the laughter the thought occurred to me that he could just be correct. The 3V23 contains a bucket-brigade circuit in the audio channel. It's used to halve the pitch when the machine is operated in the double speed mode. The result is that the double speed sound will still be fast, but the pitch will be 'normal'. In fact this arrangement works remarkably well. On putting the machine on test in the E-E mode we heard the horse-racing commentator speaking in a slurred, deep voice. He wasn't speaking slowly, but the unnaturally low pitch gave this impression. The circuit is switched on by applying 12 V at connector 35 on the audio panel and a d.c. voltage check at this point soon revealed that the circuit was indeed energized. The voltage comes from transistor X35 on the bottom mechacon board. D.c. and resistance checks proved that X35 was leaky. It's worth noting that had the engineer tried the machine in the playback mode it would have been stuck at double speed as X35 also switches the capstan servo.

Cassette jammed: The owners of these machines are reluctant to part with them. This one had been repaired by us previously. We had replaced the reel motor, the audio/control head and a part of the cassette housing known as the 'cassette bed'. The customer had asked us for an estimate first, but had accepted this. A few months later the machine came back with the complaint that a cassette had jammed. On removing the covers we found that the record safety switch lever, which was of the old type fixed to the cassette housing, had come out of its mounting and was jammed against the cassette. Refitting the lever provided only a temporary cure – after about five goes at inserting and ejecting a cassette the lever again came out of its mounting. When we made enquiries about a replacement shaft complete with lever we discovered that with the advent of the later mechacon board the lever had been deleted and was no longer available. In view of the previous repair another solution to the problem was called for – and I don't mean chopping out the old lever and forgetting about it. I ordered the later type of lever, base and leaf switch. To modify the VCR, first remove the spring, item 110 in the exploded views section of the service manual, then remove the spring hook. Select several washers to fit over the large post to keep the base from fouling the mechanism – be careful not to bring the base up too high or the cassette housing will not lower correctly. When you've found the correct height of the base, secure it with a circlip on the large post. Take out the operation board and remove the record safety LED and phototransistor, and cut away the safety switch holder. On the early mechacon board the record safe line goes low to prevent recording. On the later board it goes high, so electrical modification is also required. One side of the leaf switch has to go to chassis. It can't be taken to the supply rail because of $10k\Omega$ pull resistors on the record safe line, so an inverter is required to match the new switching arrangements to the old mechacon board. Use a BC337 transistor (see Figure 3) with its collector and emitter legs in the holes left vacant by the phototransistor's collector and emitter, leaving the base leg free. Fit a $1 M\Omega$ resistor between the base and collector to provide bias and connect the leaf switch between

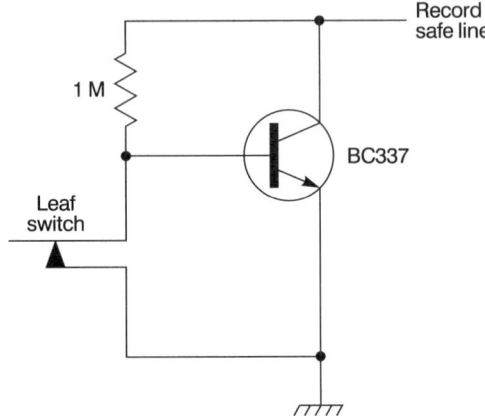

Figure 3 *Record safety switch modification for early version of the Ferguson 3V23/JVC HR7700*

the base and emitter. With the circuit modified in this way the record safe line falls to approximately 2.3 V with a protected cassette, thereby inhibiting recording.

Wouldn't memorize tuning data: Although it would tune to channels this machine wouldn't memorize tuning data. Our first check revealed that the −24 V supply at pin 9 of the memory chips was absent due to failure of X9 (2SD638). After restoring this supply there was still no channel memory. The voltage at pin 7 of the chips was 3.5 V while pin 8 was at 8 V. We changed the chips without curing the problem? Then a bulb lit up in the dim brain. The voltage at pin 8 (Vdd) should be zero. X7 was not turning on as C19 (4.7 μF) in its base circuit was leaky. There had been two faults, as the customer hadn't bothered when the tuning memory was first lost – he just continued to watch library tapes until the heads needed cleaning. . .

No tuner signal: This was due to the absence of a supply to the i.f. section of the machine. The cause was failure of transistor Q16 in the power supply.

Tape looping at stop: In my experience this fault is unusual. However, this machine would sometimes leave a loop of tape hanging from the flap of an ejected cassette. We found that the take-up spool brake was coming on after the supply spool brake because the take-up turntable tyre surface was worn to a smaller diameter than that of the supply turntable. Replacing the take-up turntable and the coil-spring that holds the brakes on cured the problem for good.

Playback colour had Hanover bars: Comparison with the conditions in another machine showed that in the faulty one the signal at TP203 was much larger. The associated adjustments R224/L203 had little effect and it turned out to be the delay line DL201 that was faulty.

Black tearing: The symptom here was a common one: black tearing to the right of black-to-white transitions while any tape was being played. This can be due to a number of things, the most common cause being worn heads. Fitting replacement heads made no difference, however, neither did adjusting the Q and damp controls associated with the f.m. preamplifier. When I attempted to set up the playback limiters as per the manual I discovered that the 'limiter balance 2' control had no effect at all. After carrying out a few further tests I decided to order a replacement for the chip that contains the suspect limiter (IC4, type HA11703). Fitting this enabled the fault to be cleared.

Wouldn't record: This machine played back OK but wouldn't record any picture at all, just noise. Someone else had recently fitted a new head. Naturally we assumed it was OK and spent a lot of time making checks and setting up the circuits. When we replaced the head with one obtained from Ferguson, normal results were obtained. The head we took out worked all right in a 3V22.

Machine half-loaded tape then stops: The drum also stopped. Rewind and fast forward were OK. To unlace the tape you had to use the on/off button. All the sensors and switches were tried, then all the panels, but still no success. We finally tried a new drum assembly, after which the machine worked perfectly. The rotor base unit in this assembly has two magnets opposite each other at the bottom. One of them was missing, so there was no pick-up pulse.

Destruction of F3 at switch-on: This proved to be the result of a faulty d.c.–d.c. converter on the display control PCB.

Tape stuck, eject had no effect: The cause was no voltage across the carriage loading motor. We found that fuse F5 had blown because C36 was short-circuit. Replacing these items put matters right.

Noisy picture, poor loading: It's not often that one of these venerable machines turns up, but this one was extremely clean. Its noisy picture was simply the result of worn out heads, which the owner thought it worth replacing. He also said that the machine didn't always load, especially with a timed recording. A new loading belt cured that. We also replaced the cassette lamp as it appeared to be the original one.

Ferguson 3V24

Intermittently go into alarm mode: If this machine was switched on and then left in the stop mode it would, after about 5 seconds, go into the alarm mode. If it was switched on and play was immediately selected the machine would work correctly until stop was selected. Then, again after about 5 seconds, it would go into the alarm mode. Intermittent alarm faults on these machines are usually caused by dry-joints on the front mechacon board. Before attacking this, however, it's as well to check the operation of the leaf switches under the deck. These are the pinch, brake, after-loading and unloading switches. In this particular case the pinch switch had lost its springiness and stayed closed in the stop position, thus invoking the alarm mode. An inspection of the mechacon board revealed that there were several suspect looking joints here. The best way of dealing with this is to carry out blanket resoldering on both sides of the board.

No picture: On examination I found that there was no drum rotation and, an added symptom, the VCR didn't enter the alarm mode. I decided to tackle the drum fault first. The cause of failure to rotate was that the drum stop signal wasn't being removed when play was selected. Tracing this brought me to IC4 on the audio/CPU board. When play is selected pin 2 of IC4 should drop to $0\,V$, removing the drum stop signal. But in this machine pin 2 fell to only $4\,V$. The reason for this was eventually traced to leakage across the board to pin 2 of IC4, between two points where the print connection is taken through the board to the component side. Carefully cutting the print prior to the feed-through links and replacing it with a length of insulated wire restored normal results – including operation of the alarm mode when there's no drum rotation.

Wouldn't load a tape: The complaint with this portable was that it wouldn't load a tape. On test we found that the capstan ran all the time while the operation switch didn't work. After much searching we discovered that the 2SC1881 $10\,V$ regulator transistor on the mechanism panel had gone open-circuit.

Initial problem with the lamp: There can't be many of these portable machines still around. The main problem with this one was the lamp, which we replaced. The machine then powered up and accepted a tape, but when play was selected the machine laced up then, after a few seconds, shut down with all the lights on the front panel strobing through. After this nothing would work until the machine was powered down and restarted. The cycle would then be repeated. As the LCD counter wasn't working I checked for pulses at the right-hand turntable sensor. They were OK here and were also present at input pin 10 of the NJM2901M chip IC5 on the front panel. This chip produced no output at pin 13, however. A replacement cured the problem.

Wouldn't power up properly: This elderly portable wouldn't power up properly. The capstan motor would spin slowly when power was applied, but nothing else would happen. Then, after about 10 minutes, the unit would spring to life of its own accord. This gave normal operation until the machine was switched off for several minutes, when the whole business would be repeated. We initially thought that one of the supplies had failed, but this was not the case. The 12 V, 9 V and 10 V supplies were all present. Wondering where to go next, we decided to check the voltages around the microcontroller chip IC4. This proved to be a good idea, as several of them were somewhat adrift. The problem was associated with the reset, at pin 7. As the voltage here was permanently high, at 7.5 V, the chip was locking up. The cause of the fault was C70 (10 μF, 16 V) which was leaky.

Ferguson 3V29, 30

Noise bars on screen at playback: This was accompanied by spaghetti, low sound and the sound-led captions as spoken. Resetting the tape guides put matters right. No capstan drive was traced to a blown Wickman fuse (CPR-D – looks like a transistor).

Intermittently refused to load: We checked all the usual things – the load switches, sliding plate under the supply reel and of course the loading motor and belt – and it was only when, in desperation, I was about to replace the complete loading block that I noticed several broken strands on one of the motor leads. Presumably it had become fatigued over the years, during successive belt changes, reducing the motor current. I find that with these machines it pays to remove and inspect, with a magnifying glass, the mechacon panel: you will usually find several ringed and crystallized joints. Resoldering these will prevent a number of confusing, intermittent fault symptoms.

Broken front panel 'hinges': Repairing broken 'hinges' on the front panel function switch operating pads has probably taxed the ingenuity and patience of us all. This latest machine to come my way had obviously led a hard life. The fast forward pad was completely detached, with very little left of the hinges following earlier repairs. I was therefore forced to try a new approach. What I eventually did was to cut thin strips from a washing-up liquid bottle, then superglue them in place to form new hinges. When set, I cut V-shaped grooves across the strips so that they would flex rather than attempt to become detached under pressure. So far the repair has proved to be satisfactory.

Ferguson 3V29

Intermittent failure to play: It wasn't the only problem! There were E-E signals but no deck functions – and the cassette lamp was on . . . A quick check around IC2 on the mechacon board revealed that the micro-computer chip was permanently reset (haven't had a reset fault since I last saw a Midway Space Invaders). The reset pulse is generated by IC3, but this and most of the surrounding components had already been changed; all except D21, which was open-circuit. We now had play but no reel drive – IC12's circuit protector had blown. I was then left with the fault I at first expected – a slipping loading belt. What a saga!

Two mechanical faults: First, when a cassette was inserted the machine would immediately go into slow rewind for a few seconds then stop. Pressing any button would then produce the alarm mode, with all the button lights flashing. The cause of this problem was a worn loading motor. It resulted in the last part of the unloading cycle being missed, so that both the after load and the unload switches were on. The second problem was very confusing: the machine wouldn't switch off when the tape came to an end in either direction. Operating the machine without a cassette in, with the end sensors blanked and then exposed to light, proved that they were working. After much head scratching we found the cause of the problem. The cassette lamp had slipped down its holder. It still shone brightly, but was too low for the light to operate the sensors.

No colour on playback of a pre-recorded tape: The machine was OK with its own recordings. The fault description proved to be correct. On playing a pre-recorded tape and tracing the signal through we found that there was a voltage drop of about 1 V across C415, the input to the main converter. On removing the capacitor to fit a replacement one leg fell off.

Capstan motor wouldn't start: The VCR would load a tape, then after 5 seconds it would unload. We found that the 'motor stop control' input from the mechacon board was high, as though the machine was in pause. Pin 22 of IC4 on the mechacon board should be at 0 V during play and at 10 V in pause. It was actually at 5.6 V, which was high enough to change the state of the following gate. Replacing IC4 restored normal operation.

Clank and judder during rewind and fast forward: The symptoms were due to the fact that the idler was broken in half – most unusual. When a replacement had been fitted the tape functions worked well except for noise bands and vertical judder during playback. Out came the carriage again. We then noticed that the back-tension band was crinkled at the tension-lever pivot: it had been adjusted by bending the band with pliers instead of adjusting the screw. The phantom fiddler had struck again – this time it was a 'helpful' neighbour.

Sound OK but no E-E video: We traced the luminance signal right up to the HA11738 chip IC201 on the bottom PCB where the E-E video was present at pin 26 but no output appeared at pin 5. Voltage checks around this chip showed that there was only 0.4 V at pin 8 instead of 2.9 V. C243 (220 μF, 6.3 V) which is connected to this pin via R244/5 was leaky.

Intermittent drum servo problem: This machine had an intermittent drum servo problem. There were no abnormalities in the waveforms on the servo boards but the tracking control had no effect. Transistor Q9 had 5.5 V at its base, implying that it was in the search mode. This switching transistor is controlled by the voltage at pin 12 of the IR2403 chip IC4, where the reading was 0 V. Pin 5 is the relevant input, which receives the search command from system control. The reading here was 6 V. When pin 5 was disconnected it was clear that this voltage was coming from within IC4. A replacement chip (M54519P) restored Q9's base voltage to 12.5 V. R10 and R207 were then readjusted as per the instructions in the manual.

Audio faults: Two of these machines have been in our workshop recently with audio faults. The first machine would record the sound, albeit almost inaudibly, but failed to erase the previously recorded sound. There was plenty of bias at T1, but it wasn't reaching the erase head. In the record mode the bottom end of RY1 wasn't being switched via pin 89 (audio dub line). In fact the voltage here varied between 5 V and 12 V. The cause of the problem was the μPA81C chip IC5 on the mechacon PCB. The second machine also failed to erase the sound, but for more bread-and-butter reasons – the erase head plug/socket was intermittent. We cut it off and rewired directly.

Loose capstan belt: We were asked to quote for fitting a new capstan motor and suggested that we should inspect the machine first. When we opened it up the cause of the problem turned out to be a loose capstan belt – it was just about ready to fall off. A set of new belts and a clean-up made the customer very happy.

Fast forward on all buttons: Regardless of which button was pressed all you got was fast forward operation. Normal operation was restored by replacement of the right-hand carriage end sensor (viewed from the front). Fine. But can anyone explain why the loading belts in these machines invariably fail the minute they arrive in the workshop? While we're on this subject, when refitting the motor after belt replacement anchor the motor assembly with one screw. Then stand the machine on end and push both loading arms forward into the V blocks. The motor can then be located with the gears meshed correctly and the remaining screw fitted.

Uncontrollable and overloaded E-E video, and no playback video: This was caused by a fault in the playback equalizing filter EQ201.

No clock display: We've had several cases of this due to the clock i.c. being faulty.

No reel drive and no supply to motor: We checked back through the junction PCB to the mechacon panel where there was no drive voltage output from the reel drive chip IC12. We soon found that there was no input supply at pin 10. This comes from regulator transistor Q1 via CP-20. The problem was that Q16, which turns Q1 on, was open-circuit.

E-E sound was OK: C243 (220 μF) which is connected to pin 8 of the HA11738 chip IC201 was open-circuit.

No deck functions: A routine check on the tape-end sensor bulb, by depressing the tape-in switch, confirmed that there was no illumination. After fitting a new bulb there was still no light! That'll teach me to be so optimistic. With the tape-in switch depressed the voltage across the lamp was found to be only 1.4 V instead of 12 V. The supply comes via the collector of the 2SB643 transistor Q1, a pnp device which is on the little PCB that holds the two-pin lamp socket. 12 V was present at the emitter of this transistor, but its collector was open-circuit. A new transistor restored the glow and the deck functions then worked normally.

No playback picture: Scope checks along the luminance signal path brought us to the playback Y level preset R201, which was open-circuit at the top end.

Ferguson 3V30

Inadequate back tension: This machine was sent in with the suggestion that the heads had failed. Sure enough the picture was all but obliterated, but the heads weren't defective. There was inadequate back tension, in fact virtually none, the clue being provided by the spooling of the supply reel. The cause of the fault was a snapped soft brake-band.

No sound in E-E, record or playback modes: A rather unusual complaint this. We soon found that the 10.5 V supply to the audio section was way down at only 0.344 V. There was no excessive loading and it turned out that the regulator transistor Q17 was open-circuit. It's a 2SD636 but we fitted a 2SD637 (60 V version) as this is a much used device we stock for Panasonic VCRs.

No E-E signals, no tuning signal: If you get this symptom, with the r.f.-through signal OK and the deck mechanics all working, check for a dry-joint at the always 9 V regulator transistor Q101.

Drum would stop after few minutes: We still see a large number of these excellent and on the whole reliable machines. With this particular one the drum would stop after a few minutes and wouldn't restart. An initial check on the supplies (we all do that, don't we?) proved to be a good move as the 12.5 V rail read 15 V. It's derived from Q1 on the power supply board. This was OK but further back the sensing transistor Q3 and its emitter zener diode D5 were faulty.

F4 blown, no rewind or fast forward: These machines just go on and on. This particular example came in with F4 (2.5 AT) blown and no rewind or fast forward. Q16 turned out to be short-circuit.

Machine would load then unload: The cause of this fault eluded us for nearly a year. Very occasionally the customer reported that on pressing play, record or timer record the machine would load then unload. We had tried the usual causes – the belts and pulse inputs to the microcomputer chip. Then finally the fault put in an appearance for us: the pinch-roller solenoid disengaged before our eyes. Transistor Q4 had gone slightly leaky from base-to-emitter. Happiness is a niggling fault put to rest! Bear in mind that the holding drive transistor is a definite possibility for this sort of trouble.

No reel drive: We found that CP2 in the supply to the 10VT05 reel driver chip IC12 was open-circuit while the chip itself, a hybrid device, had burnt up. The cause of all this was an intermittently short-circuit reel motor.

Would stop in any mode: This was because the optical reel sensor produced no reel pulses. The 11 V supply that should be present at pin 3 of the sensor PCB was missing. It comes via pin 63 on the small subpanel that's mounted to the right of the deck (viewed from above).

The track between pins 25 and 63 on this PCB was open-circuit. On another of these machines the same fault was caused by the fact that the lead from pin 63 to the sensor PCB was open-circuit.

Drum runs wrong direction (early models): If the drum runs at full speed in the wrong direction, i.e. clockwise, check the plug-socket connections to the MDA (motor drive amplifier) panel and the print continuity on the panel itself, especially around Q216. This transistor tends to run quite hot. The MDA panel was mounted vertically behind the function PCB in early models. In later models the MDA circuitry was moved on to the servo panel.

Ferguson 3V31

Intermittent colour in playback: Check for dry-joints on crystal 401 on the chroma board.

Field bounce in still frame mode: The vertical pulse control on the front panel had no effect. A dry-joint was eventually found at C75 on the servo board.

No channel storage: This was caused by loss of the −23 V supply. It should be present at pin 9 of IC205, an MN1218A RAM. Q207 proved to be open-circuit.

Noise in bottom inch of picture: There were a couple of clear lines below the noise. The tracking control, and adjustment of the guides, had no effect. It looked more like head switching point trouble. Judicious adjustment of the channel 1/2 switching positions showed that the channel 2 adjustment had no effect. The drum magnets were all present, the head hadn't been moved, nor had the drum flywheel. When I carried out checks around the drum servo I found that C53 in the pulse amplifier circuit was dry-jointed. A new set of loading rollers was also required.

Intermittent playback, record etc.: The cause was traced to the 4.435571 MHz crystal X401 being dry-jointed. It's on the colour PWB assembly board.

Wouldn't accept a tape: As the cassette loading belt was slack a replacement was fitted. But this made no difference. The next step was to check the motor voltage during the loading cycle. It was low, because transistor Q8 was faulty. A replacement restored normal operation.

Dead, but initially a failed cassette lamp: This machine had been 'looked at' by its owner before he brought it to me. He'd covered his tracks so neatly, however, that this was not immediately obvious. The original fault had been a failed cassette lamp. When the owner had removed and refitted the right-hand panel (tuning and microcontroller) the 2SC1983 transistor Q204 had become entangled in a wiring loom. As a result it had twisted and one leg had broken. The machine appeared to be dead, but actually there were no switched supplies.

Ferguson 3V32

No clock display: The display came on when the machine was plugged in, but after about half a minute it began to fade and after a minute it was completely out. Voltage checks revealed the absence of a −24 V line, which was restored when zener diode D233 on the tuner/timer subpower board was replaced.

Machine operated at about half speed: Double speed wouldn't work and the slow-speed mode was very slow indeed. After a bit of signal tracing with the oscilloscope we found that the fault was within IC6 – the waveforms at pins 3 and 4 were totally incorrect.

No picture recording: This machine recorded the sound but failed to record the picture in either the SP or the LP mode. There was no record f.m. signal at pin 5 of connector CN6 on the Y board. Checking back from here we found that the 1 kΩ f.m. record preset control R128 was open-circuit.

No fluorescent display: The 3V32 is not a common visitor to our workshop. In view of this and its complex circuitry we tend to be apprehensive when one comes along. The fault with this one was complete absence of the fluorescent display. Not suprisingly we found that the filament supply voltage was missing. It comes from a 37 kHz oscillator, which is mounted on a small subpanel behind the main display, the main items here being T201 and Q205. There should be 3 V p-p at the centre-tapped secondary winding but a scope check showed that the oscillator had stopped. It turned out that the oscillator transistor's 22 kΩ base bias resistor R281 was open-circuit.

Motors did not work followed by no functions and no display: There were two faults with this old timer. One was that the motors didn't work – because there was no 12 V output from the regulator circuit on the bottom PCB. We found that ZD34 had gone open-circuit and Q16, on the mechacon board, had literally blown up. When this had been repaired the machine worked normally. It came back three days later, this time with no functions and no display. We found that CP1 on the tuner/timer control board was open-circuit. After replacing it the machine worked for a few minutes then failed again. To cut a long story short, and three CPs later, we discovered that there's a capacitor connected between pins 5 and 10 of IC205. It's glued to the back of the PCB and is not shown on the circuit diagram. What had happened was that the leg connected to pin 5 (chassis) of IC205 was almost touching pin 4 (supply). We moved the capacitor and secured it with hot-melt. This finally cleared the trouble.

Appeared to be dirty heads: I had serviced this machine about three months previously, and had been called back because of what looked like dirty heads. Cleaning them seemed to cure the problem. I was then called back again. The heads appeared to be dirty once more, but if the machine was left to cool down the picture would be restored. The fault could be induced by going into the search mode. I took the machine back to the workshop and left it running until the fault appeared. I then scoped the drum flip-flop waveform which instead of being a square wave, was a series of pulses. As the amplified drum pick-up pulses were OK, I froze the BA853 chip IC7 and found that the fault cleared. So a new BA853 chip was obtained from JVC and fitted. Imagine my horror when the machine produced exactly the same symptoms (the chip costs over £40 trade). I subsequently tweaked R57, which sets the drum pick-up pulse level at IC7. Fortunately this cured the problem. A long soak test proved that everything was OK. It's possible that the drum pick-up pulse head may be starting to fail. Only time will tell.

Display wasn't working: After carrying out a service I found that the display wasn't working. Checks showed that its −28V supply was missing. This is derived from the timer/tuner board, where zener diode D233 was found to be short-circuit. A replacement brought the display to life.

Record and playback failure: This old-timer had been running for years with no problems. Then the ravages of time took their toll. First, the record function failed, then the machine refused to play a tape. The second fault was easily cured: a new loading belt was required. I fitted a new economy repair kit, which included the turntable rubbers. The machine then worked extremely well mechanically, and the playback picture was of excellent quality. But there was no E-E signal. Scope checks showed that video was present at the output from the i.f. board, and that it arrived at the video processing board. The signal entered pin 7 of the BA7001 chip IC201, but it didn't emerge at pin 4. Voltage checks didn't tell me much, and the control switching was OK. A new chip restored normal results.

Sometimes failed to unload and eject: When this happened the 'stop' light would flash, as if the machine were waiting for the mechanism to unwind. The loading belt and timing were OK. When I disconnected the motor I found that the 12 V was present in the unload mode. The voltage dropped to 3.6 V when the loading motor was reconnected. But a new loading motor made no difference! Q5 was OK but D19 produced a high reading. Thankfully a replacement cured the fault.

Ferguson 3V35, 36

Signal path fault and faulty playback: It was difficult to pinpoint whether the fault was in the r.f. amplifier or the converter – it was very intermittent. We decided to remove the aerial booster unit, solder all suspect joints and refit it, but the fault was still present. Taking the same course of action with the converter unit put matters right. Prior to doing this we'd checked the supply rails and found them to be correct. When we'd done all this we found that playback was faulty, the symptoms looking like the effect of misadjusted tape guides. The cause turned out to be different, however: the back-tension brake band was broken at one end. Replacement plus back tension resetting was required.

Load belt fault: A number of these machines are coming in with loading belt faults, the complaint being intermittent failure to play or record. The belt is the same as the one in the 3V29/3V30 and is just as difficult to replace. The Ferguson part number is 01X1–040–006.

Machine would occasionally stop: The customer's complaint was that the machine would occasionally stop as soon as play had started. After many hours of soak testing we found that the cause of the trouble was a lazy drum motor. With no drum rotation, the machine stopped.

Grainy picture and intermittent failure to lace up: As I'd just fitted a sensor lamp in a chargeable 3V30 I thought I'd stay with the easy ones. Things started well enough. The loading fault was due to a stretched belt. A new one was carefully fitted without removing the whole motor assembly, but there was no take-up – the belt was so slack that if it hadn't been for the flywheel bracket it would have dropped off its pulleys! This attended to, I put the machine in play and noticed that the exit guide was very sluggish and didn't reach the end of its travel. The baseplate on the loading ring had become disengaged: its pip was worn with the result that it didn't remain in its slot, consequently the spring had become undone. Rather than struggling with what was going to be a long job anyway I decided to remove the entire mechanism from the machine, a matter of unplugging and then removing eight screws including those for the carriage. To gain access to the loading rings I then had to take out six or so screws to release the top plate with the head drum and motor, the guide poles and audio/control head etc. A further three screws held the rings in place. I tried to refit everything but the exit guide baseplate was too worn. A new assembly had to be ordered and fitted after which reassembly went remarkably well.

Ferguson 3V35

Drifted off tune on BBC-1 only: My first thought was to put BBC-1 on another button and see if this drifted too. It did. The tuning voltage is obtained from a voltage doubler circuit, which according to the circuit should produce some 45 V at TP6. Ferguson told me that the voltage here should be a bit higher than this but if it gets very high, say 70 V, all the components in the voltage doubler circuit should be changed – D14, D15, C21, C22, C23, C24, C25 and R13. I did this and the machine was OK for about four weeks. Then it came back. This time I decided to change the HZT33–02 33 V stabilizer D10. The machine seems to have worked all right ever since, but I would still recommend that anyone experiencing tuning faults with these machines should look at that doubler circuit. It had gone haywire in this particular machine and no particular component could be blamed.

Tape stuck: This machine came in with a tape threaded up and the loading arms in the V blocks. We unlaced the tape manually then inserted another cassette. On pressing play we could see that the loading cycle was not being completed. A new loading belt failed to improve matters, however. After a check on the after-load switch we ensured that the gears ran freely, then suspected the loading motor. Replacing this cured the fault. Incidentally the motor used in the earlier 3V29 is identical to the one used in this model.

Machine mechanically dead: When switched on this machine wouldn't even accept a cassette, although E-E came up. After ruling out such things as the cassette lift and the power supply we turned our attention to the system control section. A cassette was manually loaded and left in the unlaced position. The relevant logic conditions around IC201 were then checked. This showed that when any action was selected the microcomputer control chip would attempt to energize the relevant motors although nothing happened. At this point we discovered that circuit protector CP1 was open-circuit. It's on the servo/mechacon panel and is in series with the 13 V supply to all the motors apart from the drum motor. With CP1 replaced the machine would load a cassette and fast wind was restored, but it still wouldn't thread the tape when play was selected. Further tests showed that the mode/lacing motor was open-circuit. A replacement restored full operation.

No playback or record colour: This fault is not uncommon with these machines, due to faults around the crystal oscillators. On this particular machine, however, the cause was the colour/monochrome/test switch at the back. It had become leaky. As the customer didn't know it was there and obviously never used it I simply removed the connections to the offending portion of the switch. Everything then worked correctly.

Video signal would fade out: This was a straightforward fault although it did involve a chase through three boards to find the cause. In the E-E mode the video signal would fade out or disappear instantly in a very unpredictable manner. A check showed that the i.f. panel's video output was present. Over to the luminance section then. The signal was present at pin 8 of the switching chip IC501 but didn't emerge at pin 4 because the voltage at switching pin 2 was incorrect. This voltage should be zero in the TV mode and 7.5 V in the video/aux mode. It was 1.6 V and varying. We eventually found that the TV/video switch itself was the cause of the trouble. The sound remained unaffected throughout.

Tapes would erase while being watched: This VCR had developed an expensive taste for pre-recorded tapes: because the bias oscillator ran in the playback mode it erased them as they were being watched. The cause of the fault was traced to Q503 on the video PCB. As it was leaky it held the Rec Start Low line at 3 V instead of 9 V in playback.

Dead: As the thermal fuse in the mains transformer's primary winding was open-circuit this machine was dead. Fortunately the pinouts on the PCB are accessible. So to avoid the cost of a new transformer we added a 250 mA fuse externally, soldering it across the pins in place of the internal fuse. The machine then worked well.

Intermittently goes dead: This machine would go dead after a few minutes or sometimes days. There would be no functions and no displays – nothing. We found that slight movement of the rear vertically mounted power supply board would initiate the fault. So we removed the board and examined it with a magnifying glass. This showed that pin 4 of CN1, which supplies the board via the mains transformer, was dry-jointed – a slight ring was just noticeable around the solder joint on the PCB. Resoldering cured the fault but for good measure we went over the whole board as some other joints looked a bit suspect.

No audio playback: The playback level potentiometer was open-circuit. The result was no audio signal but a very noisy output of crackles, pops and white noise.

Herringbone patterning on playback: Check for h.f. ripple on the 12 V supply, at pin 1 of connector CN4. A ripple voltage of 0.5 V here should lead to a check on C16 (47 μF) which you will probably find is open-circuit.

Intermittent loss of the signals: We found that the 33 V tuning supply disappeared intermittently, the cause being C21 (100 μF, 50 V) on the power supply panel. It's part of a voltage doubling circuit.

Appeared that one head was clogged: The cause of the trouble was actually the fact that the SW25 signal was of very low amplitude. We found that the 9 V supply was low at only 5 V because the 9 V adjustment potentiometer on the power supply board was faulty.

Timer wouldn't set up correctly: When the machine was on the bench I was unable to set the clock and the timer flag was flashing continuously. The cause of the problem was D10 in the timer UPC voltage stabilizer circuit. Because it was open-circuit the supply to pin 21 of the UPC was high at 8.5 V instead of 4.8 V.

Deck modes stop after a few seconds: The deck modes would all go to stop after a few seconds. All was well when the take-up reel optocoupler had been replaced.

Tuning area wouldn't light up: The machine behaved as though it was in the camera mode. The tuner/camera switch was OK, however. Replacing the HD552–088C chip cured the fault.

Channel selector lights but not the red power LED: Checks in the power supply showed that the switched 9 V output was low. The always 9 V output was OK. Although the relay produced a healthy click when power-on was operated, the contacts for this line were not very healthy. Carefully cleaning them with fine emery paper and slight bending cured the problem.

Ferguson 3V36

Tuning drift: When I checked the tuning supply at TP6 on the power supply panel I found that it was low at 20 V (the manual states 45 V). Various capacitors were checked but all were OK. I then noticed that an extra 1 kΩ resistor had been fitted at the factory, in series with R13. The missing voltage was being dropped across this extra resistor. Suspicion then fell on the i.f. module, where the 33 V zener diode D10 was found to be leaky. Replacing this diode cured the fault – with the extra resistor fitted the voltage at TP6 becomes 33 V.

Unthreading problem: This machine functioned correctly apart from the fact that when unthreading the supply spool wasn't driven in order to pull in the tape as the loading arms retract. Now although this model has a reel idler to drive the spools it doesn't have a reel motor. Instead the idler is driven by a pulley, which in turn is driven by a belt from the capstan. Whilst unthreading the capstan motor wasn't being driven and we found that in this mode the drive transistor Q206 was without base bias, although the 5.7 V bias was present in the play and fast wind modes. A study of the circuit diagram showed that in play the capstan drive comes from the servo chip, in fast wind it comes from the CPU chip IC201 and during unthreading it's switched by the expander chip IC202 (pin 38). This pin was found to be permanently low, all the other ports relating to capstan motor operation being correct. So, having encountered a number of faulty expander chips in this model, I replaced the i.c. This made no difference! Further checks revealed that R272 (10 kΩ), a bias resistor in the drive transistor's base circuit, was open-circuit.

Sound deteriorates when hot: We found that when the machine had been running for an hour or two the E-E sound became distorted and threatened to disappear altogether. The playback sound remained OK. A scope check showed that the audio signal coming from the receiver section was 'strangled' at TP4 on the tuner/timer panel because the interstation mute circuit was coming into operation. This is based on the action of the 15.625 kHz tuned circuit T5. Careful adjustment of T5 to peak up the line-rate waveform at the collector of Q11 overcame the problem.

Line hold appeared off frequency: This machine produced a picture that looked as though the line hold was off frequency. You could see that the head drum was revolving too fast. After a few seconds the tape stopped and unloaded. Our first check was on the drum FG pulses at pin 4 of IC404. The waveform here and also at pins 6 and 11 was correct. Moving on to IC406 we found that the voltages at pins 15 and 13 varied while there was zero voltage at pin 14. The 12 V zener diode D412 was short-circuit.

Dark grey picture with flashes: In the play mode the sound was OK but the screen showed nothing but dark grey with occasional flashes of picture content in the background. On tracing the video signal through I discovered that it became a little distorted in the luminance amplifiers (Q106/7) that follow the f.m. demodulator filters. It disappeared completely at the output of amplifier transistor Q103. Meter checks showed that the pre-playback 9 V supply was low at only 2.5 V. Switching transistor Q501 was the cause of the problem.

Weak E-E video: Check whether the 1 kΩ E-E level control is open-circuit.

No E-E or playback sound: Sound problems are rare with these machines. One that had no E-E or playback sound came along recently, however. We soon found that there was no d.c. supply to any of the sound chips on the main sound panel, which is mounted at the back of the chassis. The cause of the trouble was that the regulator transistor Q4 on the sound panel was short-circuit base-to-collector.

Playback picture/sound slow motion: The playback picture and sound were in slow motion, with tracking lines. Occasionally correct pictures and sound would appear, followed by a screeching noise then a return to the fault condition. When the capstan motor was removed I found that its shaft was very tight. A new motor cured the problem.

Own recordings, capstan slow: Playback of a pre-recorded tape was OK, but when a recording made by the machine was played back the capstan speed was slow. Checks showed that the capstan FG comparison signal was missing at pin 6 of IC408 (BA6305) though the input to this section, at pin 5, was OK. The obvious thing seemed to be to replace the chip, but this made no difference. After checking the chip's peripheral components I did what I should have done in the first place – check the amplitude of the pulses at pin 5. It was low of course (200 mV). When I checked back to pin 1 I found that the signal from the capstan flywheel FG coil was also low. An inspection of the flywheel revealed that the two screws, which hold the bracket, were chewed up and that someone had already fitted a new set of belts. The cause of the trouble was excessive clearance between the flywheel and the FG coil. I think that whoever had fitted the belts was unable to undo the screws and bent the flywheel bracket to get the new belt on.

Refuses to accept a tape: If the reel motor is turning when this fault occurs, you will find that a circuit protector on the bottom PCB, at the back next to the word 'Elna', is open-circuit. In every machine I've come across where this N15 protector has failed the cause has been a shorted loading motor. If you're stuck a motor from an old cassette housing can be fitted – by swapping the pulley over.

Ferguson 3V38

Spotty, weak playback pictures: Playback of a known, good, pre-recorded tape was OK. A scope check on the f.m. record signal (at TP122 and TP124) showed that it was very low. The record level preset R213 (1 kΩ) was found to be open-circuit.

Wouldn't power up but the display lit: There was also loss of the loop-through to the TV set. Checks in the power supply produced readings of 0 V at pin 1 (ALL 12 V) and pins 3 and 4 (16.5 V a.c.) of CN1. The pins had arced and gone open-circuit. Cleaning and resoldering put matters right. Another of these machines was totally dead because pin 4 of CN1 had failed in the same way.

No recorded picture: The E-E was OK. We found that C197 was short-circuit. Intermittent lockout with a dim operation LED is the symptom when the relay on the power supply board is faulty. Clean or replace it.

Ferguson 3V39

Capstan motor wouldn't operate in any mode: The relevant power supply lines were correct, but the base of the emitter-follower Q235 in the motor drive amplifier circuit was permanently low. The cause was quickly traced to the inverting gate at pins 3 and 14 of IC4 – resistance checks showed that there was an internal short-circuit at the output pin 14. I mention this because it's not the first time we've had an internal short in one of these IR2403 chips, which are used in a wide range of models. If while fault tracing you start to move towards one of these buffer chips it's well worth while checking that the input and output logic conditions are correct for the mode concerned. Another of these machines came in because the 'line hold' ran off very intermittently – in other words the drum would slow down. When I eventually managed to make the fault appear I found that there was no sample pulse on the ramp waveform at TP4. Before condemning the servo chip I decided to make a few d.c. checks and discovered that there was no voltage at pin 10, the tracking control input. The cause of the fault was traced to a broken lead inside the sleeving at plug connector 1/4 (i.e. pin 4 of plug 1) on the servo panel. We're beginning to get a number of intermittent faults on these machines due to breaks such as this.

Intermittent sound muting: Recording and E-E depended on the picture content – dark scenes produced sound muting and bright scenes brought the sound back, while in between the sound fluttered and caused a disturbance to the picture's luminance level. The cause of the fault was traced to L7 on the tuner/i.f. panel. It should be tuned to 15.625 kHz but was found to be off frequency. Normal operation was restored when L7 was peaked, using an oscilloscope, but at resonance the core was fully tightened. An unnumbered tuning capacitor within L7's can was thought to be defective. Adding an externally mounted 330 pF ceramic capacitor remedied the situation.

R.F. section damage due to thunderstorm: We had quite a succession of VCRs and TV sets with damaged tuners and aerial amplifiers after the succession of thunderstorms. This one was different, however. When operating with its internal tuner the gain was so low that the signal could hardly be seen. A new tuner was fitted but this made no difference. As the voltages around the tuner were correct attention was turned to the SL1432 SAWF driver chip. Voltage checks at pins 3 and 4 produced 10 V instead of 5.3 V readings. A new SL1432 restored normal gain.

Wouldn't accept a cassette: On investigation I found that protector CP1 (0.6 A) was open-circuit. So I removed the cassette carriage and tested the loading motor, which drew about 850 mA off load. Under the same conditions a new motor draws about 25 mA. Once the motor and fuse had been replaced, cassettes loaded normally.

Ferguson 3V42

Cassette jammed: The job card read 'cassette jammed', but the customer had released it. On removing the top we found that the deck was in the half-loaded state. When we switched on, the loading motor tried to return the loading poles to the unloaded position but was prevented from doing so because the tension pole jammed the supply loading pole. When the loading/mode control motor bracket was removed, so that we could inspect the nylon gear assembly on the underside of the deck, we found that the cogs were badly worn, as was the loading gear. Replacing both these parts and setting up the loading timing cleared the trouble.

Won't accept tape: The cause was failure of one or both of the cassette detect microswitches on the cassette housing. We decided to replace them both.

Poor end to loading cycle: This one could apply to other models which use the same deck. We found that during the last part of the loading cycle, as the pinch roller starts to pull in, the loading motor would slow down and struggle to complete the loading sequence. The cause was traced to the grease on the lower drum casting – the bit with the V blocks. Although it wasn't as hard as in some models it was nevertheless very sticky, impeding the slant pole base. To cure the problem, clean the grease off with solvent and then apply a smear of graphite grease.

Slow at lacing up for play/record: These machines are now becoming slow at lacing up for play or record. Cleaning the drive gear loading cog assembly provides only a temporary cure. The permanent cure is to obtain a complete assembly, ready lubricated and including the motor and new plastic circlips, from SEME (part number DECKR5). The motor can also be faulty. Don't forget to clean the spindles of the two small white cogs (connect gear 1 and 2) and the two larger cogs (loading gear 1 and 2) at the bottom of the deck.

Poor audio: A piece of sticky label was found attached to the AC head. I removed this and cleaned the head but the fault persisted. Audio was fine at the AV socket and was present at the input to the modulator, which was the cause of the trouble. I sent it to MCES for attention. The returned unit fixed the problem. Full marks to MCES, as usual.

Picture was 'a complete mess': This machine came in for the routine job of fitting a new carriage assembly. It didn't take long to do this, but my heart sank when I tested the machine and saw the playback picture. It was a complete mess, consisting of two very wide horizontal dark bands that resembled hum bars, with a colourless, spotty picture in between – and not a straight vertical line in sight. The display was so ragged that I had to check the tape's label to find out what I was supposed to be looking at. 'Don't panic', I told myself, 'go for the power supply'. As the E-E picture was normal, a scope check on the switched playback 6V rail (test point TP2 on the regulator board) seemed to be a good idea. The display consisted of a 2V square-wave sitting on 4V d.c. The likely culprit was C23 (2,200 μF, 16V) which measures open-circuit. A replacement restored normal playback. Naturally as far as the customer was concerned playback had been perfect before the carriage broke!

Ferguson 3V43

No E-E or playback sound: A look at the circuit revealed an awful lot of things that could have gone wrong! A great deal of the circuitry in this machine is for audio processing. The protection fuse CP3 was found to be open-circuit but replacing it produced no improvement. Many checks were made before we found (stumbled across?) the fact that there was no 9 V supply at pin 8 of the switching chip IC5. There was voltage at pin 8 of IC17, however, which the circuit showed as being the same section of print (linked by a 220 Ω resistor). As a quick check these points were cautiously linked across. This produced low, distorted sound and only 3 V at pin 8 of IC5. Mmm! A regulator transistor, Q16, is associated with CP3. When this was checked it was found to be short-circuit base-to-collector. Replacing it restored normal voltages and sound and my link could also be removed.

Intermittent failure to make a timed recording: We confidently changed the loading belt and sent the machine on its way. It was very soon back on our bench with a note to say 'same as before'. This time we checked the loading process more thoroughly, and found that there was a stiff point in the mechanism at about the half-travel point in the progress of the loading arm. It turned out that loading gear 2 (under the deck) was very stiff on its shaft. It was removed, cleaned and lubricated, and after that the customer didn't report any further timer trouble. Why it never gave trouble on manual record and playback remains something of a mystery.

Shut off by mains plug only!: When the fault occurred the only way to shut the machine off was to pull out the mains plug! Whether the machine was off or in standby, the capstan motor and the take-up reel would whiz up to a high speed, with the clock display extinguished. This strange combination of symptoms was due to loss of the unswitched 12 V line. The cause was a dry-joint on the lowest wire link between the two PCBs that make up the power regulator section, isolating the base of Q1 from pin 8 of IC1. The other joints are also worth resoldering.

Playback normal, no E-E audio: The cause was traced to circuit protector CP3 on the FMA board being open-circuit. It's in the 12 V feed to the f.m. circuit and seems to fail for no apparent reason.

Failed to read reel tacho pulses: The amplitude of the TU FG signal was low at pin 41 of IC202 because D248 (1N4148) was leaky. It's fitted on the sub-board, by IC202.

Ferguson 3V44, 45

Clock display failed after an hour: The cause was the TL066CP reset chip which changed the state of its output when warm, holding the clock chip in the reset condition.

Would stop during playback: This machine might run for an hour or it might refuse to run for more than a few seconds. In the fault condition the rewind and fast forward functions would also cut out. Suspicion immediately fell on the take-up reel sensor, which has given trouble in the past, and sure enough the 6 V peak-to-peak pulses were missing when the fault was present. A replacement sensor was ordered and fitted but the problem remained. The small PCB on which the sensor is fitted was unscrewed, and I stared at it in disbelief. With the machine in this condition I selected play and as I watched the mirrors on the take-up spool turn I saw the letter D pass through the chassis aperture through which the sensor is activated. It was one of those stick-on letters that cassette makers supply with new tapes, and appeared to be catching on the edge of the chassis aperture from time to time, thus blocking out the sensor. I've often fished these letters and numbers out of VCRs, but this is the first time that one of them has caused a fault.

'Intermittent won't play or record': We're now getting a lot of these machines with this complaint, and of course the loading belt has stretched. Unlike the 3V29/3V30 series in which this was also a common complaint, belt replacement in these models is very straight-forward, just remove the top cover, hinge up the top PCB, remove the screening can and there it is right on top. No circlips etc. to remove!

No red light on the power switch: Circuit protector CP4 (ICP-10) being open-circuit has on several occasions been found to be the cause of no red light on the power switch and no drum rotation. It seems to go open-circuit for no apparent reason. When it's open-circuit the switched 5 V supply is removed.

Ferguson 3V44

Intermittent playback colour: Check BPF301 for open-circuit or dry-joints. If the on LED doesn't light but the capstan runs when the on switch is pressed, check whether the 3.9 V zener diode D3 in the power supply is short-circuit.

Clock, timer and timer problems: There were several problems with this machine. The clock couldn't be set, there was no tuning and the timer couldn't be set being amongst the most obvious ones. A case of a very large brain failure. After perusal of the circuit diagram we suspected the microcomputer chip IC601. As its −5 V supply and clock were OK we tried a replacement. This cured all the symptoms. It's type M50730–607.

Jammed cassette: This VCR came in with a jammed cassette, a bent cassette housing and the tape still partially wrapped around the drum, but the machine did its best to eject the cassette. To start with we straightened and retimed the cassette housing, then reset the deck timing. As we didn't trust the VCR completely we inserted a dummy cassette and watched the operation of the deck very closely. Everything worked well until eject was selected, when it was obvious that the cassette was ejected much too early – had it been a proper cassette the tape would still have been partially wrapped around the drum. My first thought was to replace the mode control switch. This machine uses an LED and sensor assembly, however, with the light shining through holes in the cam gear. When this item was replaced the deck mechanics worked correctly.

Problems with own recordings: Because of the extremely intermittent nature of the fault with this machine it unfortunately bounced. Operation with prerecorded tapes was perfect, but with its own recordings there were occasionally tracking errors and an interference bar would roll through the picture. As a complete repair kit had already been fitted I decided that a mechanical cause of the trouble was extremely unlikely. Eventually scope checks revealed that the machine didn't always record a control track on the tape. After some time had been spent checking around I found that C430, which couples the control pulses to the head circuit, was dry-jointed. Resoldering was all that was required.

Drum and capstan running slowly: The drum and the capstan were both running slowly. A check on the servo reference signal, using a frequency counter, showed that it was running at only 2.5 MHz. The cure was to replace the 4.433 MHz crystal in the chroma circuit.

Machine wouldn't accept cassette: As the power supply circuit protectors were intact we turned our attention to the carriage assembly. The cassette could be loaded manually, after which all functions such as fast forward, rewind and play worked normally and the cassette was ejected correctly. We found that the cause of the problem was the leaf switch at the right-hand side of the carriage assembly. All was well after fitting a replacement.

No clock display: This machine had no clock display although the function display worked correctly. Scope checks on the timer/display board showed that there was no output from IC401. The supply to this chip was correct, but there was no clock signal either at this chip or where it enters the board at pin 6 of CN1. The missing signal was traced back to broken print on the power supply module. Repairing this print restored the display.

Drum motor running backwards: For once the symptoms displayed agreed with the fault description on the job ticket. This said dirty heads. Cleaning them didn't help, however, and the reason for this was soon apparent: the drum motor was running backwards at full speed. D408, a 5.1 V zener diode, was short-circuit.

Loss of colour: A tap on the top panel of this machine would restore it temporarily, but even after a mass soldering operation the colour was still tap-sensitive. So methodical checks had to be carried out. This brought me to filter BP301, which must have had a poor internal contact. Anyway a replacement restored reliable colour.

Blank raster on playback, sound OK: The E-E picture and sound were present but on playback there was just a blank raster, the sound being OK. Scope checks showed that there was no video output at pin 9 of IC102 and no sync output at pin 2. Voltage checks on the chip were inconclusive, the voltages at pins 2, 24, 27 and 33 being incorrect. I finally had to change the chip, thus proving that it was the cause of the fault. It's a small 'end-on' PCB, part number PU22031A.

Ferguson 3V45

Intermittent won't play or record: We're now getting a lot of these machines with this complaint, and of course the loading belt has stretched. Unlike the 3V29/3V30 series in which this was also a common complaint, belt replacement in these models is very straight-forward, just remove the top cover, hinge up the top PCB, remove the screening can and there it is right on top. No circlips etc. to remove!

No drum rotation: Although this machine would accept a cassette and the fastforward and rewind modes were OK there was no drum rotation and the function LED wouldn't come on. The cause of the problem was an open-circuit fusible link, B3, which is located in the power supply.

During playback and record machine would enter the stop mode: This machine would sometimes play almost to the end of a tape before shutting down: at other times it wouldn't even enter the play mode after lacing up. We noticed that slight pressure anywhere on the main PCB would make the machine shut down. Voltage checks were of no use in this situation, so we used a magnifying glass to carry out a careful scan of the board. This revealed that R501 (680 Ω, 0.5 W) was dry-jointed at one end. IC404 in the servo section receives its 17 V supply from R501. Resoldering provided a complete cure.

Odd symptoms when D408 is leaky: D408 is a 5.1 V zener diode. In this machine the drum immediately took off backwards at high speed, and there was just a quick burst from the capstan motor. A replacement diode put that right. But the remains of the back-tension band were jammed in the carriage. This prevented the arm coming into contact with the tape. I know that the machine had been used like this. What had the owner been watching?!

Couldn't be tuned: I found that the 30 V supply was missing at pin 5 of CN1, though 45 V was present at pin 3 of CN2. Nor was there 13 V at the collector of Q13. Lifting the tuner panel revealed all. Water had come down the aerial lead into the r.f. converter/booster and flooded under the nearby tuner/i.f. panel, corroding away the link between the emitter of Q12 and the collector of Q13. Hence the fault – the booster was undamaged.

Ferguson 3V48

No take-up: This machine nearly drove us to desperation and certainly had us all questioning its parentage. On the fourth visit to the workshop the fault showed up as no take-up: the capstan motor would stop after an indeterminate time. The merest suggestion of heat or cold on IC202 (M50742614SP) had the desired effect so we replaced it. But the fault was still present. A scope was connected to pins 8 and 7 to monitor the capstan drive while a meter was connected to the unswitched 5 V supply at pin 9. We waited, and waited! Eventually the fault showed up as a distortion, reduction and final disappearance of the capstan drive. IC403 (BU2710) proved to be faulty, again in a thermal manner.

Timer problem: The following fault caused us some sleepless nights. The machine came in with a damaged timer door and PCB. Putting this right was quite easy but when we tested the machine we found that while instant and normal recording worked fine all we got with a timer recording was garbage. The mechanism sprang into action correctly, and checks showed that all the record voltages were the same as with an instant or normal recording. A clue was that when coming out of the timer record mode the E-E display had swirling interference on it: this cleared when the machine was set for another function. We eventually traced the cause of the fault to C10 which is connected to the switched 5 V rail. It allowed parasitic oscillation to get into the f.m. modulator with a timed recording only, destroying any chance of a successful recording. The display gave the impression that there were the remnants of a picture behind the scrambled garbage.

Required a new loading belt: A 3V48 and a JVC HRD140 machine, which have similar mechanisms, had the same fault – they required new loading belts (remember the 3V30?). The 3V48 came back, however, with the complaint that it would only rewind/wind fast forward in the picture search modes. A plastic lever on the loading block had come adrift – it's used to free the brakes for rewind/fast forward. I can't quote a reference/part number for it as it's not listed as being separate from the loading block.

Intermittent drum rotation: This was caused by a dry-joint at the 2SB1052 transistor Q1.

Ferguson 3V53, 55, 57

Intermittent mechanical failure: On a number of occasions we've found that intermittent failure to carry out mechanical operations such as rewind/play etc., even when the display shows that the command has been received, has been due to an intermittently open-circuit or very tight loading/mode motor. Attend to the usual regreasing at the same time.

Intermittent loss of capstan lock: After replacing a worn audio/control head and a broken front flap I was rewarded with intermittent loss of capstan lock, the symptom being noise bars rolling through the playback picture. The fault cleared when the main PCB was hinged upwards. Careful checks around the servo circuitry revealed that there were dry-joints at both pins of the control head connector CN401.

Red LED lit no functions or display: The red LED was lit but there were no functions and no clock display. Integrated circuit protector CP3 in the power supply had failed, causing loss of the unswitched 12 V supply. Failure of one or other of the four ICPs in the power supply circuit for no apparent reason is not uncommon. The engineer can be responsible, however, when the screening can is removed from the mechanism with the unit powered – switch off first.

Ferguson 3V53

'Fine net curtains' on playback: Scope checks narrowed the sources of the noise to the MSM6989RS delay chip IC2, which had obligingly cleared the fault when it was frozen.

Machine wouldn't switch on: It would accept a cassette, but wouldn't eject it. The clock was flashing away and could be set, but basically the machine was dead. No faults could be found when initial checks were carried out in the power supply. Then I found that the on-line didn't go low when the on button was pressed. The switch itself was OK, but the power-on output from the microcontroller chip IC202 didn't change. If a shorting lead was used to take the line low the machine made an attempt to start and various supply lines became active. A replacement microcontroller chip put matters right.

Ferguson 3V54

Would play for 5 seconds and stop: The head was spinning and the wheels were going round, so it appeared that an input signal to the mechanism control chip was missing. We found that the drum FF signal was correct but the take-up reel sensor signal was irregular. It would start at the correct amplitude and then diminish. In addition, when of the correct amplitude there was a ripple on the lower edge (see the Figure 4). A check on the signal from the TU sensor to the buffer transistor on the deck terminal board proved that this was correct, with no ripple. The ripple appeared at the collector of this transistor. It's supplied from the switched 12 V rail, which also had a ripple on it. A voltage check revealed that the voltage was low at 10.5 V when the tape was running. Time to investigate the switched 12 V supply more thoroughly. The rail also supplies the tuner/i.f. board, and when this was unplugged the ripple disappeared. Was the fault due to this panel drawing excessive current, or was the power supply regulator unable to supply sufficient current? As the E-E picture was good, a fault in the tuner/i.f. panel was ruled out and attention was turned to the power supply. Cold checks on the semiconductor devices here revealed that Q10, which switches the switched 12 V regulator on, had high-resistance junctions while the 3.9 V zener diode D3 which biases Q10's base was short-circuit. After replacing these two components the VCR worked correctly.

No colour: This was traced to a faulty low-pass filter, LPF301.

No colour: Sure enough a test tape played back in monochrome. Checks revealed that the playback chroma was leaving the colour module for the delay line at the correct amplitude, but was distorted

Figure 4 *Ferguson 3V54 take-up reel sensor signal with ripple*

when it returned. This distortion can best be described as a colour-bar waveform with two sets of bursts and chroma in the white bar, seen at the composite video output socket. As this waveform was also present at the output from the delay line the latter was replaced. This did no good and a lot of time was then spent checking the components around the delay line, all to no avail. Replacing the chroma module didn't improve matters either. At this point I did what I should have done to start with – I made a recording and played it back. As expected there was no colour, but the head-switching point must have been some 30 lines before the field sync. A double check with a pre-recorded tape proved that the playback-switching point was correct, so now we had two problems, no colour and an incorrect record head switching point. Attention was next turned to the servo circuit, where the capstan servo was found to be unstable. When the shield over the drum assembly was removed the drum was seen to be running slightly too fast. Assuming that the machine was working correctly before the colour was lost, something was causing four symptoms: no colour, an unstable capstan (not audible though), the drum running too fast and the record head-switching point incorrect. In the hope that these were all related we decided to tackle the servo fault. Now the only common factor in the drum and capstan servos is the 4.43 MHz reference signal, which is derived from the chroma module. Ah! A check with a frequency counter revealed that the frequency was 4.4764 MHz instead of 4.43 MHz. Replacing the 4.43 MHz crystal cleared all the faults. Why was the chroma waveform coming out of the delay line so distorted, with two sets of bursts? My theory is as follows. With the drum rotating too fast the time the heads took to read the video tracks was shorter than the delay line's preset delay. Since the direct path signal was faster than the delay period, the direct and delayed signals would not overlay exactly in the adding circuit. Has anyone any other theories? On the circuit diagram the delay line is referred to as a comb filter, with no details as to what goes on inside.

Intermittent play and record, and chewing tapes: When the fault was observed tape movement ceased, the tape being left in the fully laced position. The capstan motor had stopped due to a dry-joint at the emitter of Q603. I've often resoldered the connections to this transistor and Q604 when servicing these machines as the joints have always looked suspect.

Head drum rotated clockwise: As we had a loan machine of the same type to hand the drum and motor drive amplifier were swapped over. This proved that the fault was on the main board, where we found that D408 was short-circuit.

No timer display and no operation: The job card said 'dead' but the supply lines were OK. The actual symptoms were no timer display and no operation, with a cassette still in the machine. Checks showed that the display grid drive was correct but there was no segment drive. We then found that the microcontroller chip was not receiving the a.c. clock reference pulse though it was arriving at the timer board. C214 was found to be faulty with a resistance of 330 Ω.

Capstan motor appears to run through with no control pulses: If the capstan motor appears to run through with no control pulses present at the control amplifier, check the condition of C405 on the main PCB. You will probably find that it's either dry-jointed or open-circuit.

Clock worked, no other operation: None of the front controls had any effect. As the circuit protectors in the main body of the power supply were all in order we turned our attention to the main PCB where we found that the switched 5 V supply was missing. The cause of the fault was C605 on the top panel – it was leaky, a replacement restoring normal operation.

Damaged tape on eject: On this machine it seemed that the slant poles didn't fully retract. Replacement of the loading sensor assembly, part number PU35632A3, fixed the problem. I've also had this failure with Model 3V55, which uses the same part.

Ferguson 3V55

'Dead, no functions, tape jammed': We extracted the customer's tape by rotating the tape loading motor manually. After switching on there were still no functions. A check on the power supply section revealed that there was no voltage at pin 3 of plug CN1, nor at pin 2 of CN3. The cause of the missing supply was the fact that CP1 (F10) was open-circuit. Replacing this restored normal operation.

Intermittent, no E-E or record vision: For intermittent no E-E or record vision check that R105 on the tuner/i.f. panel has been correctly inserted in the PCB.

Loading motors run after ejection: As a result of this fault the loading belt squealed loudly. We found that the cause of the fault was the mode optoswitch assembly. It's available from CPC at a very reasonable price under part number TNPU35632A3.

Channel selector would not move: We all drop clangers from time to time but this was a beauty. The complaint was that the channel selector couldn't be moved from the auxiliary input position (position 0). As there had been several heavy thunderstorms we came to the conclusion that either the timer or the mechacon microcontroller chip had succumbed. But substitution checks showed that neither was at fault. We decided to force the machine into the timer mode by making the auxiliary line go low. This produced snow on the screen in the E-E mode and we were able to tune in stations in the auxiliary position. We then found that we were also able to tune in the other positions and when the auxiliary line was released everything worked normally. This whole business wasted several hours. What had happened of course was that the electrical storm had wiped the tuning memory clean and it then refused to select any channel other than auxiliary until this position contained information.

Intermittently the drum would begin to go too fast and the machine would shut down: In the fault condition the drum FG signal at TP414 went missing, although it was present at pin 6 of IC404. The supply to this chip was intermittent because R501 was dry-jointed.

Would stop in play: Watching the tape counter in the fast-wind mode led me to the reel sensor – the counter worked in fits and starts instead of providing a steady count. As cleaning the underside of the reel disc and the faces of the opto device made no difference, a new sensor was fitted. But the fault was still present. A scope check on the sensor's output showed that it was intermittent: so a new reel disc was tried. The results were no better. The cause of the trouble was actually down to me: when I first removed the reel disc to clean it one of the height setting washers had been lost. Thus when the disc was refitted it was too close to the sensor. Matters were put right when a new washer was fitted beneath the reel disc. So the cause of all the trouble had been a dirty reel disc, but carelessness had resulted in lost time. I should be shot at dawn!

Ferguson 3V57

Display fault: The part of the display that shows fast forward, play etc. didn't work although the clock part did. IC301 (MN1250) drives the faulty part of the display. Its serial data lines (pins 14 and 17) had the same signals on them, which seemed odd until we found that they were shorted together by a solder bridge across pins 7 and 8 of CN304.

Clock alight but no switch-on: The following note could apply to any of the machines in this range. The symptoms were not uncommon – clock alight but the machine wouldn't switch on. This is usually due to the absence of one or more of the switched supplies because of an open-circuit ICP in the power supply. In this case, however, the ICPs were intact, but the switched lines were missing because the power supply CRTL line at pin 9 of CN3 remained high at 3.2 V. This high came from pin 1 of the syscon microcomputer. The power switch input to pin 17 of the microcomputer chip was very low but was being earthed as the switch was pressed. The same applied to the timer input at pin 20. Separate pull-up resistors are used, so the fault had to be in the supply. In fact the unswitched 5 V line was missing. It's obtained from the unswitched 12 V supply via a regulator on the main panel (not the PSU). The 12 V supply was present here but the regulator transistor Q602 was not being biased on as there was a 90 Ω leak from its base to chassis. We found that the culprit was not the zener diode, as expected, but the parallel 0.022 μF ceramic decoupler C6050. The interesting point was that the fault had been reported as intermittent some days earlier. We'd returned the machine with a 'no fault found' note under the assumption that a mains lock-out had occurred, something that's not uncommon with these machines.

Intermittent stop: This was due to either reduced amplitude or missing take-up reel pulses. The usual causes of this problem with these machines are dry-joints or a defective reel optocoupler. This time, however, I found that the switched 12 V supply to the reel sensor board was low and varying. When the supply and the reel pulse waveform were monitored you could see that the latter disappeared when the supply dropped to 9.5 V. The cause of the problem was on the power supply board: D3 was leaky with a reading of approximately 600 Ω.

Dried grease around mode cam: For some time now we've experienced problems associated with dried and hardened grease around the mode cam, guide bases etc. In recent cases of mode problems, usually showing up as the mechanism being out of sync and the motor and belt running (and usually squealing) against the clutch, it has been necessary to replace the motor in addition to carrying out the grease treatment. It tends to become lazy. It's best to replace the motor as an assembly, with the cam and bracket pregreased. This saves a lot of time and mess and the assemblies are reasonably priced.

Mode timing out of sync: When you get one of these machines in this condition – for example, the arms are laced, the carriage is up and the pinch roller and post are engaged as if in the play mode – do the usual regreasing/replacement of the mode motor/cam assembly. Then replace the loading belt and retime the mechanism. Also expect the mode sensor photo-interrupter fitted around the black cam on the underside of the mode cam assembly to be faulty. It's not always the case, but if you want to avoid having to retime the mechanism again replace it as a precaution. The cost is negligible.

No playback colour: The cause was traced to IC301 on the main PCB. Part number is PU22046A. Chroma was present at pin 24 of the chip but there was no output at pin 22.

Ferguson 3V58

Dead with no display or mechanical functions: We found that the supplies from the three regulator circuits were missing. The main one is the switched 12 V supply which should always be present when the machine is connected to the mains. The bias for this circuit is taken from the 45 V supply on the mains transformer board. It was missing because the 10 Ω safety resistor in series with the rectifier diode was open-circuit. This in turn was due to the fact that the associated reservoir capacitor C5 (47 μF, 63 V) was short-circuit. When the resistor, capacitor and diode had been replaced the machine worked fine.

Completely dead: After checking the obvious fuses I decided to check for an open-circuit primary winding on the mains transformer, having had this fault previously in one of these machines. It was intact, as were all the secondary windings. A check at the plug that connects the regulator PCB to the transformer then showed that the unregulated 45 V supply was missing. The safety resistor R2 was open-circuit, but the rectifier diode D4 was all right. The actual cause of the fault was the reservoir capacitor C5 (4.7 μF, 63 V). It read about 10 Ω! I've had this fault several times now.

Wouldn't respond to remote control commands: We found that D501 on the infrared receiver panel was open-circuit.

Ferguson 3V59

'Time remaining' function failed: We found that there were just dashes in the display when the machine was put into play and the function was selected. So we waited for it to calculate from the two reel speeds and display the results, but nothing happened. We knew that the take-up sensor was OK otherwise the machine would have cut out. So we checked the supply sensor input to the syscon chip IC601 (pin 35). There was no high/low cyclic change, just a steady high level. A check was then made, with the bottom cover removed, at the sensor's output connection. There was a high/low switching output here, and within a few seconds a time remaining display. The cause of the fault was a loose earthing screw – the one that earths the optocoupler's LED and optotransistor emitter connections. The pressure we'd applied when connecting a meter to its output was enough to re-establish the earth connection.

Both fluorescent and LCD displays would go off intermittently: Despite this fault the machine still worked. A look at the Ferguson fault diagnosis pocket book suggested that the timer chip IC101 was suspect in this event, so a replacement was fitted. Unfortunately the fault persisted. Interrupting the mains input would bring back the supplies, so it seemed that something was causing the timer chip to freeze – but what? A scope was left connected to the 5 V supply line. This showed that the voltage dropped momentarily. We traced back to the 12 V supply and found that this also dropped occasionally. The STK5481 regulator chip was faulty. I suppose I had to come across an intermittent one sometime!

Machine's drum rotated in reverse: Replacing the VC2023A chip IC2 cured the fault.

Ferguson 3V65

Fusible resistor open-circuit: We've had two almost identical cases of the 220 Ω fusible resistor that feeds the 30 V regulator being open-circuit. The cause in both cases was severe corrosion and shorting on the underside of the tuner/i.f. panel in the vicinity of IC3. No leaky capacitors were found, and the cabinets showed absolutely no signs of spillage. Strange . . .

Smeary E-E picture: Playback of a test tape was good but there was a smeary E-E picture. Replacing the luminance module IC101 put this right.

Defective upper drum?: This machine displayed all the symptoms of a defective upper drum but remained the same after a new one had been fitted. When the head pre/rec panel was pressed the fault suddenly cleared. All ten pins of connector CN2 were dry-jointed!

No colour on pre-recorded tapes: This machine came in for head replacement. When we'd done this we made a recording and found that the picture/colour etc. were normal. The machine was therefore returned to the subscriber who subsequently complained that it played his tapes only in monochrome. When we tested the machine in the workshop with a pre-recorded tape, sure enough no colour. The reason for this was that the upper drum had been fitted 180° out of its correct position, something that's very easy to do with this model as the 'relay pins' are both made of the same colour plastic.

Failure of motor drive chip: Failure of the M54644AL or M54644BL motor drive chips on the mechacon panel in these and clone models is not unusual – they control the capstan and reel motors. Recently we had one of these machines in which IC603 had died (all inputs and supplies OK but no output), so we confidently quoted for a replacement. When this was fitted the reel motor remained still and the new chip started to overheat. The reel motor was short-circuit through 300° or so of its rotation! It carries a high price tag for a brush motor, and a difficult telephone conversation with the owner prefaced the fitting of a replacement.

Tried to rewind a non-existent tape: This machine thought it had a tape inside and was trying to rewind it. When I've encountered this fault on previous occasions the cause has been loss of the switched outputs from the STK5481 chip. But in this case the supplies were OK. There was no voltage at the cassette lift, however. On inspection the motherboard was found to be suffering from dry-joints at the connectors, especially at CN1.

Ferguson FV10B

Power supply chip (STK5481) failed: The now infamous power supply chip IC801 (STK5481) had failed, with the 5 V line rising to 9 V as a result of an internal short-circuit. We replaced this and the clock display returned to normal. When a tape was loaded it would wind and rewind all right but if play was selected the machine would lace and then unlace as soon as the pinch roller engaged. On investigation we found that there were no drum flip-flop pulses at pin 34 of IC601 (M50731–623SP), although the drum rotated normally. When we checked the servo chip IC401 (HD49712NT) there was no pulse output at pin 62. Replacing this i.c. provided a cure. We've had several machines since with similar problems, including a JVC HRD170 which has more or less the same circuit with different chip types.

Dead, no display, no functions: The repair ticket said 'dead', and the machine was very dead indeed, with no display and no functions.

A check at pin 10 of the voltage regulator chip IC801 showed that the 13.2 V which should have been present here was missing. A glance at the circuit diagram showed that there were several possibilities for this. The culprit turned out to be R1 (10 Ω) which had gone open-circuit. A replacement brought the machine back to life.

Servo fault suspected: This fault could probably apply with any VCR but serves to show that things aren't always what they at first appear to be. Because of regular (1 second) drum speed variations a servo fault was suspected. We noticed, however, that there was quite a lot of graphite around the base of the audio/control head. Closer inspection showed that the head had been screwed down almost to its limit and was damaging the top edge of the tapes. Head realignment restored normal drum operation and of course improved the sound quality. The customer denied that the machine had been tampered with although he did admit that his son had removed the cover – 'he's usually quite good with electrical things'. Now where have I heard that before!

Missing lines in playback picture: This machine's playback picture was marred by what appeared to be three or four lines missing every inch or so. The effect consisted of thin, horizontal black lines that varied in intensity with adjustment of the tracking control. We found that C801 (47 μF, 25 V) in the motor 12 V supply had gone low in value. The problem was not present in the record mode. The relevant capacitor in JVC HRD170 series machines is C10.

Ferguson FV11

Intermittent playback picture: L15 is faulty (off pin 13 of IC101).

Poor or intermittent tracking: IC2 (VC2023A) faulty.

Low gain on high channels: Try repositioning the aerial lead that connects the tuner to the aerial splitter.

Sound but no playback picture: From this description I suspected clogged video heads, so before trying a tape I took a look at them. There was some oxide present, but it wasn't too bad. The drum was next warmed to remove any moisture and a tape was played back. The picture was fine! It was only when I moved the luminance/chroma board – I had to fold it up to the vertical position to remove the video head screening can – that the fault occurred. The effect was similar to an off-tune monitor TV set – just off the higher end of the r.f. output.

The fault could be made to come and go by slightly twisting the panel. Close examination revealed a dry-joint on a small luminance coupling capacitor (C134), which is next to the right-hand subpanel.

Intermittent eject and rewind: A couple of faults are becoming very common with these machines. First is intermittent failure to eject, or other intermittent mechanical faults, caused by a loose earthing screw on the mecha connection PCB. The other is intermittent going into rewind from play during a tape. This is caused by dry-joints on the supply photosensor. Both faults can be exceptionally intermittent.

No tuning voltage: The temporary field engineer accused the tuner of being the cause of no signals. He ought to have known better! There was no tuning voltage – the regulated 30 V supply was being lost across R53 as there was a short from the tuner's BT pin to chassis. We found that the pin hadn't been trimmed and was shorting to the bottom cover. When this was corrected the machine still didn't tune as there was now no load on the BT rail. The control transistor had no drive from the frequency-synthesizer tuning chip IC3 as there was no 5 V supply. A break was discovered in the print between IC3 and the 5 V regulator IC1.

Very bad hum in both E-E and playback: A scope check proved that this was being caused by the 5 V supply from the STK5481 regulator chip. The input at pin 2 of this regulator is smoothed by C4 (2200 μF). A replacement capacitor cured the initial problem but on test F3 (1 AT) failed. It took us some time (and some fuses) to establish that the STK5481 chip had an intermittent fault.

Capstan motor slow, drive i.c. hot: The capstan motor was very slow and its drive chip was very hot. Unlike more modern DD motors, the chip is not part of the motor and doesn't cut out when it overheats. A new chip got the motor running but was getting hot. It's not an easy motor to dismantle, but once the bearings had been cleaned in alcohol and reassembled with fresh oil the motor ran at full speed without the chip overheating.

Ferguson FV11R

No clock and no timer functions: This machine would otherwise accept tapes and perform deck functions normally. IC1 (UPD75208CW-097) on the front panel seemed to be a logical place at which to start. Supplies arrive at CN1: the −30 V and filament supplies were both correct. Unswitched 12 V was present at pin 1 of CN3 and was being converted to unswitched 5 V by IC2 to feed pin 64 of IC1. The reset line at pin 39 of this chip twitched normally at switch-on, and the clock oscillator across pins 30 and 31 was OK – there was little to show by way of data output, however, despite various requests being made of the chip. A new chip restored normal operation.

VCR kept stopping after about an hour's use: The complaint with this VCR was that it kept stopping after about an hour's use. On test we found that the tape counter stopped counting. The cause of this turned out to be the famous loose deck earthing screw that provides the earth return for the take-up optosensor and the mode switch. What puzzled us was that in every previous case we've dealt with this screw it has caused a loading fault when loose. Presumably a high-resistance connection was enough to upset the optosensor but not the input to the microcomputer chip from the mode switch. Incidentally the official cure is to fit a small shake-proof washer beneath the screw, not just tighten it.

Loss of capstan servo control: The symptom was noise bars, which ran through the picture. The control track pulses were OK at pin 20 of the servo chip IC2, and someone had already changed this i.c. So further investigation was required. We eventually found that C25 (4.7 µF, 25 V) was the cause of the trouble. It tested OK but a replacement cured the fault.

Drum would not rotate: Two of these machines came in with the same fault. They would accept a tape and the front controls operated. There were no functions, however, because the drum wouldn't rotate, and there were no E-E signals. The tape would be ejected after a few seconds. In both cases replacing the STK5481 power supply module cured the trouble.

No video on playback: E-E was OK. On playback there was sound but only a blank screen (no video). Checks showed that the E-E 5 V line was always at 5 V. The DTC144W digital transistor Q503 was faulty.

Ferguson FV12L

Machine operation 'haywire': Another easy job I thought – the customer complained that the machine's operation was haywire and my tests confirmed this. I replaced the STK5841 chip and the machine seemed to be all right, but on soak test it went haywire again. This time attention was directed to the mechanics, which were found to be dry and stiff. After removing the mechanism baseplate and cleaning and lubricating the mechanism the machine worked well.

Intermittent play and record: This was a JVC HRD230, equivalent to the FV12L. The play or record would be suddenly interrupted at random times, reverting to the stop mode. It was because the reel pulses disappeared. To obtain a reliable earth connection for the optocoupler we had to fit a tiny shake-proof washer to the reel sensor PCB assembly's fixing screw. There's a well-known weak spot in certain JVC VCRs. It's cured by fitting a shake-proof washer to the underdeck PCB fixing screw to ensure a good electrical earth connection.

Large hum bar in E-E and playback modes: A check on the switched 5 V line showed that a distorted 500 mV, 50 Hz square wave was present. We initially suspected C14, which decouples the switched 5 V output at pin 3 of the STK5481 chip IC1. It was OK, however, the cause of the fault being the chip itself.

Ferguson FV20

No sound: These machines are very popular and have excellent reliability. The fault with this one was no sound. In fact there was just a buzz in the E-E and playback modes but when going through the scart socket into my monitor the sound was OK. A new r.f. converter put matters right.

No playback, low luminance on E-E: There were two fault symptoms with this machine, no playback picture (blank screen, sound OK) and what looked like a low luminance level in the E-E mode. Both problems were caused by the same faulty component, the low-pass filter LPF102, which is connected between pins 10 and 21 of IC101 and is used for both E-E and playback.

No colour: Use of another machine proved that the faulty VCR was recording the chroma signal, so the fault was only on playback. From the head amplifier circuit the chroma passes to the playback processing circuitry via a couple of filters. There was an input to the first filter LPF201 but no output. As a check, the filter was removed and a 0.1 μF capacitor was used to couple its input and output connection points. This restored colour, so a new filter was ordered.

No record colour: We found that the machine played back pre-recorded tapes all right but it wouldn't record any chroma. As a first step we scoped the down-converted chroma output at pin 5 of IC201. There was only a very small signal here, about 10 mV peak-to-peak. This signal is fed to an emitter-follower buffer stage via a low-pass filter, LPF202, so the input to the filter was disconnected to see if the problem was due to excess loading. There was no change in the signal level. All the inputs to the chip were then checked and found to be correct, suggesting that the chip itself was defective. As we had another FV20 in the workshop we borrowed its main converter chip. Still no improvement. After spending some time going round in circles we decided to replace the low-pass filter with the one in the second machine. To our relief a healthy 900 mV peak-to-peak chroma signal then appeared at IC201's output pin.

Intermittently rewind then turn off: This machine would, with or without a cassette inserted, intermittently go into rewind or fast forward then switch to standby. If there was no cassette present when this happened the machine would light the cassette symbol in the display. Checks showed that the end sense condition at pin 43 of the syscon microcontroller chip IC6001 was incorrect. The d.c. pull-up was low because R628 (120 kΩ) had gone high in value.

Wouldn't load a cassette: We found that the right-hand deck infrared sensor was open-circuit.

Ferguson FV21

No cover over mains fuse, beware!: Note that early machines have no cover over the mains fuse which is always live.

Dead machine: Now we've had so many of them in with the STK5481 hybrid regulator chip faulty that I didn't bother to carry out any checks, I simply fitted a new STK5481. Guess what? The machine was still dead! Checks around the STK5481 regulator then showed that the voltage at pin 5 of connector CN801 was missing. This voltage is applied via resistors to pins 7 and 10 of the chip. When the source of this voltage was traced back to the mains transformer PCB we found that R4 and D9 were OK but the 10 Ω surge limiting safety resistor R1 was open-circuit. Replacing this restored full operation. Maybe the STK5481 had been the cause of this, but I wasn't going to find out and left the new one in!

Parts of playback picture missing: There was an interesting fault with this machine: intermittently part of the playback picture, sometimes all of it, would be missing. The fault was different each time the machine was put into the playback mode. A scope check showed that during part of each field there was a complete absence of signal. When we looked at the FG signal from the lower drum we found that FG was there but no PG. The cause of the fault was a 3.3 µF capacitor on the lower drum PCB.

Ferguson FV26D

Accepted tape but no deck functions: When eject was selected the cassette was returned but a loop of tape was left caught around the left-hand roller guide. Quick checks at TP801 and TP802 showed that the 12 V and 5 V supplies were missing. Circuit protectors CP601 and CP602 were OK, the culprit being the STK5481 regulator chip. This chip is prone to failure and it pays to check the voltages around it before carrying out any other work on the VCR. As a guide, the following voltage readings were obtained with a working machine, using a 10 MΩ/V DMM. Pin 1, 3.4 V; pin 2, 8.4 V; pin 3, 5.2 V; pin 4, 12 V; pin 5, 0.2 V; pin 6, 12 V; pin 7, 13.2 V; pin 8, 18 V; pin 9, 11.9 V; pin 10, 13 V; pin 11, 18 V; pin 12, 0 V.

No playback video: This machine came from another dealer. On test we found that the E-E picture seemed to be slightly overmodulated. There was a good f.m. envelope at pin 39 of the PB20166C chip IC101. Further checks on this i.c. showed that the output at pin 10 wasn't passing via LPF102 to reach pin 21. As we didn't have this filter in stock we had to order one from CPC. It arrived next day. When we'd fitted it the machine was back in good working order.

Machine would shut down: When the on button was pressed the deck mechanics shuffled back and forth for a couple of seconds and the channel indicator appeared briefly then the machine shut down. The cause was loss of the 12 V and 5 V supplies because the STK5481 chip was faulty.

Tape spillage: This was because the brakes didn't come on in the rewind and fast forward modes. A check with an identical machine showed that in the faulty machine the windmill, item 54 in the exploded view, turned both ways while in the good machine it turned only one way. The cause of the fault turned out to be a broken clutch spring, item 55 part number PQ42002. Note that in this machine the capstan motor has to be removed before the loading block assembly can be lifted out.

Half-loading arm sticky: It was very sticky, to the extent that most of the time during play it was outside the tape path. At other times the loading sequence would be aborted because the arm jammed with the guide poles. When the cam gears were stripped down I found that the grease was hardening. So a complete clean and relubrication was carried out. After this the machine played faultlessly. Unfortunately the audio/control head was so badly adjusted that the machine would play only its own recordings. It's OK now, but there's a pile of tapes that are of no use!

Dead machine: A check in the power supply of this dead machine revealed that R1 was open-circuit. It's a safety resistor, which is mounted near the mains transformer.

Ferguson FV30

Dead: This machine was only about five months old and came in a box covered with labels giving the owner's name and address. There were further labels on the machine, leads and remote control unit. These also had messages like 'do not drop or scratch'. I'm glad he told me – I'd never have thought! Anyway the 1 AT mains fuse had died and although the bridge was intact the start-up supply rectifier D11 was found to be leaky. A new fuse and BA159 diode put matters right.

Weak or no E-E signals: If the off-air (E-E) signals go weak or are lost in a snowstorm, check for dry-joints at the pins of the modulator/booster module. It's mounted on the main PCB and it seems that the pins can be strained by plugging and unplugging aerial cables.

Went into stop mode during play: This machine went into the stop mode after a few seconds of play, record, rewind etc. The fault in this machine was caused by the supply spool optosensor.

Rolling on pre-recorded tapes: This machine worked all right with its own recordings and could be set up for near perfect f.m. with an alignment tape, but on all other pre-recorded tapes there was rolling as though the TV set was tuned to the wrong channel. If the micro-controller chip IT01 has been changed you will have to follow the instructions in the manual on resetting the head switching pulses. BT08 has to be shorted to chassis for 1 second: this starts an internal programme within IT01 to set up the chip for the deck. It's quite hard to find BT08, which is at the right-hand side of IT01 on the component side. There's a convenient chassis pin next to it. Surprisingly enough, doing this cured the fault.

Intermittent shutdown: A recent case of intermittent shutdown, with the tape remaining laced up and all the motors stopped, was cured by fitting a new reel optocoupler and mode switch.

Playback OK, no E-E or record: Playback was OK but there were no E-E or record signals just snow. Checks showed that the 12 V u.h.f. band switching voltage at pin 8 of the tuner was missing. TT06 (BC558) was open-circuit.

Power supply blown up: When the kit of power supply parts had been fitted the 12 V line could be set up correctly, so the power supply was connected to the rest of the circuitry. But there was no clock or mechanism activity. Checks showed that the voltage on the 7 V line was low, the other supply voltages being correct. The chopper transformer LP40 was faulty.

Switching-mode power supply failure: After we'd replaced all the usual items that fail when the switch-mode power supply dies the voltages were pulsing and the machine was still dead. Replacing CP38 (470 µF) cured the fault.

Ferguson FV31

No-go, no light-up condition: If you come across this type of machine with a no-go, no light-up condition, check the supply voltage at pin 32 of the microcontroller chip IK60. The chances are that you'll find little or no voltage here because the BC337 regulator transistor TK44 has failed. Also check its 1 Ω, fusible series feed resistor RK44, which is a safety component.

Mains fuse blackened, chopper OK: Replace the mains rectifier's 150 µF, 385 V reservoir capacitor. It tends to flash over when the mains voltage is applied. A cold test will suggest that it is OK.

All functions OK except stop: Playback and record were fine, but when stop was selected the machine might carry out any function. It would sometimes switch off, and at other times perhaps go into reverse picture search. If a cassette was loaded and the machine was left in the stop mode, it might set off by itself after a while. A replacement HD614081S microcontroller chip cured the fault. We later learnt that the fault had started after a storm during which the power supply to the house had been struck.

Ferguson FV31R

Power supply fault: The installer complained that this new machine was dead. On the bench that ever so nice power supply pumped and whistled at me. With the covers removed I switched the machine back on at the mains and it started up correctly. My next fault-finding step may seem odd to those of you who don't know these VCRs: I tapped the tuner/i.f./signals PCB that sits across the top of the machine. It could then be made to stop and start. My experience has been that, however strange the symptoms, this is very often the area in which the cause of the fault lies. After looking at thousands of perfectly good joints I eventually found that there was an intermittent short in the tuner/r.f. amplifier. I should have looked there first of course.

Mechanism problems: I'm not impressed by the mechanism in these machines. Extensive use is made of plastic components and this can lead to trouble. The initial problem with this one was that a cassette had stuck in it. It was the owner's fault – he'd stuck stickers all over the cassette rather than in only the specified positions. Our field engineer managed to extract the tape but then found that the mechanism was out of sync, something that afflicts these machines. Because of the design, in the event of even a small snarl-up there's a chain reaction of breakage in the rest of the mechanism. This machine required a new pair of loading rings and retiming.

Dead VCR, power supply whistling: The card said that this machine was dead. When checked on the bench the power supply was pumping and whistling at me. With the covers removed I switched back on at the mains to find that the machine started up all right. Now for a bit of highly technical fault-finding: I tapped the tuner/i.f./signals PCB that sits across the top of the machine. Doing this would stop and start the machine. In my experience this is very often the area in which the fault lies, however strange the symptoms. After looking at thousands of perfectly good joints I found an intermittent short in the tuner/r.f. amplifier.

Appeared to have damaged heads: However, a new drum produced no improvement. A look at the circuit showed that the outputs from the heads enter the TA7772P preamplifier chip IQ80 at pins 2 and 6, the output is appearing at pin 10 where there should be a 0.3 V peak-to-peak f.m. waveform. In fact the output from only one head was present, a straight line being displayed where the other head's output should have appeared. So we had a head-switching problem or a lower drum fault. The head drum flip-flop signal from the servo panel is connected to the signals panel at pin 8 of connector BW04. There's a test point, BW11, and the amplitude should be 3.6 V peak-to-peak. The waveform was missing, however. So it was back to the servo panel, where the drum FF square wave is generated by the microcomputer chip IT01. The output, at pin 14, depends on pin 8 receiving a pulse from the drum optocoupler via IM01. As there was a signal at pin 8 but not at pin 14 we replaced IT01. To our relief this cured the trouble. Unfortunately the manual provides no details of the voltages or waveforms around this chip. Note also that two types have been fitted in these machines. If, as in this one, there are two small subpanels mounted vertically on the servo/power supply panel, use type ZC93168P. The other type is EF6801U4DTD243. Conclusion: if a badly damaged head is suspected, check the drum FF signal.

Breaking back-tension arm: This machine had a nasty habit of breaking its back-tension arm as the deck mechanics mistimed themselves, no matter how carefully the instructions in the manual were followed. We noticed that when the machine set off in play the drum motor didn't rotate. This turned out to be a vital clue. The drum stood still because the 5 V supply to pin 2 of chip IM02 was missing. From a look at the circuit diagram this appears to be totally unrelated. The PCB layout holds the clue: the link that supplies 5 V to IM02 also supplies the pull-up resistor RT67 in the mode-sensing circuit, the cause of the trouble being a dry-joint on this link. With the dry-joint resoldered and the deck mechanics realigned yet again everything worked correctly. All that was left to do was to fit a new back-tension arm.

Picture disturbance and momentary loss of sound: These symptoms were caused by a very intermittent loss of drum sync. After an extensive investigation of the motor drive and servo circuits a chance brush against the ribbon cable connector immediately behind the upper drum produced the fault. Remaking the ribbon cable connection to the free socket cured the trouble.

Wouldn't tune in to local channels: On test we found that although all other functions worked correctly there were problems when the machine was put in the channel preset mode. The tuning display counted up through the channels numbers, but the monitor didn't display any sound or vision and the tuner wouldn't lock on to any of the channels as it passed through them. After removing the case and raising the top PCB we took the screening cans from above and below the tuner/i.f. section and examined the area carefully for print cracks, dry-joints etc., something that's common in this range of machines. Everything seemed to be OK this time, however. So checks were made at the tuner's base pins where we found that the tuning voltage (pin 11) was missing. We moved back to the tuning control transistor TT12, whose collector voltage was at 0 V, then to pin 9 of the TD6316AP tuning control chip IT20 where the voltage was 3.2 V. This was turning TT12 fully on, hence the absence of its collector voltage. A new TD6316AP chip restored correct tuning.

Machine dead but fuse OK: The characteristic squeal was heard as the power supply started up. Failure to operate was due to loss of the 5 V supply to the timer microcontroller chip. Fusible resistor RK44 (1 Ω) was open-circuit.

No or unstable playback: Check whether the insulating washer beneath the head of the screw that secures the top PCB is missing. If it is, tracks short-circuit to chassis.

Dead: No functions worked and there were no displays. Checks showed that the switched 5 V supply was missing – all the unswitched supplies were present and correct. The culprit was TP73, which was open-circuit. It's on the timer/display board.

Dead, no clock display: Checks on the supply voltages revealed that the on/off monitor wasn't working. We traced back to the timer display board and found that RK44 (1 Ω fusible) was open-circuit. A replacement restored normal operation.

Dead, power supply appeared to be OK: We found that the machine would power up via a variac at 235 V or less. For the machine to get going, a start-up supply must be developed across CP14 before the 12 V supply is developed across CP38. If not, the start-up oscillator TL16/17 quickly dies. Zener diode DP21 had changed its zener point from 9.1 V to 11 V. As a result the voltage across CP38 was incorrect, locking up the power supply. As a replacement we used the heftier BZX61 1.3 W type.

Would jam intermittently: The machine would then turn itself off. It uses optocouplers as mode switches. A subpanel mounted on the power supply/servo PCB feeds the pull-up resistors associated with these optocouplers. The cause of the fault was a dry-joint on this panel.

Dead, fuse OK: All the transistors in the power supply were OK. The start-up oscillator signal could be seen at the junction of RP28/29 but not at the base of TP28. The diodes were all OK except for one – DP16 was very leaky! Incidentally, never leave the base of TP28 disconnected when power is applied.

Failure to load: The loading fault was cured by removing the debris that jammed the entry guide. I then gave the machine a good clean-up and tested it. Playback produced normal sound but no picture: the display consisted of white diamond shapes, about the size of a large postage stamp, on a black background. Using the AV connector output produced the same result. I hinged out the top board and attached the scope probe to test point BW10. Normal video was present. Refitting the board produced the same fault symptom. The cause of the trouble was traced to the board fixing screw at the right-hand side, where it screws into the cassette housing. It was shorting an adjacent PCB track. An insulating washer beneath the screw cured the problem. You could get the same trouble with Models FV30 and FV32.

No E-E or V-V signals: I found that LW03 (22 μH) was open-circuit, removing the 12 V standby supply (U4) to the power splitter. Unfortunately the owner had 'looked at it' and had neglected to refit the insulating washer under the right-hand PCB's mounting screw. The TA8607P FM amplifier chip IQ40 had objected and died.

Wouldn't store any channels: The cause was traced to the BC337 transistor TK44, which showed signs of distress (it had bubbled out). Standby battery XK59 was responsible for TK44's distress. There should be 2.4 V across it but the reading was 1.2 V.

No results and no display: We found that there was no 5 V supply because the TIP120 transistor TP73 was open-circuit.

Ferguson FV32

Machine mechanically out of sync: This machine was confused: it continually tried to unlace, even though it was fully unthreaded. It wouldn't accept or eject a cassette. The clock and signal circuits worked all right and all the supplies were present, although it looked as if work had been carried out on the power supply PCB recently. Metering between the mode select switch and connector BT04 revealed that the ribbon cable from the deck to the system controller was open-circuit. The ribbon cable's part number is 556 555 00.

Would only partially load tapes: This machine would load a tape and both rewind and fast forward were all right. But when play was selected the machine would start to thread then suddenly stop in its tracks, leaving the tape partially loaded. We found that the timer i.c. was supplying a servo stop signal to pin 26 of the system control/servo chip

IT01. So a new HD614080S chip was ordered, but when fitted made no difference. Maybe the machine was entering an alarm/stop mode, although the drum and capstan motors still rotated. We decided to get out the scope to check waveforms and found that there were no drum PG pulses at test point BK02. So we replaced TK25, still to no avail. At this point we decided that it had to be the drum motor. But have you seen the price?! We carefully dismantled the direct-drive drum motor and found that the PG pick-up consists of two single printed circuit coils, with a small surface-mounted electrolytic capacitor (C6) connecting one end to the driver chip. This capacitor was open-circuit. Its value is $3.3\,\mu F$, rating $50\,V$. We used a $10\,\mu F$ subminiature type from an Amstrad 4600 i.f. can as a replacement. The result was perfect – and a fraction of the cost of a new DD unit.

VCR would load up, start to play, then stop: This machine, a Telefunken VR4940, is similar to the Ferguson FV32. The one we had would load up, start to play, then stop. I suspected the reel sensors, but replacing them made no difference. Replacing the mode switch put matters right.

Tape chewing: When I tried it none of the functions worked. The cause was a wire broken off the loading motor, which made me think that someone had been there before me. After reconnecting the wire I selected play. A strained sound came from the tape, so I removed it and inserted an empty cassette. When play was selected I could then see that the take-up spool was rotating at the full fast forward speed. The faulty component was the MC14094 chip IT62.

Ferguson FV32L

No r.f. output: Now dry-joints in the r.f. amplifier/modulator unit can cause this, and I was thinking about it when I picked up the job. But this time the cause of the trouble was quite different. There was no tuning as the pulse-width modulated drive from IT20 was missing. Furthermore, the following BC558 transistor that supplies pin BT of the tuner was short-circuit all ways round. The tuner had a short to earth from pin BT and all three devices had to be replaced.

Dead machine, mains fuse F1 blown: As no fault could be found and the fuse wasn't blackened a replacement was fitted. At switch-on the new fuse flashed violently. There was still no readable short-circuit, so the chopper transistor was removed from the board and another new fuse was tried. It again blew straight away. The only likely causes of the fuse blowing were the mains rectifier diodes and the 150 μF, 385 V reservoir capacitor CP07. The latter was our first suspect and when it was removed a new fuse held. When tested on our capacitance meter the old capacitor seemed to be all right, but a replacement cured the fault. The old one must have been flashing over internally when the mains voltage was applied.

When first switched on the drum motor hunted and there was picture tearing and line slip: The picture began to stabilize as the machine warmed up, but it never became stationary. Since the fault was worse when the machine was cold I got out the freezer and hairdryer and began my attack. Any heat in the servo area lessened the effect of the fault, but no amount of freezer made it worse. So I removed the bottom cover and applied gentle heat to the drum motor PCB. The fault then disappeared. When a shot of freezer hit C6 (3.3 μF) the drum almost spun off the deck. After replacing this subminiature capacitor perfect control at all temperatures was restored.

Irregular jumping between LP and SP: Finally the machine creased a tape. It had already received attention from someone who had resoldered several connections on the servo PCB. After a bit more resoldering around the capstan chips the machine seemed to work all right. But after an hour I gave the board a push and the fault returned. This time I was able to localize the source of the fault to the area around the LM393 capstan speed comparator chip IT45. Checks here showed that there was a jumping voltage at pin 2. When a spotlight was trained on the area I found a very small crack in the print between pin 2 of IT45 and CT48 (1 μF). Repairing this finally cleared the fault. So you not only have to look out for soldering problems with this range of VCRs but also keep a watch out for cracked tracks.

Incorrect tracking: The reported fault was incorrect tracking. When we tested the machine we found that playback of its own recordings was poor. Replacement of the upper drum produced little improvement, but replacing the lower drum cured the fault.

Ferguson FV41

Roll in pause or search mode after a few hours: If pause or a search mode was selected when the machine had been running for a few hours the picture would roll. A check on the video signal showed the reason for this: in these modes there was a field sync pulse only on alternate fields. The pseudo-V sync pulse that's used in these trick modes is generated by the servo microcomputer chip and is fed to pin 28 of the luminance processing chip IC26. Even when the fault was present the 50 Hz pseudo-V signal was correct. Checks around IC26 showed that the supply voltage was high at 6.4 V instead of 5 V. A small regulator circuit in the servo area of the main PCB produces this 5 V supply from the U7 8 V line. The regulator circuit isn't the same as that shown in the provisional circuit diagram. We found that the faulty component was the 2.7 V zener diode DT53 which had nearly 4 V across it.

Chroma fault after 2 hours: After playing back for a couple of hours a chroma fault would appear. The symptoms were as follows: on the left-hand side of the screen the colour remained correct but further across to the right its phase changed, e.g. blue changed to orange. The fault would clear if the tape was stopped and then restarted or the cue and review functions were used. As the conditions around the chroma signal processing chip IC08 were correct we decided to swap it with one in another machine. The fault moved with the chip, proving that the latter was defective.

Rolling pictures with some tapes: The complaint was rolling pictures with some tapes. A look at the f.m. waveform at BF14/6 revealed all the switching points were incorrect. Adjustment does not involve presets and an oscilloscope with these machines. You simply play back an alignment tape and press and hold 0 and 8 simultaneously. Then, after about 2 seconds, press stop. This completes the alignment. A tweak on the left-hand guide to straighten up the f.m. waveform completed the 'repair'.

Ferguson FV42L

Poor record: This new machine could be tuned in to the TV set and played back a pre-recorded tape, but with record and E-E operation the picture was terrible. Checks showed that there was a very distorted video signal as coil FW11 was open-circuit, in fact physically damaged.

Noisy picture in playback of VCR's own recordings: The complaint was of a noisy picture during playback of the machine's own recordings. After many hours we saw the fault. White dots washed across the picture, at times looking almost like a test pattern of vertical lines of dots. The symptom was also present in the E-E mode whilst a recording was being made. We noticed that there was a slight delay between the machine starting to record and the appearance of the dots. The cause of the problem was spikes produced by the audio bias/erase oscillator appearing on the 12 V supply to the tuner etc. The culprit was CS29 (10 nF) which is connected between the base of the oscillator transistor (TS26) and chassis.

Dead, but OK without top cover: This machine would work when the top cover was removed. All then seemed to be well – until our trusty PCB whacking tool was brought to bear on the main board. We soon established that the regulated 5 V supply would go missing. This comes from TT64 (BD435), which is fed with a 7 V supply. TT64's three legs were all dry-jointed.

Ferguson FV61LV

Various faults, some intermittent: Here's the list – excessive fast-wind torque (to the extent that once-wound tapes were too tight to play!), no tape end sensing, failure to accept a cassette, failure to eject a cassette, no capstan operation and switching off with no mechanical movement. Just about every operation is controlled by the ST90T30-QFP80 servo microcontroller chip IT01. It's a many-legged, four-sided flatpack device mounted on the lower PCB. We found that it was thermally faulty.

Loud hum or drone from mechanism: This machine's mechanism produced a very loud hum or drone. The cause was sticky dirt in the capstan motor bearing. It was generating so much friction that there was burning – the capstan was discoloured at this point. Stripping the motor, cleaning and lubricating provided a complete cure.

Poor erase: Because the strength of the erase bias varied, bits of the previous track were left superimposed on bits of fresh recordings. The cause of the fault was the BC337–40 transistor TL01, which is mounted on the PCB beneath the deck. Replacement set everything to rights. We discovered that the transistor's gain was varying – it was in fact heat sensitive.

Ferguson FV62

Dead, no display etc: The cause of the trouble was in the r.f. converter/tuner, where the SDA and SCK lines were shorted. A replacement module cleared the trouble.

Wouldn't accept tapes: The cause was simply dust in the timing slots at the bottom of the drum assembly. A thorough clean restored the normal operation.

'No E-E sound': The job card said 'no E-E sound', but on test there seemed to be plenty. After a while the sound disappeared. The repair shop engineer had replaced the tuner, and was surprised that this hadn't cured the fault. I traced the sound path from the main board to the little PCB that has the audio record/playback amplifiers on it. There was sound at the input to this board, but not at the input to the chip. C017 had never been soldered at one end. Resoldering it restored permanent sound.

Ferguson FV68TX

Playback sound switches to mono: This is becoming a very common fault with modern VCRs. Because the entry and exit guides had worked loose the playback envelope was very poor. Setting them up and locking them cured the fault.

No hi-fi recording, replay OK: There was no hi-fi recording, though playback of pre-recorded hi-fi tapes was OK. We suspected the TEA5712 chip in the head amplifier, but fitting a replacement made no difference. Luckily another VCR of the same type was in the workshop. Swapping the drum assembly over proved that this item was the cause of the trouble.

Dead, mains fuse open-circuit: The chopper transistor was short-circuit. The cause of the failure was attributed to diodes DP91 and DP92 which are wired in parallel. When they'd been replaced the chopper control chip IP01 still didn't work, however. To cut a long story short, we had to replace FP01, TP34, TP35, DP16, DP11, RP27 and, again, IP01. To my relief the machine then worked.

Ferguson FV71

Old sound track not erasing: The bias oscillator was of course responsible. Its circuit, consisting largely of surface-mounted components, is on the small PCB to the right of the deck. We found that the transistor, IT001, was short-circuit and its 18 Ω feed resistor RL02 was open-circuit. A bell began to ring about a modification in this area. Solder up the oscillator coil LL01, and change the value of C002 from

1.4 nF to 2.7 nF (part number 20136340). This modification improves the oscillator's ability to start. We also soldered the erase head wires directly, as failure in this area is common with other machines we've had in. The fault can be intermittent.

Failed to come back on after power cut: This machine had failed to come back on after a power cut. There were no dried-up capacitors this time: TP91 (2SA1020) was short-circuit collector-to-emitter. A replacement restored normal operation.

Refusal to start: Try replacing DP06.

Ferguson FV71LV

Fuzzy picture: The picture gave the appearance that there was no output from one head. Cleaning the drum made no difference, nor did a replacement preamplifier chip (IR01). We finally found that CR05 was leaky.

Dead machine: Fortunately replacing IP01, TP01, RP18 and RP21 in the power supply restored operation without the blow-ups you usually get with switch-mode power supplies of this type.

Power supply tripping: We were told that this machine was found to be dead after a thunderstorm. When I tested it the power supply was tripping. The cause of this was RP18 (1.5 Ω), which had risen in value to approximately 8.8 Ω. A replacement stopped the tripping and brought the machine back to life.

Power supply dead but ticking: We replaced CP007 (10 μF) and the 10 μF capacitor on the print side of the board. But the fault was still present. The culprit turned out to be CP008 (100 μF).

Finlux

Finlux VR3724

Wouldn't play and/or damaged tapes: On inspection I found that the capstan motor was not turning. The cause was Q5102 which was open-circuit.

Heavy black horizontal lines on playback: When we carried out scope checks around IC101 we found that the waveform at pin 41 was OK but the waveform that returned to the chip at pin 42 was noticeably faulty. The cause of the fault was traced to the $1\,\mathrm{k}\,\Omega$ resistor R1007, which biases buffer transistor Q1003, having gone high in value. A new resistor cured the fault.

Loss of channel memory: This machine wouldn't memorize or change channels or tune in properly. A replacement memory chip (IC602) put matters right.

Colour flashing with pre-recorded tapes: The cure is to replace the video processing chip IC101.

Fisher

FISHER FVHP5100
FISHER FVHP520
FISHER FVHP615
FISHER FVHP710
FISHER FVHP906

Fisher FVHP5100

No remote control functions: Try fitting a three-core mains cable. If this doesn't cure the fault replace IC151.

Failure to play with auto-ejection of the cassette: A new idler assembly cured this. But there was occasional colour dropout when playing a known good tape, leaving a white-spotted monochrome picture. Scope checks showed that one head's f.m. signal was missing at the output from the BA7253S three-head amplifier and switching chip. By the time the head-switching signal at pin 1 could be measured, the fault had cleared and didn't return for a month – in the customer's home, of course. Replacing the upper drum assembly cured the fault. The original head assembly appeared to be in perfect condition. There must have been an intermittent break in a head winding or head ferrite, possibly caused by the tape brought along by the customer – it had been spliced using Sellotape.

Jammed loading mechanism and no capstan rotation: This machine had two faults. First, it would shut down after six seconds because the loading mechanism was jammed. I found that the flat of the mode switch spindle had turned 180° then presumably jammed. Securing the switch and clearing the jam was all that was required. Secondly, there was no capstan rotation. This was caused by a leaking electrolytic capacitor (C1) on the motor plate. It had leaked electrolyte, removing the supply to the SA3001 motor drive chip. Replacing C1 and fitting a wire link (after cleaning) saved the cost of a new motor.

Fisher FVHP520

No operation: A no-go fault in a VCR can be caused by a multitude of things, from a blown mains fuse to some fiendish problem or other. This no-go machine's loading motor was not being driven. Our investigation showed that one of the motor-drive transistors, Q868, was burnt to a blob – for the very good reason that the control microcomputer chip was simultaneously giving load and unload commands! As well as the microcomputer chip and Q868 we felt it prudent to replace both the multi-switch chips IC862 and IC864. The loading motor itself was unscathed and cheerfully did its stuff when the repairs were complete.

Fast forward OK, no other functions: The cause was at first thought to be a keyboard or system control fault, but subsequent checks ruled these out. A mechanical fault seemed likely therefore. Turning the machine on its side and pressing play proved fruitful as it then worked. The cause of the trouble then became apparent: the loading belt was slipping. Fitting a replacement restored normal operation.

Low reel torque: The cause of this can be difficult to find on these machines. We've found that it's usually due to reel spool tyre wear. Cleaning with alcohol and drying will prove the point.

Fisher FVHP615

Chewing tapes and poor rewind: This is a common fault with this model. We replaced both idler assemblies and the machine was returned to the customer. A few weeks later it came back with the symptoms of wow and flutter and tape chewing, but only at the very start of a tape. This had started to happen after replacing the idlers. When we played a fully rewound tape in the machine we found that the capstan speed varied and that as the tape was being taken back into the cassette it slipped down the guide and was crushed on the lower edge of the housing. This indicated high take-up torque, which was found to be over 200 grams. The cause of the problem was the take-up reel. These reels have a clutch at the bottom and this one had virtually seized.

No record fault: The switching signal at pin 16 of the LM6416E-239 chip IC503 goes low in the record mode, turning on the REC 9 V switching transistor Q513. The REC 9 V line was low because Q513 was faulty. A BC640 gave good results.

Juddering E-E pictures, strange video waveform: Here was another case where the cost of spares pushed us into mending rather than replacing. The trouble was juddering E-E pictures and a very strange video waveform from the tuner module, which includes the u.h.f. tuner, the i.f. section and the demodulators. The source of the trouble seemed to be in the vision a.f.c. section, but the service manual gives no circuit details – the black-box approach. Sanyo, who handle Fisher spares, quoted us £93.73 plus VAT (trade price) for this mystical module. We went into it with a hairdryer and freezer and found that C12 (0.47 μF, 50 V) was the cause of the trouble. A 0.47 μF, 35 V tantalum bead replacement put all to rights.

Fisher FVHP710

Fuse blows for no reason: If you find that fuse F902 (500 mAT) on the mains transformer PCB has blown for no apparent reason, check for dry-joints at the PCB connections of pins 10 and 12 of connector PV903 on the regulator panel. Considerable current passes through these pins, and some of it will be diverted through F902 and C928 in the event of a momentary open-circuit. The other joints on the plug/socket connections to this board are also worth checking.

No display or channel indication: This machine worked to an extent but lacked any trace of a display or channel indication. Checks at IC101 (LC6502B633) showed that the supplies were present and the clock oscillator worked, but the reset line (pin 19) was in a permanently low state. The reset pulse is produced by Q009/Q010 on the preset/tuning board. The transistors were healthy but C017 (0.01 μF), a disc ceramic capacitor across the reset line, measured 5 kΩ. Replacing this cured the problems as the chip could now initialize.

Intermittent symptoms: There were various intermittent symptoms with this machine as follows: failure to accept tapes, the counter zeroing, and no functions. The cause was dry-joints at plug PV903 on the power supply panel.

Loading arms problem: Although rewind and fast forward were normal, when play was tried the loading arms began to move forwards then stopped, returning almost at once to rest. A check on the supply lines showed that the 5 V rail to the micro/syscon departments was high at 8 V. This supply comes from the STK5431ST multi-regulator chip in the PSU. As the other sections of the chip were clearly all right I decided to carry out a modification. Pin 1 of the chip was disconnected from the print and taken to the input of a standard three-pin 5 V regulator whose output was connected to pin 1 of PV904, the 5 V line. The regulator was then screwed to the flat pad at the top of the power supply heatsink. This course was adopted because the customer required the machine in a hurry. I recall being told in my training days that 'a bodge is only a bodge if it can be considered unreliable and/or unsafe – otherwise it becomes a modification'!

Fisher FVHP906

No playback colour: A test recording played back on another machine showed that it didn't record colour either. All relevant l.t. rails were correct and d.c. checks at the pins of the ICs involved proved inconclusive. Scope checks revealed that the oscillators were all running, at the correct frequencies, and that while chroma was going into the main processing chip IC203 it wasn't coming out. At this stage of chroma signal processing only two chips are involved, IC203 and IC204. We replaced both in turn and found that IC204, an LC7342, was the culprit. The same chip was responsible on another occasion, with a Model FVH-P907.

Clock worked, nothing else did: The job ticket read 'not playing' and the truth was that you couldn't switch the machine on. We found that the power control input to the main microcomputer chip was correct, but nothing happened at the power control output. Replacing the chip cured the fault.

No functions: At switch-on the power light lit then, after 6 seconds, the machine went to standby. In addition there was no loading motor shuffle. If you get these symptoms check for the presence of the 12 VSW supply at pin 2 of PA902, and the power control condition at pin 4. If the 12 V line is low and pin 4 is at 4.5 V, change the STK5466ST chip IC901.

General

General VGX520

Tuning problem: The VGX520 was a complete stranger to us. Its problem lay in the tuning department, which would not self-seek no matter how much the tuning-up button was pressed. The culprit was the tuner/clock/display microcomputer chip, type μPC7519, which lives near the fluorescent display panel. It's a surface mounted chip which can be removed only by cutting off all its legs (use a very sharp scalpel to do this) and clearing up all the stumps afterwards. When the chip had been replaced we found that we could only seek as far as the top end of Band I, as indicated in the display panel on the 'VL' bar graph. To progress to Bands III (VH) and IV/V (U) we had to fit a switch across the vacant key connections marked 'band' at the bottom of the key switch PCB. Once all available programme settings have been assigned to u.h.f. in this way the switch can, if required, be removed since the new chip is now programmed.

Intermittent tape chewing: This machine uses the Panasonic NV430 tape deck, so the fault could apply to either model. When the stop button was pressed during playback the tape would stop but not unload. If the tape was then ejected the loading poles would unload but the tape would not be wound back into the cassette. The cause of the fault was traced to high-resistance contacts in the mode control switch.

Hum bar: There was a slight hum bar in the E-E mode and a stronger hum bar with colour reversal in it in the playback mode. Electrolytics C1002 (4.7 μF, 40 V) and C1003 (47 μF, 40 V) in the power supply section were the cause. As the panel is difficult to get at, replace both capacitors – either can cause this fault, and it would be annoying to have to take the panel out again.

Moving band on playback: The playback picture had a moving band, like a hum bar, in which the colour was missing. The cause was traced to C1003 and C1007 on the power supply PCB.

Colour for only 5 seconds: This machine worked well in the E-E mode, but when playback was tried the colour was present for only about 5 seconds after which the picture became black and white. Call it intuition if you will, but I suspected the power supply and was rewarded when the colour returned for a few seconds after a shot of freezer on capacitors C1002 and C1003. Replacing these two 47 μF capacitors restored good pictures in all modes, but as I was in the mood I decided to replace all the electrolytics on the board.

Goldstar

Goldstar GSE1290IQ

Poor playback picture, tape chewing: We replaced the pinch roller and arm assembly and gave the deck a good clean/service. After a soak test the machine was returned to its owner. A couple of days later it came back, again because the playback picture was poor. After several tries at loading and unloading we found that the back-tension arm sometimes stopped before it reached the play position. As it is mechanically linked to the main cam we decided to replace the mode switch. This cured the fault for good.

Won't accept or eject tape: Remove diode D521 and fit a wire link in its place.

Intermittent refusal to accept a tape and intermittent switching from SP to LP: I suspected the mode switch but decided to replace the whole loading block, which includes the mode switch and loading motor. The speed fault was cured by replacing the pinch roller.

Goldstar RQ504I

White spots and lines on playback picture: The static discharge brush had obviously fallen off. Wrong again! As numerous checks on the earthing failed to bring anything to light, attention was turned to the bottom of the machine. I temporarily connected another earth wire to the shaft of the drum spindle. This made a difference to the spots, but only for an instant. So I took the motor apart and found that one of the stator turns was loose and had been rubbing on the rotor/magnetic flywheel. Redressing and fixing it with some superglue kept it away from the flywheel, clearing the discharge effect.

No drum rotation: It was soon apparent that the motor was faulty, as the chip on the motor PCB had a hole in it. This is the second faulty motor we've had in these comparatively new machines. When the cause of the fault is not as obvious as in this case, the important motor connector pins to check are 4 earth, 5 always 12 V and 6 where the control voltage should be present. When this reaches 1 V the motor should be turning. The other pins are for the PG and FG pulses, playing no part in starting the motor. Pin 1 of the connector is identifiable by the black wire. The part number for the lower drum/motor assembly is 413–220A.

Poor colour with pre-recorded tapes: This problem occurred after fitting a new drum. So we set the machine up as per the book, but the results were the same – the colour in the playback picture was still out of phase. Advice was sought from GoldStar technical, who are very helpful. 'Try another drum' was their recommendation, but again the results were the same. Time for some head scratching. We had already checked the chroma signal path and had found no faults. The only thing that didn't match up with the measurements given in the manual was the FM envelope, which was smaller than specified although beautifully formed. Fortunately a similar machine came in for repair, so we were able to carry out comparison checks. With the good machine the FM was not appreciably larger and wasn't as well formed as with the faulty machine. So we decided to change all the subassemblies, again without any improvement. When a third new drum was fitted the machine worked perfectly. So was this two faulty new drum assemblies? Well, not actually faulty but incorrectly assembled. The upper drums were on the wrong way round. Correcting this produced perfect results. Should you get this sort of problem, note that the upper drum is coloured green on one side and white on the other. The upper face of the lower drum is coloured green. White goes to green, not green to green as with the faulty drums.

Grundig

Grundig 2 × 4 super

Wouldn't accept cassette or carry out any function: We found that the relay didn't stay in when power-on was selected. The cause of the trouble was that C446 (470 μF) had lost capacitance.

Stopped playing after a few seconds: This was because the head drum was rotating very slowly. A quick check on the supply lines revealed a 5 V ripple on the 15 VR rail. C451 was low in value. It should be replaced with a Grundig part (from GCAS) – if a standard capacitor is used there is still some ripple and the servos take a long time to lock in.

No or poor braking: Check the +12 VR line at pin 18 of the power supply. In the machine we had the supply was intermittent because the relay contact was poor.

Handset intermittent: Flexing the unit had some effect, and we envisaged dry-joints. During the fault we found that the battery voltage fell from 6 V to 2 V, which we assumed to be due to poor internal contact between the cells. We've since had one other case. Both these units still had the original battery fitted. Replacement batteries of different manufacture do not seem to have this problem.

Cassette-in switch wasn't making: The customer brought this machine along in a great hurry as he wanted to record a programme. He said it wouldn't load a cassette. The cause was quickly found – the cassette-in switch CL wasn't making. As a temporary measure to enable the customer to make his recording we shorted the switch out by linking pins L1–6 and 7 on the switch board. This restored normal operations including unloading. To load it was necessary to insert a cassette then press 'tape', after which the machine would load. Note that if the CL-closed signal is not present no functions are available (play, wind etc.) even with a loaded tape.

Wouldn't record: This machine would play back, search and wind/rewind, but it wouldn't initiate record or thread in. The problem was that the left-hand keyboard didn't respond as button 1 was permanently 'on'. In fact the pushbuttons jammed in when pressed – buttons 4 and 7 exhibited this to a lesser extent. The keyboard is part of a large board that's secured to the front panel by plastic blobs and contains all the other front panel pushbuttons. Grundig Service told us that a repair kit is available but that relieving the front panel cutouts around the buttons is usually adequate. This provided a perfectly acceptable cure.

Power supply fault: After removing the 25A car fuse and replacing the intermittent on/off switch I eventually found the real cause of the fault – the coupling capacitor between the TDA4600 chopper control chip and the chopper transistor. I rarely see one of these machines.

Grundig VS180

Reel motors running continuously: Check C301 or C305 in the motor drive circuit for being leaky.

Both spools were running fast and there were no other functions: After threading up manually the machine would unthread and eject the cassette, with both spools continuing to run fast and not switching off. This would tend to suggest something wrong with the cassette eject switch, but the threading motor stopped after eject so the micro-computer chip did that all right, indicating that this chip was not faulty. After a word with Grundig Pete we decided that the most likely culprit was the M722 series-to-parallel interfacing chip, which indeed it was. We came to the conclusion that erroneous data was being sent to the microcomputer chip. The faulty chip can cause other symptoms depending on which data bit is corrupted.

Transport damage: This machine had suffered transport damage while being brought back from abroad. While sorting this out we found that the machine would sometimes initialize by winding forwards and backwards very slowly. The cause was eventually discovered to be the on/off switch. It switched off the 33 V supply but not the 5 V supply, leaving the clock on and the machine partially operating.

No clock/counter display: Checks around the clock chip revealed that the 256 Hz clock pulses were missing. They come into the keyboard panel from the control panel on two matrix lines designated K4 and K8, and are generated by IC245 and the associated 32 kHz crystal. We found that the 32 kHz oscillator had stopped due to shorting vanes in an associated trimming capacitor.

Tape path fault: The off-tape f.m. at the start of the rotation of the heads was very low, although the last half was OK. The paint seals on the guides were unbroken, the tape tension was normal and a replacement tension arm had no effect. A new head drum was finally tried and this restored normal operation.

Would play but didn't unthread every time: The switch situated on the threading ring FA1 worked correctly but the FB switch was dirty.

Noise bars on the picture every 30 seconds during playback: A lot of time was wasted checking the tape tension etc. before the scope came out. We then found that the control track pulses were missing because the control head was open-circuit.

Failure to record, no keyboard LEDs: Sure enough the record button had no effect, also the tape counter didn't work. The common item was IC235 (4066) on the control board, where its inputs from IC240 were being held down. To be on the safe side I replaced both IC235 and IC240. The problem with another of these machines was that sometimes at switch-on the tape would be spooled very fast for 20 seconds. If the tape was near the start the end sensors didn't work and the tape would be broken. If record was selected the machine would thread up then immediately unthread and the pause light would flash. The common clue was that when either fault occurred the +12 F supply was missing. The power supply was sensing an overcurrent on the +12 F line because C769 (100 μF) on the video module was short-circuit.

'Spots on play': The spots were confined to the centre of the screen and gave the impression of a tape-path fault. When the off-tape f.m. waveform was scoped, however, it showed that this was not the case.

Adjusting the tracking had no effect of course, so I suspected a faulty drum motor. When the motor baseplate was removed the cause of the trouble could be seen – grease had built up on the brushes. When this was cleaned off the spots had gone. The first fault report in this section found that leaky capacitors in positions C301 or C305 would result in the reel motors running all the time. In the case I had the capacitor was very slightly leaky and the reel motor rotated only when the on/off switch at the front of the machine was in the off position!

Stuck in pause: Grundig VCRs are rare visitors to our workshop. What happened with this machine was that it would have a brief go at any function selected and would then shut down with the pause LED alight. This is an 'emergency' indication, drawing attention to lack of reel sensor pulses, and in this case came on because the reel brake solenoid didn't operate. A pull-in current is routed via a switch (behind the solenoid) which opens when the armature has moved over. Its contacts were dirty, leaving only the hold-in current in the solenoid. Giving the switch contacts a good clean restored normal operation.

Total picture dropout at top of the screen: This clearly indicated that there was a tape-path problem at the entry point to the drum – bear in mind that his model uses the Beta format U-wrap. Scoping the f.m. waveform confirmed this diagnosis. The fault cleared when slight pressure was applied to the tape at the entry point. The entry height and angle are set by two guide poles that are located on either side of the full erase head. They were still tightly secured by the original locking paint. Having cleaned everything to no avail I inspected the entry guide poles closely and noticed very slight wear on the one closest to the drum. The pole consists of a collar placed over a fixed pillar and secured. Removing the collar and turning the worn side away from the tape cured the fault, although I must add that realignment of the entry poles called for some patience.

Machine was dead: There was just a momentary twitch of the capstan motor at switch-on. The power supply is commendably simple and I soon found that D425 was short-circuit. A BY299 was used as a replacement. This made it possible to play back a tape to check the deck functions. No recording was possible, however, as there were no signals. The main 33 V supply was missing at the relevant power supply pin, a short to chassis being recorded at this point. It was due to the 33 V zener diode on the front panel.

Grundig VS200

Wouldn't accept a cassette: The customer's complaint was actually that it wouldn't record! In this machine there's a safety system in the syscon circuit to prevent activation of the mechanism if the brake release solenoid hasn't been energized. After removing a diode to override this safety arrangement a cassette could be inserted. Then, intermittently but gradually becoming permanent, the brakes wouldn't release, causing a terrible noise when the machine tried to lace, and stopping rewind and fast forward dead. The cause of the trouble was soon traced to the BC876 solenoid coil driver transistor, which is a Darlington type.

Tape-path problems: This type of problem, such as the sound going off after a few seconds, is often caused by a faulty pinch roller. No sound or no control track recording/playback can be caused by an open-circuit head winding. Intermittent failure to initialize, especially from cold, is often caused by poor connections on the ribbon edge connectors to the sequence control module.

Failure to mechanically initialize: We see a lot of these machines. They seem to be very reliable electronically but the mechanics are not so hot. This problem is often due to a faulty brake switch which is located between the reel motors. If it has a plastic cover this must be discarded and the switch changed, not just cleaned. The reel motors are suspect if the machine chews tapes or shuts off. A light tap on the top of the reels will sometimes free a tight motor. The feed motor causes the worst problems – when it becomes tight it will jam the mechanics. Changing the motors over will often cure the problem. We've also had a few sound heads go open-circuit recently. You need to take note of the pink slip that comes with the new head as the wiring is different. If you get this wrong you'll still have no sound.

Various symptoms: The symptoms with this machine were as follows: no E-E or record sound, a rolling picture with its own recordings, and the search tuning didn't stop when a station was found. I couldn't see how there could be a common cause for these varied symptoms and therefore decided to check the ribbon cable connectors to the sequence control module. The contacts were OK but someone had fitted plug CP1 on back-to-front. Yes, the phantom fiddler had struck again! For problems such as loss of tuning information or cannot memorize the clock or date check the CMOS battery on the tuning board. It can go open- or short-circuit. When it's short-circuit you can't store information even when the machine is powered up.

Machine loaded, unloaded and ejected: There was an F3 fault indication. This means that there were no tacho pulses, i.e. there's a deck or electronic fault giving no reel or drum rotation. Thankfully the cause was a simple one. The capstan drive belt was very slack and sometimes slipped under load. When this was put right the machine worked OK on test for an hour or so. It then again failed to load. Adjustment of the deck load micro-switch finally put matters right enabling the machine to complete its full play cycle.

Defective audio/control/erase head: We've had several of these machines with this fault. The last one that came in had another problem – tape creasing when 'back editing control' was working, prior to a recording being made. It seems to be essential to bend the leading roller on the threading ring backwards or forwards, despite perfect alignment of the ACE head, to ensure that the tape path is exactly central on this roller. Beware of this one.

Occasional tape damage and loading problems: With a little perseverance we found that the switch behind the brake solenoid didn't make contact every time. Thus the machine locked up. Cleaning cured the problem.

Modifying modulators: The owner of this machine bought a second-hand satellite unit to add to his existing two VCR and three TV set-up. As all the signals had to travel around the house the connections were made at u.h.f. This, together with our local group A transmissions, didn't leave much room for another modulator! My solution was a small modification to the only varicap modulator in the set-up, the one in this VCR. The 9 V zener diode that regulates the supply to the tuning preset was replaced with two zener diodes, providing 5.1 V and 5.6 V, connected in series. This increased the tuning voltage sufficiently to provide an extra three channels, releasing enough room for the satellite receiver. I've since modified several Grundig modulators in this way, in each case obtaining enough room for pattern-free reception.

Failed clock: The owner of this machine had only recently arrived in our area and called me in to retune it to the local transmissions. Would I set the clock while I was about it? The tuning memorized all right, until the power was disconnected when all memory was lost. It then emerged that the clock hadn't worked for a long time. So a quick tune-in turned into a workshop repair. The fact that the clock could be set but wouldn't run led me to the power supply, where I found that the 50 Hz pulses were weak and distorted. This was a complete red herring, however. The memory and clock functions were both restored by replacement of the 1.2 V power pack. It had shorted internally, one result being an excessive load on the circuit which buffers the 50 Hz pulses that trigger the battery-powered crystal oscillator when the mains supply is available. This battery is quite hard to find: it's on the board immediately to the left of the deck. A Philips-type memory cell worked all right when the mounting had been 'fine tuned'.

Wouldn't record: The owner was more used to Far Eastern models. All that was wrong was that he didn't select an input before pressing record.

Grundig VS200, 220

Intermittently stops playing: When this happened it would unthread and show F2 in the display. It would also sometimes stop during wind or rewind, but the display would then show F6. Both of these fault codes indicate that the microcomputer chip thinks one of the spools has stopped. As there was no tape spillage scope checks were made on the spool optocouplers. With the machine playing or winding the optocouplers' output should consist of a signal of about 9 V peak-to-peak. The output from one of them was only 2 V pk-pk. A new supply optocoupler was required.

Special record function: If you look in the front of the service manual for these machines you'll see that there's a special function which allows you to put the machine in play without a tape. I tried this on several occasions and it didn't work. Then, while looking through Grundig's technical tips one day, I found out why – you have to lower the cassette tray first! Do this by pressing the cassette-in switch by hand for a few seconds. It isn't necessary with models such as the VS340 as with these the machine lowers the tray for you.

Dead machine: If the machine is completely dead check whether D410 (ZDl6) is short-circuit. While you have the panel out look at the centre pin of C407. When this electrolytic capacitor begins to swell it pushes the positive pin through the board, cracking the solder.

Grundig VS300

Would not accept cassette: If an attempt was made to load a cassette nothing happened except that F1 flashed on the display. On the basis of past experience with older models I homed in on the brake solenoid switch. This had dirty contacts, but when a new one was fitted it still didn't pull in. Its driver transistor T2141 (BC876) was open-circuit. For anyone not familiar with these machines, when the on/off switch is operated with no cassette inserted the brake solenoid should pull in for a moment. Sometimes the driver transistor goes short-circuit as a result of which the solenoid's thermal fuse blows. This is available as a spare part from Grundig (GCAS).

Would not accept cassette: If a cassette was wound in by turning the loading motor by hand, however, it would be ejected when asked. We found that the brake switch had a high-resistance contact. It's a two-way switch, so check both ways! When you fit a replacement be careful not to strain the switch pins – they can move if handled roughly and you then end up with the same fault!

Wouldn't initialize or play: This machine would accept a tape but wouldn't initialize and wouldn't play or wind. F7 was flashing in the display. The capstan motor had a dead spot, but before fitting a replacement I turned the motor a few times and tried a test recording. This revealed that the recorded sound was weak and that the colour from the previous picture showed through. The customer had had a quick look, had accidentally pulled off plug L14 and had then fitted it back to front. This meant that the erase head wasn't connected and the sound bias was excessive.

Tape damage: The customer brought two cassettes along with the machine, both of which were creased towards the end of the tape. When I tried fast forward then stop, tape spilled out as the brakes didn't come on. The cause of the problem was that the BC876 brake solenoid driver transistor T2141 was short-circuit.

Grundig VS310

Intermittently damaged tape: This machine intermittently damaged tape. Grundig Pete spotted it by chance while we were discussing other things. He put a tape in (mine) and it scrunched up! The small, flat copper-coloured guide spring fitted to the top of the audio head had broken off.

Poor reverse search: This can be caused by faulty or worn heads.

Wow on sound: Check whether the capstan belt has gone hard or has split. You should fit a Grundig replacement (from GCAS). To get to the belt you have to hinge out the electronics chassis – don't remove the deck or you'll never dress the wires in the right place again. If you have a different fault when it's back together check all the plugs – they are of very poor quality.

Dead: There was no clock, no booster, nothing at all from the power supply. The chopper transistor was OK, with 30 V across it, but there was no drive. D425 was short-circuit.

Displayed F1 and wouldn't take cassette: A check can be made on the cassette-in switches at connection 3 of plug L12. The voltage here should be 5 V until the switches close, when it should drop to zero. But the voltage never rose above 1 V. A new M722AB1 chip was required.

Warble on sound: This machine had already been somewhere else for repair. When we tested it the sound was indeed poor. We were told that the capstan belt and motor had been replaced. After checking that the power supply voltages and ripple levels were OK we removed the chassis screws and hinged it out so that we could watch the flywheel and belt running. There was a noticeable vibration with the belt, which looked new but didn't seem to be as tight as one would expect. When a new, original Grundig belt was fitted everything was all right. The old belt had a lot more stretch than the new one and was obviously not a proper Grundig spare. The usual fault with these belts is that they split and break, maybe due to the hard rubber of which they are made.

The display said F1, no loading motor movement: The loading motor movement – it wasn't being driven as R2155, the PTC thermistor in series with the output to the motor from pin 3 of IC2150 was dry-jointed at one end.

Grundig VS340

Intermittent failure to play, or leaves a loop on eject: If the machine intermittently fails to play or leaves a loop of tape hanging out of the cassette on eject, suspect the capstan motor of having a dead spot.

Loop in ejected cassette: When the tape was ejected there would occasionally be a loop left out of the cassette. This is usually due to a dead spot on the capstan motor, but a scope check on the motor seemed to show that this was not the case. Just then the capstan stopped. Some careful waggling of the wires showed that there was a dry-joint where the capstan motor wires are soldered to the small PCB at the back of the deck.

Tape shreading: Early versions of this model tend to produce black spaghetti after white when playing pre-recorded tapes. The cure is to get an exchange panel from Grundig (GCAS). It will have the following changes: C1406 220 pF; L1406 will have two green and one blue paint spot; R1501 100 kΩ; a 560 Ω resistor will be added in series with C1420. If you modify an early module the DOK control will need to be reset. Make a recording then play it back. Connect your scope to pin 16 of IC1430 and set the Y amplitude so that the waveform takes up six vertical divisions. Then adjust the DOK control so that the waveform occupies eight vertical divisions.

Capstan servo fault: In play the capstan would stop, then go too fast, then stop and so on. Checks showed that the capstan tacho signal at pin 2 of IC700 was erratic while there was no input at pins 17 and 18 of this chip. The wire to the tacho coil was dry-jointed where it connects to the threading motor's PCB.

Sound warble: It was caused by tight capstan motor bearings, a new motor curing the trouble. Unlike some who would strip the motor down and lubricate it, 'to save the customer some money', I prefer to work to manufacturers' standards.

Loss of control pulses: Check for dry-joints where the cables connect to the audio/control/erase head.

'Buzz on sound': There was a loud, 25 Hz beating noise during playback of any hi-fi tape. This is usually caused by worn hi-fi heads, but a scope check at the hi-fi f.m. test point (pin 35 of the hi-fi sound module) showed that the signal here was fine. So we started to check through the signal path. Everything was OK up to the output (pins 9 and 3) of IC1135, where the audio signal had 20 msec chunks cut out of it. IC1135 is part of the extrapolator circuit. It should be switched on for 12 μsec periods, but was being switched on for 20 msec ones instead. The timing is set by the 4070 quad exclusive-or chip IC1140, which was faulty. A new 4070 chip restored normal sound.

Damaged tape during ejection: A neighbouring dealer had recently given this machine a deck service. It now damaged the tape as the cassette was ejected. Tape guide 112 failed to return to the eject position because it caught on the pinch roller: the pinch roller return spring 48 had been fitted upside down.

Grundig VS400

Loop of tape after eject: The only way in which I could make the fault occur was to press stop with the machine in rewind search. The tape then wasn't drawn back into the cassette. Slight readjustment of the mecha-state switch was required.

On another machine, I had just fitted a new mode control switch because the machine looped the tape on eject. When play was selected, however, noise bars ran through the picture as there was no output from the control track. A scope check showed that the control head's earth connection was floating, although the fault would clear if the chassis was disturbed. The cause of the problem was eventually traced to a poor connection at plug A3 on chassis board 1.

No tuning: The BT supply at pins 15 and 16 of the tuner was permanently high at 32 V. The bus lines to the tuner appeared to be OK, and disconnecting the link between pins 15 and 16 then injecting a varying d.c. voltage proved that the tuner itself worked. So there was a fault in the tuner's PLL/synthesis circuit. We sent the tuner to MCES who speedily put matters right. Incidentally this machine uses the Panasonic D1 mechanism.

Wrong functions: When fast forward was selected the FF symbol showed in the display but rewind was what you got. Play, rewind and forward search were all OK. This ruled out problems with the end sensors. A new mode switch was tried, solving the problem.

Grundig VS440

Intermittent no erase or sound bias: The problem with one of these machines was intermittent no erase or sound bias. It would clear if the PCB on the back of the audio/control/erase head was pressed. When the PCB was removed we found that there was a solder bridge from one of the pins to the screening can.

Played OK but only snow in E-E and record: If the search button was pressed once, the correct channel number could be seen to be stored OK but no signals were tuned in. The +A (record) supply was missing as transistor T485 was open-circuit base-to-emitter.

Display shows A0, but child lock not on: This is often caused by a flat memory battery with the VS500 range, but the VS440 doesn't have one.

I cleared the child lock and gave the machine a test. It would tune in but wouldn't memorize. There was a faulty RAM in the tuner, so a new tuner was required.

No E-E signals only snow: By using the direct tuning method rather than search tuning the channels could be set in the tuner preset positions but there were still no signals. Checks on the tuner module showed that all voltages were present and correct. The other thing to check here is the I²C data and clock lines. These were at the correct d.c. levels, and pulses were present. Inside the can there's an SDA.1202-2 PLL chip. We found that it had no output at pin 18. A replacement restored correct tuning.

No vision in the E-E mode: Playback was OK. The tuning worked in that the channel numbers were right, but one of the 12 V supplies was missing. Transistor T685 (BC548) was open-circuit.

Lost ability to tune in stations: A defective tuner was the first thought that occurred to us, but a closer study of the circuit diagram revealed a cheaper possibility – the SDA3202-3 PLL chip. It's housed within the tuner assembly, and is fairly easy to replace. After doing this normal operation was restored.

Grundig VS500

Chewed tapes: When a cassette was inserted the deck went as far as the half-load position then the tape spilt out of the cassette and off the guides. The cause of the problem was that the left-hand spool wasn't being braked as spring 26 had come off. When the spring was refitted in its hole in the chassis normal operation was restored.

Mains fuse blown: This was because the BUZ90A chopper transistor was short-circuit. When d.c. checks were made to find out why we discovered that R1318 (300 kΩ) was open-circuit. The f.e.t. and resistor were replaced, but the BUZ90A again failed at switch-on. We had to fit a new TDA4650 control chip as well.

Power supply problem: The voltage at the power supply pin of the sequence-control microcomputer chip CIC230 in this machine was low at 2 V instead of 5 V. The dealer the machine had come from asked us to replace the chip as he didn't have the tools to complete the task. Fair enough, at least this would save damaging the print. But let's see, is the chip getting hot? It wasn't. So perhaps the 5 V regulator was faulty. A replacement made no difference and I then noticed that the 12 V supply was also affected. I checked both supplies by powering them from an external source: the 5 V rail took 330 mA while the 12 V rail took 125 mA. No shorts then. The power supply was either not regulating or not providing sufficient power. The power supply was actually switching in bursts, so it could have been working in a heavy-load mode. In fact this was not the case. The cause of the problem was an open-circuit capacitor (C1326) in the regulator control feedback circuit. Grundig tell me that the fault is not unknown to them.

Inoperative tape start and end sensors: Check the drive to the tower LED. R285 (47 Ω safety) can go open-circuit, as can CT285 (BC848C). If the clock display flickers (the flicker gets worse if you put your hand near the display) change the fluorescent display itself.

Dead machine symptom initially: There were very low voltages at the secondary side of the chopper transformer. Over the course of half an hour, however, the voltages gradually increased, creating a chatter from the deck solenoid and eventually normal operation. The cause of all this was C1409 (33 μF) in the primary side of the chopper circuit. It had gone low in value. The replacement must be a 105°C type.

Dead, no output from power supply: The mains side was OK, but the primary side wasn't oscillating. A check at pin 6 of the chip produced a reading of 10 V, which should have started the drive but didn't. So we replaced the chip. This was the wrong thing to do, as it made no difference. The cause of the problem turned out to be C1326 (47 μF, 25 V), which was low in value. It's the reservoir capacitor for the chip's d.c. supply.

Grundig VS510

Pulsating picture: A tape that accompanied the machine showed the fault. The best way to describe it would be a jitter on vertical lines of the picture every 2–3 seconds. It was most noticeable with the machine's own recordings, though it could just be seen on playback of a pre-recorded tape. The sound didn't seem to be affected. Following our usual practice with Grundig machines the first step was to check the supply lines for correct voltage and absence of ripple, but there was no fault here. The drum and capstan signals were rock steady. To break the deadlock I made the ultimate sacrifice: I brought in my VS510 that I'd won in the Engineer of the Year competition, and swapped over the decks. This proved that the fault was a mechanical one. A long screwdriver was then used as a stethoscope to listen to each moving part in turn. A noise could be heard coming from the right-hand rotating guide P3. So I fitted the one from my deck, but the noise was still present. Then I noticed that the speed of the impedance roller P1 was varying. If it was stopped, the noise and the jitter stopped. Dismantling the roller carefully and applying minimal lubrication cured the problem. The tape tension must have amplified the noise, making it appear to come from somewhere else! Don't remove guide P3 unless it's absolutely necessary to do so. It has a marked effect on the tape path. I know as I had to set it up on both decks! The guide is not only adjustable up and down, as usual, but also from side to side.

No teletext: When this mode was selected the page number appeared but there was no clock while the no teletext active message was present at the bottom of the screen. Tests around the SAA5231 chip on the DOS/TEXT board (DOS means display on screen) showed that there was no video input at pin 27. The BC848 transistor CT1655 was open-circuit.

Dead, clicking noise: When this machine was powered all that could be heard from it was a clicking noise. It was otherwise dead. Whenever you get an odd fault like this with one of these VCRs head straight for the electrolytic coupling capacitor in the chopper transistor's base circuit (usual value $47\,\mu\text{F}$). In this case replacing it stopped the clicking and restored the power to the rest of the circuitry.

Blank raster in play and E-E modes: If the sound, the display and the deck are all OK, inspect the DOS (display on screen) board visually for damage. The place to look is just behind where the front, right-hand side top retaining screw goes. If the wrong length of screw has been fitted, CR1550 can be damaged, producing the above symptom.

Machine totally dead: I removed the power supply and soon found that there was just 2 V at the 12 V output. Moving to the primary side of the circuit, I connected the scope to C1326 ($47\,\mu\text{F}$, 25 V) and found that excessive noise was present. C1326, which decouples the 10 V feed to the TDA4605 chopper control chip, was in fact open-circuit. A replacement brought the power supply back to life. I also replaced C1325 ($1\,\mu\text{F}$, 100 V). After refitting the power supply the VCR worked perfectly.

Hinari

HINARI VTV100
HINARI VXL5
HINARI VXL6
HINARI VXL8
HINARI VXL9

Hinari VTV100

Intermittently refuses all commands: This would occur during play, record or any other mode. Spraying the microcontroller chip with freezer proved that this was the culprit. It's a Sony type CXP5058H-118. Fitting a replacement and cleaning the tape path put the machine back in working order.

No playback colour: I checked the VCR section of this TV/VCR combi with another TV set and everything was fine. I then connected another VCR and found that the fault was present. After a lot of component replacement and adjustment checks I replaced the 4.43 MHz crystal XTL301. This provided a complete cure.

Another unit was brought in with the same symptom. The cause of the fault turned out to be in the TV section. We noticed that unless the off-air signal was perfect the colour lock was poor (good job we've some dodgy flyleads to help with tests!). So attention was turned to the chip that seems to do almost everything, including colour decoding. A check at pin 38 showed that the line pulse input was incorrect. C481, an $0.47\,\mu F$, $50\,V$ electrolytic, was low in value. A replacement restored good lock with both off-air and off-tape signals.

Hinari VXL5

Operates in play or record for about 20 seconds and then shuts down: The take-up reel sensor was faulty. Here's a tip: switch to counter, press play and observe the erratic and irregular number changes.

E-E and playback unstable: There was also very poor contrast. There was hardly any vertical or horizontal lock and no colour. A scope check at the video input pin on the r.f. in/out converter module showed that whilst the luminance signal was normal there were no field or line sync pulses. Tracing back from this point – we'd no manual – we came to a 47 μF, 16 V electrolytic capacitor (C354) which is connected to the collector of a transistor. At this transistor's base the video signal was correct. It was also correct at the collector once C354 had been disconnected. The capacitor checked out all right with our scope tester but we decided to fit a replacement. This restored normal operation.

No E-E or playback audio: A scope check at pin 9 of the i.f. module showed that the audio signal was present here and wasn't being muted by Q301. Audio was present at two of the pins of IC309, an electronic switch, so this device was ruled out – it's a one-pole, two-way switch, although the manual doesn't say so. The audio signal was then traced to the servo board. It was present at pin 13 of the LA7096 audio chip IC103 but there was no output at pin 18. We got low-level E-E audio by connecting a 47 μF electrolytic capacitor between pins 13 and 18 of this chip, thus ruling out anything that followed it. A new chip failed to cure the fault (Sod's law), so further checks had to be made on the components around it. The 12 V supply was present at pin 27, but there was no voltage at pin 3. A small 47 μF decoupling electrolytic capacitor between this pin and chassis was short-circuit. Fitting a replacement cured the fault. This is not the first time we've had capacitor troubles with this particular model. A fault we've had many times is that the machine loads then immediately unthreads. This is caused by loss of the drum PG pulses. In every case we've found that C145 (100 μF, 10 V) in the PG amplifier circuit had gone short-circuit.

Drum ran very fast and tape unlaced: Two non-working, ex-rental machines we'd purchased had the same fault – when play was selected the tape laced, the drum ran very fast then the tape unlaced. The cause of the trouble was that the 6 V supply to the drum feedback amplifier IC104 was very low as C145 (100 μF, 10 V) was short-circuit. We used a replacement rated at 16 V.

Video signal low: It was low, smeary and contained crushed sync pulses in the playback and E-E modes. Sound was not affected. As we didn't have a manual we carried out scope checks back from the r.f. modulator. Things were clearly wrong around Q306, where we found that C353 (47 μF, 16 V) was short-circuit. It appears to be a video signal coupling capacitor.

Recorded sound low with some tape: This problem will probably be experienced with any VCR that uses the same deck, for example the Amstrad VCR4600. The machine worked perfectly with some tapes, but with others the recorded sound was very low. After much searching, soldering and component checking I soldered all the connections on the audio/control head. This provided a complete cure.

Hinari VXL6

No drum rotation: All the other functions were in order. Voltage checks in the power supply disclosed that the P-on 5 V rail was low at 2.8 V. The culprit turned out to be Q504, although the device read OK when checked on a transistor tester. As we didn't have a direct replacement we fitted a TIP42C. This restored normal operation.

Intermittent stop, counter erratic in play: We also noticed that the counter continued to count even in the stop mode. Replacing the take-up reel sensor cured the fault.

Video signal crushed and distorted: This occurred in both the E-E and playback modes, and the video signal was of low amplitude. Not having a manual, I was forced to follow the print. This brought me to Q306 (2SC1740), whose base voltage was too low for it to switch on properly. The cause of the trouble was C353 (47 µF, 16 V) which was short-circuit. We've had problems with other 16 V electrolytics in these machines. Symptoms have included no drum rotation and excessive capstan speed.

No erase and no record: This machine started to dislike the 47 µF, 16 V capacitors that are sprinkled throughout its circuitry. The cause of no erase was C160 short-circuit, with R164 (22 Ω fusible) open-circuit. There was no record because C212 in the head preamplifier can was short-circuit (record 12 V line). Just for variety C145 (100 µF, 10 V) went short-circuit, with the result that the 6 V supply to IC104 (drum feedback amplifier) was very low. The symptom this time was that the machine unthreaded immediately after threading.

Hinari VXL8

Capstan wouldn't run: We'd no service information but we had a bit of luck: when we touched the ribbon cable to the capstan PCB the motor started to work. There was no further trouble after stripping this back and resoldering it.

Switching between SP and LP modes in both record and playback: In the past we've found that this has been due to tape wrinkling, usually as a result of a faulty pinch roller or incorrectly adjusted back tension. This causes distortion of the signal from the control track, a problem from which the Amstrad VCR6000/6100 series machines also suffer – they use the same deck. The tape path was carefully examined, but was blameless. The control track pulses can be monitored at pin 7 of IC103. They were seen to be of reduced amplitude and varying, but adjustment of the control head failed to produce any improvement. This was in no small part due to the large amount of hum on the 'ucom +5' rail. You can measure this at pin 1 of the power supply plug. The cause of the trouble was bridge rectifier D506, which had an open-circuit diode. We also replaced the associated 2000 μF, 16 V reservoir capacitor C505.

No fast forward/rewind although motors can be heard: This fault is often intermittent. It can be caused by seizure of the operating levers at the front of the deck. Figure 5 shows these levers in closer detail. There

Figure 5

are two white nylon gear levers operated by slide bars located centrally between the supply and take-up spools. These gear levers are slotted to locate on to pins on the slide bars. The fault occurs when the pin in the left-hand gear lever sticks in the slot. We used to rectify the fault by dismantling the gear assembly and widening the slot on the left-hand lever fractionally, using a needle file. More recently, however, we've noticed that the pivot on the left-hand gear lever (D in the figure) has a very small amount of play when this fault is experienced. We now feel that filing the slot simply counteracts the effect without removing its cause. So we've taken to replacing the left-hand gear lever which we obtain from CPC under part number HN62D085909305. Should you encounter this fault condition you should inspect this whole area, looking for signs of wear.

Another cause of this fault is the distortion of the rubber grommet on the stop lever. It sits upright next to the slide bars/operating levers which rest against it in the eject mode. Where this is the case a replacement grommet provides a complete cure.

No operation due to mains-borne transients: The symptoms were no EE operation, no channel changing, cannot programme etc. with just the letter E in the display. Unsolder the back-up capacitor for 30 seconds then reconnect it. Switch on and the microcontroller chip should recover from its crash. We've had this more than once and the routine has worked each time!

Virtually no capstan drive: There was no play, rewind or fast forward. Control of the capstan motor centres around the BA6219 chip IC106. Its supply at pins 7 and 8 was OK at about 15 V, but pin 4 was at 0 V. This could have been because Q105 was short-circuit or biased fully on, but what we actually found was that the 9.1 V zener diode D110, which is connected to pin 4 of IC107 (another BA6219), was the cause of the fault. As IC107 is the loading motor drive chip this calls for some further explanation! D110 straddles the print run from pin 4 of IC106. It's glued down, with just the PCB varnish for insulation. Lifting the diode clear and insulating it provided a complete cure. Just to add to the fun, the fault was initially intermittent.

Playback speed was faster than the search mode: Although this machine was several years old it had hardly been used. The complaint was that the playback speed was faster than the search mode – in fact selecting fast search slowed the capstan down. There was obviously something wrong in the servo department. Several chips can influence the capstan speed, but voltage checks in this area were inconclusive in relation to the figures given in the manual. The gods must have been smiling at us on that day, however, as an identical Orion machine came in for service. After making some comparison voltage checks with both machines in the playback mode we replaced the digital servo chip IC102. All was then well.

No rewind or fast forward: Play and record were fine. We soon spotted that the brakes didn't come off in the rewind and fast-forward modes, but it took a lot longer to find the cause. The brake trigger mechanism, items 259–262 in the exploded view in the manual, was suspected. Trigger lever 260 didn't latch properly. The cause of this was traced to a square rubber pad that sits over a pin on the mechanism. In the stop mode it rests against brake plate 261. This rubber bumper holds brake plate 261 to the right, enabling trigger lever 260 to latch. As the rubber bumper wears down, the trigger stops latching and the brakes stay on in fast forward/rewind. As to replacing the rubber bumper, you can either get one from a scrap machine or turn it round by 180°, which will cure the problem. Alternatively it is available from SEME under part number VPAR6833 (rubber damper).

Channel display goes to E, or pause or record lights up, or jumps between LP and SP modes: Before carrying out any checks replace C509 (220 µF, 10 V) in the 5 V supply. A scope check on his supply will probably show that a 50 Hz ripple is present. These problems tend to be very intermittent. So give the machine a long test before returning it.

Hinari VXL9

Machine would load and unload straight away when play was selected: Fast forward and rewind were OK. After a very unsuccessful search with the scope the machine was put to one side while a manual was obtained. It was not up to the usual Hinari standard: the circuit diagram had to be viewed through a magnifying glass as it's all on a single page. Checks around the TD6364NPAL digital servo controller chip IC102 showed that drum switching pulses were present at pin 18 but there was no head-switching square wave at pin 9. Since the 5 V supply was OK at pin 38 it seemed that the chip was faulty. As a check a Matsui VX820 that was waiting for a new carriage was pressed into service – its head-switching signal was fed to the Hinari VCR, which then played but with the drum rotating too fast. Normal operation was restored after obtaining and fitting a new TD6364NPAL chip.

Wouldn't tune: The BT line was permanently high, and altering the channel number (FS tuning) had no effect. The clock and data lines at pins 53 and 52 respectively of the flat-pack, surface-mounted micro-controller chip IC601 seemed to be OK but the supply 'load' at pin 51 was low as it was dry-jointed.

Intermittently selected external input mode: This machine caused us a series of problems, one after another. I finally got down to the last two faults, which seemed to be linked. The machine would intermittently go into the external input mode of its own accord – replacing the channel number with an E. Even more intermittently it would for no apparent reason go into pause. This happened only in the play mode, never during record. As scope checks showed that the spurious commands weren't coming from the local keys, checks were carried out around the IR amplifier. The supply was found to be slightly low at 4.7 V and had a 1 V p-p ripple on it. This supply also powers the microcontroller chip. So over to the power supply where both C505 (2200 μF) and C507 (220 μF) were low in value. Replacements restored a clean supply at the correct level and the mysterious happenings ceased.

No E-E video or audio: Playback was OK. You will find that there is no or very little video output from the machine's i.f. can, because C312 (470 μF, 16 V) is short-circuit or leaky. It couples the video signal to the switching chip IC208.

Hitachi

HITACHI VT11
HITACHI VT120
HITACHI VT130
HITACHI VT150
HITACHI VT17
HITACHI VT220
HITACHI VT33
HITACHI VT410
HITACHI VT430
HITACHI VT520
HITACHI VT530
HITACHI VT63
HITACHI VT64
HITACHI VT65
HITACHI VT8000
HITACHI VT9500
HITACHI VTF770
HITACHI VTF860
HITACHI VTM622
HITACHI VTM722
HITACHI VTM822
HITACHI VTM830

Hitachi VT11

Drum motor running continuously: This appeared to be due to no supply to the motor start transistor Q601, possibly because D614 was open-circuit. Replacing this diode didn't cure the fault, but replacing Q601 did – despite the fact that a cold check-out of circuit didn't show any leaks.

Machine would only select one mode: It didn't matter which button was pressed. We found that D453 was leaky.

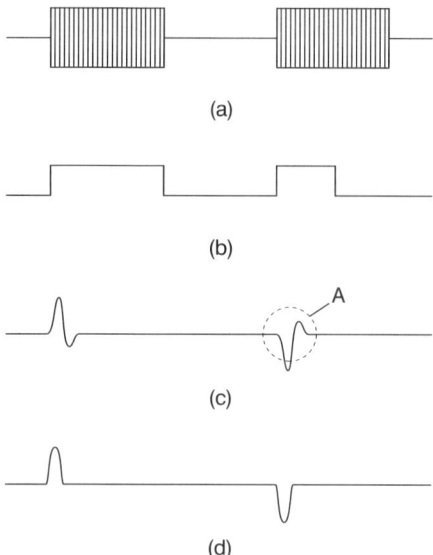

Figure 6 *Waveforms associated with the Hitachi VT11 fault described below*

Appeared to be dirty heads: The fault with this machine looked very like dirty video heads but after several attempts at cleaning them it was obvious that we had to look elsewhere for the answer. A check at the playback f.m. test point produced the waveform shown in Figure 5(a). Assuming that the heads were OK, we decided to check the drum flip-flop waveform which was as shown in Figure 6(b). This signal is produced by the drum pick-up pulses which were then found to have the correct mark – space ratio. So the servo chip which contains the multivibrators that generate the drum flip-flop signal was accused of being faulty. Unfortunately replacing it made no difference. Time to look closer at the drum pick-up pulses. The waveform was as shown in Figure 6(c) – Figure 6(d) shows the correct waveform. The small positive-going overshoot pulse circled in Figure 6(c) was of sufficient amplitude to make the positive monostable trigger, thus creating the incorrect drum FF signal. Replacing the Hall i.c. on the drum motor cured the fault.

No E-to-E, playback sound or vision: The fault was traced to a faulty coil, L2, inside the r.f. converter.

Drum intermittent high speed: The picture would dissolve into a large number of lines. The cause was a dry-joint on the drum PG head connector. To locate and repair this we had to remove the DD unit. A new cassette housing damper to prevent any more cassettes going into orbit completed the repair.

No response from front panel: There was no play/fast forward/rewind etc., although the machine worked normally when asked to make a timed recording. The remedy was to replace the BA6304 chip that interfaces the front panel to the microcontroller chip.

Hitachi VT120

Tapes were being damaged during cassette loading and unloading: This was because the left-hand tape guide was not fully retracting during the tape unload phase. In fact the loading motor was shutting down prematurely. The problem was cured by slight adjustment of the mecha-position switch.

No fluorescent display: A faint glimmer could, however, be seen if all the workshop lights were extinguished. Checks around the display driver/timer chip showed that the −30V supply was missing. This comes from the visual search board where we found an open-circuit circuit protector.

Picture intermittent: On test the machine displayed a noise bar that rolled through the picture, as though there were no control pulses. Both own recordings and pre-recorded tapes were affected. We found that the control pulse input to the servo i.c. was present and correct, but a check on the other waveforms around IC601 revealed that the capstan phase control pulse at pin 11 was missing. As all the inputs were correct the chip was replaced, clearing the fault.

Reduced tuning range: The machine wouldn't tune up to the h.f. end of the band. In the search mode the varicap tuning line reached only 15V before sweeping back to the low end. The cause of the fault was the tuner/i.f. unit − it was not a synthesizer fault as at first suspected. Excessive current was being drawn when the tuning voltage tried to rise.

Clock display random, no functions: The customer complained that there had been intermittent problems for some time. Now none of the operation keys worked and the clock display was random. We found that there was no 5V supply at pin 32 of the timer control and operation (key scan) chip IC751. This supply follows a very devious route: we eventually found that circuit protector IC805 on the VS tuning PCB was open-circuit.

Drum motor would slow down: The drum motor would start when play or record was selected but as soon as the tape was laced up it would slow down and eventually stop. The machine then unlaced and switched itself to standby. On many VCRs with digital servos the 4.43 MHz chroma subcarrier oscillator signal is used as a servo reference/clock signal. On this machine it was missing. Replacing the HT4539B hybrid chip solved the problem, restoring the 4.43 MHz signal at pin 27.

Would half load tapes: When a tape was inserted this machine would sometimes load it very slowly half way then just sit there and switch off, leaving the tape half laced. On one occasion the machine accepted the tape normally then switched off, again without ejecting the cassette. Suspecting a faulty loading motor, we applied an external 9 V supply to its terminals. The carriage operated normally. As the M54649L carriage motor/loading chip IC902 has given trouble in the past we checked the voltages here while trying to load a tape. There was only 2.3 V across the motor, measured at pin 10, so we checked the 12 V supply at pin 9. This dropped to 2 V when the chip was asked to drive the motor. As replacing the chip made no difference attention was turned to the source of the 12 V supply, at pin 7 of the STK5471 regulator chip. A new chip restored normal operation.

Threaded-up tape stuck inside: If the power switch was pressed the machine would power up for about 3 seconds then revert to standby. The first thing to do was to check the power rails. They were all OK. After several attempts to switch the machine on we noticed that a slight thud occurred when the power button was pressed. It appeared to come from the underside of the deck. On further examination we found that it came from the capstan motor. When its pulley was touched the capstan motor began to rotate and the tape unlaced. Subsequent checks showed that the capstan motor had a dead spot.

Mode switch problem: The mode switch in this VCR has given us trouble in the past. The symptoms are numerous and varied. Should you suspect the mode switch, check the following voltages at pins 47, 48 and 49 of the HD614042SD37 system control chip IC901.

Mode	47	48	49
Rec/play/forward search	0	0	0
Fast forward/rewind	0	5	0
Stop (carriage up or down)	5	5	0
Rewind search	5	0	0

Very intermittent vision overloading in the E-E and record modes: This was traced to the HT4757 hybrid chip.

Low E-E sound: The complaint with this machine was low E-E sound. We traced the cause to a leaky 4.7 μF, 35 V capacitor, C08, in the i.f. block.

Play and stop didn't work: All the other functions worked perfectly. The play and stop buttons had to be held or pressed repeatedly before they would operate – sometimes. Suspicions that there was something sinister in the system control or timer microcontroller circuit turned out to be unfounded: both switches were faulty. I wonder why?

No functions, only clock display: There were no functions at all, only a clock display that randomly changed from bright to dim and light from the operate LED. Checks in the power supply indicated that the STK5471 regulator chip was faulty. A replacement restored the machine to life.

No colour: This machine wouldn't record or play back in colour. The playback picture also had what can best be described as an 'orange peel' effect which was more noticeable in light grey to white areas. In addition white flaring occurred when the picture contained sudden light to dark areas, this symptom being more pronounced when the sharpness control setting was advanced. Slight finger pressure on the HT4539 hybrid decoder module would restore normal colour for an indefinite period. We wasted a lot of time soldering in the area around the module and the module's pins, but in the end it was a new module that restored the colour. The part is available from Charles Hyde but is expensive, the trade price at the time of the fault being nearly £40.

Low E-E and record sound: We traced the cause to C08 (4.7 μF, 35 V) in the i.f. unit. The fault seems to be turning up quite often now that these machines have seen a fair bit of use.

Cassette-in symbol shown at all times: Also the carriage wouldn't stay up. All functions were normal, including eject. But you had to be quick about getting the cassette out before the carriage took it back in again. For once the sensors were OK. I followed the wiring from the plug (socket on the main board) and found a dry-joint at R906. Resoldering this cured the problem.

Hitachi VT130

Tuning problems: According to the job card the machine couldn't be tuned. In fact the whole receiver section was out of action, although the tuning signal was present and a good tape played back correctly. We found that the not-PB 12 V supply was missing due to switching transistor Q504 being open-circuit. A BC328 turned out to be a suitable replacement for the 2SA952. The machine was a later model with an ICPN5 (200 mA) protector in the 12 V feed to Q504's emitter. This protector had also failed and had to be replaced. Later versions of this model have several changes and modifications which are covered in the Hitachi supplement manual number 2705E.

Very bad audio flutter: When an alignment tape was played a scope hooked to the audio output socket produced a display with marked wobble at the right-hand side. We found that the capstan motor drew about 140 mA off load and that the waveform generated across its input connections was rough. A new motor drew 85 mA and produced a much smoother waveform. When it was fitted in the machine the sound returned to normal.

Would chew bottom edge of tape: This was almost certainly a one-off fault. I relate it here to show how careful you have to be when diagnosing VCR faults – even mechanical ones! The machine would intermittently chew the bottom edge of the tape. It was not immediately obvious that the problem stemmed from the fact that the cassette was not going down fully: the tape was fretting on the bottom edge of the plastic housing. This was because the vertical spacing pole (centre, front of the cassette) was bent, or rather the plate that it's anchored to was buckled. The backward leaning pole rubbed on the inner wall of the cassette shell and stopped it short of its correct position, although a gentle push would force it home. Each cassette inserted into the machine came out with a tell-tale scratch mark on the plastic face behind the tape. Sounds obvious doesn't it? But it was the devil to find . . .

No tape transport: Suspicion fell on the A5 V supply to the syscon section of the circuitry. Sure enough it was missing. We traced the circuit back to the 9 V circuit protector which was open-circuit. Before fitting a new one a resistance check was made between the 9 V line and the deck. The reading was only a few ohms. Further checks showed that there was a short-circuit inside the A5 V regulator IC802 on the VST board. Replacing this restored normal operation.

No capstan rotation in play: This machine accepted a tape all right and fast forward/rewind worked. When play was selected, however, the drum rotated and the machine laced up but there was no capstan rotation. So stop was initiated and the machine unlaced. Slight flexing of the servo panel would enable the machine to enter the play mode but after the slightest movement it would shut down. Much time was spent checking the voltages around the servo chip (IC601) whilst trying to flex the panel in an effort to note any changes in the readings in the go/no-go states. This didn't work so I decided to let the machine shut itself off then carry out checks, hoping that it was still in the fault condition. There should be a 4.43 MHz signal at pin 32 of IC601. This was missing. I traced it back to pin 27 of IC301 on the luminance panel where it was also missing. Slight movement of IC301 made it come and go. Resoldering IC301 provided a complete cure.

Intermittent stopping of capstan and drum motors: We've had this fault on a couple of occasions. Because of its intermittent nature it took us some time to get to the bottom of the problem the first time round. We found that tapping the luminance – chroma subpanel would instigate the fault, giving us the clue to its cause – the fact that the 4.43 MHz reference oscillator signal is also used by the servo circuits. A scope check at pin 32 of the servo chip IC601 showed that the 4.43 MHz signal went missing when the Y/C panel was flexed or tapped. After much searching and soldering we decided to replace the HT4539B chip IC301, which is the source of the 4.43 MHz signal. It's a hybrid chip that contains some surface-mounted components and has a crystal stuck on top. We didn't attempt to replace the crystal but think it's the cause of the trouble since no amount of panel tapping stops the oscillator when the crystal is cooled. Those without an account with Hitachi can obtain the chip from Chas Hyde.

Picture rolling in E-E and playback: This is a common fault with Model VT130. The cause is C524 (220 μF, 16 V) which is on the top PCB near the converter module. You usually find that it's swollen and discoloured. The same fault occurs with Model VT14, but in this case the capacitor is C859 (470 μF, 16 V). It's best to fit a capacitor with a higher voltage rating in both models.

Fishnet interference on picture: During playback of this machine's own recordings or pre-recorded tapes the picture was covered with a fishnet type of interference irrespective of picture content. The cause of the fault was the HT4757 chip/module on the YC panel. A replacement and a deck service brought a smile to the customer's face – until he received the bill.

No functions until play was pressed: When this machine was switched on and a cassette was inserted none of the functions would work until play had been pressed. Once play had been pressed, stop, fast forward and rewind worked all right, but when the cassette had been ejected and the operate button pressed the loading arms would start to come out and the loading motor continued to run, much to the distress of the loading belts. Extensive checks eventually brought me to pin 6 of the syscon chip IC901 – the power control pin. The voltage here was high at around 7 V. Since the supply to the chip is 5 V, the voltage at pin 6 in the high state should be 5 V. Tracing back through R952 I found that the voltage was even higher, about 10 V. Checks in the power supply revealed that D865 (IS2473) in the power on/off control system was short-circuit. With a sigh or relief I fitted a 1N4148 as a replacement and then tried the machine again. To my horror the fault was still present, and this time the machine couldn't be turned off, remaining on all the time. As the power control line was still at 10 V I lifted pin 4 of CN852 on the power PCB, but the 10 V was still present at R952. The only other component connected to the power control line is D508 (1SS119): when this was removed and checked it was also found to be short-circuit. Another 1N4148 restored normal operation, with the power control line back at 5 V in the on state.

Hitachi VT150

'Shush-shush' noise from head drum: Our warehouseman Reg staggered into the workshop with this one. It had had to be replaced a few days after delivery. The customer said that when it got warm it squawked and the picture twitched laterally. After several hours' running, sure enough it did just that. The squawk was in fact more of a 'shush-shush' noise. It was coming from the head drum. On switching the machine off and then rotating the drum by hand we could feel friction at one point in the turning circle, suggesting that there was a bearing or rotor clearance problem in the direct-drive motor. Accordingly a new lower drum assembly (incorporating the motor) was ordered under guarantee. We were apprehensive about fitting it, anticipating a long setting-up session – after all this is a dual-speed machine. In the event virtually no adjustment to the tape path or the electronics was required, although everything was checked. This is a tribute to the tolerances to which modern VCRs are built.

Intermittent severe overloading in the record and E-E modes: The picture was crushed into white. As with the best intermittent faults it would lie dormant for long periods, fooling you into thinking that either you'd cured it or that it had gone away. At length we found that although the signal entering the HT4757A hybrid chip IC203 on the YC panel was OK in the fault condition it was much too large at pins 8 and 27. The chip itself was defective, with intermittent failure of the internal a.g.c. system.

No capstan rotation in play: This machine accepted a tape all right and fast forward/rewind worked. When play was selected, however, the drum rotated and the machine laced up but there was no capstan rotation. So stop was initiated and the machine unlaced. Slight flexing of the servo panel would enable the machine to enter the play mode but after the slightest movement it would shut down. Much time was spent checking the voltages around the servo chip (IC601) whilst trying to flex the panel in an effort to note any changes in the readings in the go/no-go states. This didn't work so I decided to let the machine shut itself off then carry out checks, hoping that it was still in the fault condition. There should be a 4.43 MHz signal at pin 32 of IC601. This was missing. I traced it back to pin 27 of IC301 on the luminance panel where it was also missing. Slight movement of IC301 made it come and go. Resoldering IC301 provided a complete cure.

Tape stuck inside, no functions and no eject: This machine is almost the same as the VT130 but has long play. Whilst checking around we found that the M54649L loading motor and cassette lift motor control chip IC902 were very hot. As both motors ran when powered from a separate d.c. supply we replaced IC902. Unfortunately this made no difference. Voltage checks then showed that the J2V supply at pin 9 was very low at 0.5 V. It's worth noting that this chip has two 12V supplies, one at pin 7 for the internal logic and one at pin 9 for the high-current motor drive. Tracing back from pin 9 brought us to the power supply where IC851 had 18V at its input but no 12V output. Although the power supply panel looks the same as that in the VT130 the regulator chip is different – type STK5476. This is a 12-pin device with only pins 1–10 used. We didn't have one in stock although we did have the STK5471 as used in the VT130. When we removed the STK5476 we found that the heatsink was drilled with two sets of holes. The smaller 10-pin STK5471 was quickly fitted to the heatsink, restoring full operation. Could the STK5476 have been fitted because of a shortage of the other type of regulator?

Snow over playback picture: Although intermittent, the fault was present most of the time. The symptom gave the impression that a head had failed or was clogged. It could be instigated by touching the YC board almost anywhere. The HT4757A luminance processing hybrid chip was the cause. When this item fails the symptom is often patterning on the playback and E-E pictures. Not on this occasion, however.

Wouldn't play tapes: On inspection we found that there was no drum or capstan rotation. We also found that the SP/LP tape speed switch didn't operate. The cause was soon tracked down to failure of the STK5476 power regulator IC851 to provide a 5 V output. A new chip restored normal operation.

Hitachi VT17

No clock display: This is due to the absence of the 10 V supply on the timer panel. The cause is that Q1795 on the back-up board goes open-circuit. It's not very easy to change.

Lower half of picture noisy: The heads had worn out and were replaced. After doing this I was left with a problem: the top half of the picture was fine but the lower half was noisy and there was a definite division between the two. The cause of the fault was traced to the relay on the video drum PCB. It shorts out the trick mode heads during normal playback but was not doing its job properly. A replacement provided an effective cure.

Drum ran at full speed: This was because there was no FG signal at the servo board. We found that there were dry-joints on the FG board within the drum motor.

Appears to have faulty head: If any of these machines appears to have a faulty head, before condemning it check that the relay clicks in the pause mode. If it fails to click the fault is either in the relay (dirty contacts) or its control circuit. It's a good bet that the head is OK.

Interference on playback: Two thick bands of noise moved up or down the picture. A scope check on the playback f.m. signal showed that there was an h.f. signal superimposed on it. The h.f. signal wasn't locked to either the f.m. signal or the drum FF pulses. A scope check was then made on the 9 V PB line. This showed that a 0.5 V peak-to-peak ripple at approximately 2.7 MHz was present. It cleared when C1159 on the servo/reg board was replaced. A meter check showed that this electrolytic charged correctly. It also produced the correct display with the component tester built into our Hameg scope. Unfortunately these testers work at 50 Hz and don't show up h.f. faults.

Tape would sometimes stop moving: The intermittent nature of the fault made diagnosis a long and difficult job. Eventually the symptom continued long enough for me to see that the capstan motor stopped. Further investigation showed that the cause was dry-joints within the 5-pin plug and socket at the nearby control panel. Resoldering and cleaning put an end to this elusive and time- consuming fault.

No functions, no clock: The 'operate' button LED said that the machine was on but there were no functions and no clock display. The NiCad back-up battery should read 2.5 V: this one read 2.5 Ω! It had also killed the 10 V regulator transistor Q1795. A new NiCad battery and 2SC2030 transistor restored normal operation.

Loss of speed after a few hours: The machine would eventually stop. When we opened it up we found that all the chips had been replaced and a new capstan motor had been fitted. Someone had been beaten by this one! We found that the flywheel became stiff when the fault occurred. All that was needed was cleaning and lubrication.

Motorboating on sound in playback: This was caused by relay RL401 on the audio PCB. Removing the plastic cover and cleaning the contacts provided a cure.

Head wear problem: As with all machines that have a few years of service behind them head wear is becoming a problem with this model. This particular machine's symptoms were poor playback with a lot of tracking noise evident with pre-recorded tapes: reproduction of the machine's own recordings was poor, but not quite as bad as with pre-recorded tapes. We tried fitting some second-hand heads from a scrap machine but the fault persisted. Replacement of the relay behind the head drum assembly also made no difference. The clue to sorting out the problem was the fact that the tracking control seemed to be loose while we couldn't find its centre position. The control turned out to be open-circuit, a replacement restoring normal results.

Hitachi VT220

Failure of STK5471 regulator chip causing various symptoms: This machine uses a multi-legged slab type chip (STK5471) to produce the regulated power supplies. We've had several of these fail, giving the following symptoms: operate light comes on, no deck functions and a noisy 'r.f.-through' picture to the TV set, this picture disappearing altogether when the VCR is switched on. It's the 12 V line that fails. Use plenty of heatsink compound when replacing this device.

Intermittent shut down, tape fully laced: This machine is similar to the VT120 and others. The problem was an elusive one: at very rare intervals the machine would shut down with the tape fully laced. With the fault present, switching on would bring up the on LED and function displays for a second or two, then the syscon would turn them back to standby with just the clock display remaining. A long soak test and much button stabbing established the fact that the fault showed only when the pause and frame-advance facilities were used – and then only rarely. The cause was a minute dead spot in the capstan motor.

Machine wouldn't switch on: The syscon didn't receive data from the timer, though the operate button produced the correct results at the timer chip. Further checks around this chip showed that it wasn't receiving data from the search-tuning board. Replacement of the search-tuning chip brought the machine back from the dead.

Poor colour: In fact on playback the chroma produced a diamond pattern. A scope check at pin 16 of IC301 (playback main converter output) showed that the signal was virtually the same as that at pin 8 (4.43 MHz chroma input). Replacing the bandpass filter CP302 seemed to be the logical thing to do but the fault remained. After pursuing several other red herrings we cured the fault by changing the chip.

Machine dead, no 5 V or 12 V supplies: The display lit up and the 'operate' LED was on continuously. We found that IC901's reset pin 49 was high at 5 V instead of being at 0 V. The reset pulse is generated within IC802 on the VS tuning BB. A check at pin 9 of this chip showed that everything was in order. The collector and base of transistor QR804 were both at 5 V; however, the collector should have been at 0 V. Replacing QR804 restored normal operation.

Machine almost dead: There were no clock or deck functions. But the power supply voltages were all present at the output connector. Circuit protector IC405 (N5) was open-circuit. It's not easy to find, as it hides close to the side of PG604. There was no apparent cause for its failure, and the machine worked perfectly after fitting a replacement.

Hitachi VT33

Dead, clock OK: This one belonged to a girl who lived in Zambia. She brought it back on a visit after a local service engineer had said he couldn't repair it due to lack of information and spares. The player was dead, although the clock functioned. Checks around the power supply chip IC151 revealed that the always 12 V line was present but the switched 9.5, 12 and 16 V outputs were missing. The power-on signal from the microcomputer should bring in the 16 V regulator via pin 13 of the chip, which in turn switches on the 9.5 and 12 V lines via R153 and R151/2. Replacing the 16 V output with an external supply restored all the power rails. Normal operation was obtained when IC151 had been replaced, but before returning the machine to Zambia I decided that a new idler and set of belts would be prudent.

Wouldn't play or record when hot: The machine would thread up, but the head speed would then begin to vary, making the micro think that the head had stalled. If you went into play – pause first the machine would sometimes play all right, but not always. A look round the head servo revealed that the 9.5 V supply started to oscillate when threading was complete. Replacing the many electrolytics associated with this supply had no effect. A new STK5421 finally cured the problem.

Generated a noise in most modes: The symptom was similar to that produced by a dry capstan flywheel bearing in the later VT64 series. As suspected, the flywheel bracket was fitted with the nylon bearing. Further pushing, prodding and listening revealed that the noise was coming from a pulley on the clutch assembly, item 301.

Intermittent E-E and playback picture: The problem with this machine was intermittent loss of the E-E and playback picture. The video signal was present at IC202's output but was missing at pin 8 of plug PG233. It was being lost at Q210 whose base voltage was low at 1 V instead of 6.7 V. Tracing back we found that Q224's base voltage was also low. The culprit was C292 which was leaky.

Poor sound on own recordings: This machine played pre-recorded tapes reasonably well. So, as cleaning the audio control head failed to improve matters, a replacement head was fitted. This cured the problem.

Tapes remained looped in the machine at eject: Whilst removing the covers I noticed a label on the baseplate saying that another company had repaired the machine some months earlier. Considering this fact I was amazed at the amount of dust inside it. On running the machine with an old 'test' tape I found that there was no fast-forward or rewind action. This was because the rubber drive wheels and the clutch assembly had all been coated with a layer of oil, which appeared to be due to overgenerous application to bearings during the previous service. This meant a complete mechanical stripdown, clean and rebuild, something that could have been avoided if the previous 'service engineer' had been more careful in cleaning the machine properly and had not been so generous with the oil.

Shuts down after 5 seconds: The drum and capstan motors both worked but a hunting sound came from the drum. At start-up a scope check showed that there were square waves on the 9.5 V supply to the servo: the supply was also trying to rise to 10.5 V. Replacing the STK5421 chip IC151 on the regulator board cured the fault.

Tape threads, then unthreads: It's quite common with many Hitachi models to find that the tape threads up then almost immediately unthreads. What happens is that the drum slows down because the servo reference signal is missing. It comes from the 4.43 MHz chroma subcarrier oscillator section of the HT4539B hybrid chip IC203. I changed IC203 on this machine as a matter of course, but this time the fault was still present. Doing what I should have done to start off with, I then checked the voltages around IC203. The 9 V supply was missing at pin 9 because choke L216 was open-circuit. The questions now were whether IC203 had been faulty and whether L216 had burnt out? I broke L216 open and found that it wasn't obviously burnt: I guess that one day I'll fit the old HT4539B to another machine to see what happens. It's the unpredictability of it all that keeps me going!

Would only work for a few seconds: Simple idler replacement and service I thought. But when I inserted a cassette the lift moved so quickly that it almost pulled me in. Then the power LED went off. When I switched on again and ejected the cassette it came out very fast. I decided to check the power supply to the cassette lift motor and found that the voltage was high. A check around the STK5421 power supply chip then showed that the 9.5 VA and 9.5 VB lines were both at 17 V. A new STK5421 restored normal operation but a new idler and deck service were required.

Wouldn't eject tape: No matter what mode the machine was in, pressing the eject button wouldn't release the cassette. In fact pressing the button would sometimes change the function, for example if the machine was in the play mode pressing eject might put it into the fast-forward or some other mode. Usually, however, pressing the button had no effect at all. The cause of the trouble was the eject switch itself. It measured all right out of circuit when checked with an ohmmeter, but a replacement from a scrap machine cured the fault.

Wouldn't accept a tape: The display and the E-E mode were OK. When a tape had been inserted manually there were still no deck functions. Checks showed that the switched 9 V supply was missing, because the SKT5421 regulator had failed. A replacement restored normal operation.

Cassette would eject on play: The cassette would be ejected a few seconds after being loaded. On investigation I found that the supply side was being loaded but the take-up side wasn't going down. The cassette flap wasn't opening because the flap lock release arm had lost its tension. I was able to retension the arm after removing the cassette holder, saving the cost of a new carriage.

Hitachi VT410

No colour during playback: The machine actually recorded good colour. The chroma signal path was present and correct throughout the up-conversion circuitry, but the signal was lost at R3U7 which was dry-jointed to the PCB. This resistor provides matching for the 2 H delay line.

Wouldn't accept tapes: The cassette-in indicator was on, and if the cassette housing was removed the play mode could be selected. The cause of the trouble was no supply to the cassette LED – it was shorted to chassis. We found that the ribbon connector between the deck PCB and the main PCB was displaced at the main PCB.

Would load a tape then shut down: If the operate switch was pressed the machine would come on for another 30 seconds and then switch off. No deck functions worked. The power supply was working correctly and checks around the syscon microcomputer chip failed to reveal anything amiss. A clue was given by pins 8 and 9 of IC901, the outputs that control the loading motor: pin 9 was high at 4.7 V while pin 8 was at 0 V. The system control was trying to unload but the loading motor wasn't doing anything. A quick check on the loading motor drive chip IC902 (BA6209U2) showed that its inputs were correct but there was no output. Replacing the i.c. cured the fault.

'Sound slurs': Many customer descriptions are merely the first clue in the guessing game that occupies an engineer during the first part of his diagnostic session. 'Sound slurs' is one of the better ones, usually indicating problems in the capstan department. On this machine the capstan ran slowly and erratically. A meter connected to the speed control line to the motor showed that the servo was working flat out to speed up the motor. The motor itself was faulty, although it ran very freely when turned by hand.

Machine would unload and shut down: This machine would unload and shut down almost immediately after the load sequence had been completed. It seemed logical to start by checking the flip-flop signal to the microcomputer chip. Careful observation of the screen during the very brief period of time when this signal was visible gave a further clue. Lines on the screen indicated incorrect drum speed. Replacing the lower drum cured the fault.

Vertical black and white bars in playback: The complaint was of vertical black and white bars instead of a picture during the first 15 minutes of playback. On test we found that a tap on the panel would clear the fault. The signal appeared to be getting lost within IC202, but a replacement made no difference. We then found that connecting a scope probe to pin 14 of IC203 would remove the black and white bars, leaving snow on the screen. It seemed that the f.m. demodulator wasn't operating correctly and that as a result the dropout compensator was working overtime, causing instability in the form of the bars. Replacing IC201 cured the fault but it was impossible to prove which part of the chip was defective.

Intermittent audio recording: The problem was due to a loss of bias. We traced the cause to a switch within IC401 closing. This occurred because the voltage at pin 5 fell below 11.3 V, thus activating the switch. C431 was found to have a 2 MΩ leak. Note that according to the diagram pin 5 has to go high for the switch to close: in fact it has to go low.

Poor or very noisy eject: In this model the cassette carriage is driven by the capstan motor via a series of cogs and gears. The cause of the fault was the worm on the right-hand side of the carriage, part number 6435571. As a precaution I also replaced the worm wheel assembly, part number 6896971.

Machine would lace up, play briefly, then unlace: This was an obscure fault. The counter was working, and the waveforms around the syscon chip IC901 were all OK except for the 25 Hz signal at pin 23 which was found to be of low amplitude when a comparison check was carried out with a working machine. The signal comes from the servo chip IC601, where a comparative resistance reading to chassis revealed that the faulty machine had a 400 Ω leak The cause of this was C23, a surface-mounted capacitor on the video head preamplifier PCB!

Sound but only a white raster: I inserted our colour-bar test tape and checked at the video output socket with a scope. Only a chroma signal was present. This was strange, as one would have expected to see flickering colour on the monitor's screen. The scope was next used to check the luminance signal path. The signal was present at pin 7 of the HT4848B chip IC202 but didn't emerge at pin 1. After confirming that its supply was present we replaced the chip. This restored normal operation.

Slow rewind and fast forward: We found that this was being caused by a power supply fault: the 12 V output from IC851 (STK5372) was low.

Drum might kick rather than rotate: Very intermittently when play or record was selected the drum would just kick rather than rotate. The cause was traced to failure of the thick-film chroma processing chip IC301.

Hitachi VT430

No E-E sound, playback OK: When we removed the plug-in tuner/i.f. module we found that the cause of the problem was a stray splash of solder on a surface-mounted resistor to the audio output line.

Damage by mains-borne lightning: The mains input PCB had to be repaired. This got the machine working, but in the play and E-E modes there was a blank raster. Replacing Q3301 – we used a BC640 – restored the playback luminance but we now had weak E-E. Scope checks proved that the i.f./demodulator and p-in-p modules were OK. The video was traced to pin 7 of the LA7016 chip IC1501, but there was very little output at pin 4. Replacing this chip cleared the final fault.

No tuning and the clock display was dim and flashed 'on/off': When the top was removed there was a terrible smell. Thinking that there had been liquid spillage we checked the boards thoroughly but everything seemed to be in order. We eventually found that the 1SS130M-T –40 V supply rectifier D917 had burnt up – it turned out to be the source of the very toxic smell. To gain access to the PCB to replace D917 we had to remove the 100 µF, 63 V electrolytic C918, which is the reservoir capacitor for the +40 V supply to the voltage-synthesis tuning PCB. While we had it out we thought we'd test it –

and found that it was open-circuit! After replacing these two items everything worked normally.

Machine stuck in pause: When a cassette was inserted the pause indicator lit up. Other functions could be selected, e.g. play, but this produced only a still picture. A check at pin 56 of IC901 produced a reading of 3.2 V instead of 0 V. This led us to I1581, a 100 µH choke that's in series with the camera pause socket.

Wouldn't eject tapes: The cause was simply a slack reel drive belt – in the eject mode the belt drives the carriage. A new belt put matters right.

Hitachi VT520

Poor playback sound and vision: This quite new machine had an unusual fault: the playback picture was very poor and was rolling while the sound was low and very muffed. With a fault like this the audio/control head is the first thing that springs to mind, but in fact the symptoms were caused by insufficient back tension. A check on the back-tension post showed that it was about half an inch away from the tape. The reason for this was not at all obvious. Nothing was jamming it. The mechanism that controls the arm looked OK, but it didn't move the arm far enough. Curiously, if the loading motor was given a few more turns the mechanism and back-tension post moved to their correct positions and normal operation was obtained. Of course when the machine was stopped and play was again selected the fault was back. We made a note of the number at which the arrow on the mode switch pointed. In play or record it should normally be 6, but the arrow pointed to 5. This is the position for reverse play, in which no back tension is needed. A new mode switch from Hitachi, part number 5610702, cured the problem.

Capstan motor would stop and start: When this machine had warmed up the capstan motor would stop about every second then restart a second later, as though someone was pressing the pause key at 1 second intervals. This produced a very curious symptom. If a recording was made in the fault condition the playback picture would appear to jump and skip frames. When freezer was squirted on the chip on the capstan motor assembly the fault cleared. Fitting a replacement capstan motor assembly, part number 5571454, provided a permanent cure.

Pre-recorded tapes' chroma noisy: This machine played its own recordings perfectly but with pre-recorded tapes the chroma was at best noisy and at worst non-existent. Checks around the chroma processing chip IC301 showed that the chroma signal was of about the right amplitude but very noisy. Matters weren't helped by several mistakes in the circuit diagram: the playback not the record voltages are the ones in brackets while pin 18, the search switching line, should be at 1.5 V or so in play and 0 V in search. I felt that the f.m. entering the chroma processing was poor but it looked fine. Luckily I had another head amplifier, but substitution made no difference. And so to the video heads. Strange though it may seem, the cause of the problem was the lower drum assembly/rotary transformer/drum motor.

Capstan motor failure: We've had the direct-drive capstan motor fail in a very misleading way in a couple of these machines: the on-board drive chip gets too hot to touch after running for a few minutes. You might find that the unstabilized 16 V line is as high as 23 V. Even so the motor itself is the cause of the fault, which often shows up as intermittent stopping and starting of the capstan.

Machine would load, then drum stopped: The capstan never started – and the deck would remain still until the mains supply was switched off and on again. The deck would then unload and the machine would switch off. When the voltages at the mode control inputs (pins 11, 12 and 13) of IC901 were checked, I found that 3.3 V was present at pin 11 whatever the position of the deck. The cause was a leaky mode switch. A new switch restored normal operation – note its position before you remove it.

End sensor problems: Printed flexible ribbons are used to link the tape-end sensors to the main PCB in this model. A common cause of problems, mainly concerning the end sensors, is poor contact with the edge connector at one end or other of a ribbon. The usual symptom is failure to accept a cassette or retraction of the cassette after ejection, or alternatively deck shutdown a few seconds after entering the forward mode. The cure is to clean the connectors and ribbon ends. We are now starting to find worn audio/control heads in these machines. The first indication of this is loss of capstan servo lock with a machine's own recordings.

Capstan motor would stop: We've recently had three of these VCRs in which the capstan motor would stop momentarily every few seconds when the machine had been in operation for between 30 minutes and an hour. In each case the motor-mounted coil-drive chip was too hot to touch and a new capstan motor had to be fitted.

Tried to load cassette without one being inserted: This machine tried to load a cassette without one being inserted and the wording 'Code 1' appeared in the clock display. We suspected the start and end sensors: fortunately both pins of the rewind sensor were dry-jointed. After soldering this up the machine worked perfectly – and the wording 'Code 1' disappeared as well. Phew!

Low sound and jumping/critical intermittent tracking: This indicated that the audio/control head was faulty. At last we seem to be getting some sensibly priced alternatives.

Hitachi VT530

No playback, E-E picture appeared to suffer a.g.c. overload and the search tuning didn't stop when channel was found: All these symptoms were caused by one component, however, CP205 (LC delay) on the YC board.

Cassette housing needed to be retimed in main deck: This machine kept coming back to the workshop because the cassette housing needed to be retimed to the main deck. When this had been done it would work perfectly, but it would return with the same problem about three months later. We finally traced the cause of the trouble to the start and end sensors on the cassette housing and replaced the IR emitter.

Sound wow in playback and record: The capstan and pinch roller were both fine. I traced the source of the wow to the clutch assembly.

Hitachi VT63

No recording: From the initial recordings we made with this machine it looked as though the heads were faulty. But the customer's complaint was of no recording. Making further recordings proved that the heads were not faulty, as good pictures were obtained. Oscilloscope checks revealed that the video f.m. envelope was varying – and could be varied by applying pressure anywhere on the board. The f.m. signal passes from the chip via a 1 kΩ resistor to the f.m. record level preset. The waveform was stable at the chip, but varied at the other side of the 1 kΩ resistor. Various filters were accused, but all proclaimed their innocence. After a lot of pushing and prodding the fault was eventually traced to the preset. It was broken, one leg having snapped close to its body.

Failure in display: This was an unusual one. The fluorescent display worked except for the third block from the left (3 and Sunday) which remained blacked out at all times. If a finger was firmly pressed along the tube's lead-out pins this block, amongst others, would light up, so the tube was exonerated. The 3/Sunday section is designated grid 8 (tube pin 11) and should be fed by a pulse/strobe signal from pin 46 of the timer control chip IC101 (M50757–681SP). This signal was missing and replacing IC101 restored the full display.

Overloaded video: A quick scope check revealed that the cause of the fault was at the front end, on the tuner/i.f. board. C832 (470 μF) was short-circuit.

Screaming and clanging when the tape was ejected: The machine then switched off. When we checked out the deck functions we noticed that there was no auto-stop at the end of fast forward. Both problems were caused by the fact that the left-hand tape sensor wasn't working. When the voltage at the EST pin on the carriage PCB was checked it hovered at around 4 V instead of 9 V, but the slightest movement of the carriage produced the correct 9 V and normal operation. Careful positioning of the end sensor holder monitoring the EST voltage and securing it with a drop of superglue solved the problem.

Tracking problems: With a pre-recorded tape the picture was unwatchable when the tracking control was in its centre position. If it was moved to either end there was a stable, watchable picture. We found that the f.m. envelope varied and couldn't be corrected by operation of the tracking control. Replacing the audio/control head provided a complete cure.

Wouldn't accept a cassette: The tape-in indicator was permanently lit and the carriage loading motor was trying to eject the carriage from the machine! This was all because the carriage- mounted tape-end sensors were slightly leaky. The leak couldn't be measured with a multimeter, but the scope component tester displayed the tell-tale waveform. This is a very useful device!

Hitachi VT64

Would load, play for about 5 seconds, then unload: We found that the drum flip-flop signal from the servo i.c. was of reduced amplitude. Checks were made around the servo and syscon microcomputer chips but nothing we did restored the flip-flop signal to its correct amplitude. The flip-flop signal is also fed to the Y/C panel, and although the circuit diagram gave no clues as to what could be wrong here all became clear when the panel was removed – a liquid had been spilt into the machine at some time and was loading the flip-flop signal. The odd thing is that no other traces of liquid spillage could be found.

Noisy rewind: However, after running two or three tapes through the machine nothing unusual could be heard. We decided to leave the machine on test on the soak bench, and after a few minutes the fault developed but it was in playback, not rewind. Back on the work bench the machine worked normally when switched on. Time to leave it running in playback and make the tea. When I came back it was droning away nicely, but when the machine was tipped backwards to inspect its underside the noise went. Lowering it brought the noise back. The help of a colleague was summoned: he held the machine while I looked up from beneath. Pushing the capstan flywheel up would clear the noise, and on inspection we saw that there were signs of rust deposits on the nylon plate in the middle of the flywheel support bracket. A smear of Vaseline on this plate cured the fault.

Drum motor had stopped: Voltage checks around the drive chip IC601 revealed that the supply voltage at pin 10 was low. This i.c. is supplied via D642 and a cold check suggested that the diode was OK. Replacing it restored drum motor operation, however.

Intermittent eject: This machine intermittently thought that eject had been pressed when in fact the operate button had been selected. There's nothing very complicated about the switch matrixing circuit. Most of the switches go to the syscon chip via the matrixing resistive ladder network RA9001, which is on the function switch PCB along with the switches themselves. When a switch is pressed a series of resistors is connected to chassis and the syscon chip and its decoding i.c. works out which key has been selected. The faulty item was in fact the operate switch, which had gone resistive. This fooled the syscon by altering the resistance in the matrix ladder. Come back the piano- key Ferguson 3V00!

Machine kept stopping: The problem was extremely intermittent. Various mechanical items had already been changed, including a faulty capstan motor, which we've known to cause this problem, but the fault persisted. The cause of the problem turned out to be totally electronic – that plague of all electronic equipment, dry-joints. Several were found around the 9 V regulator Q605 on the servo board, also on R699. After re-soldering these, the problem had been cured permanently.

Drum and capstan failed in record: Playback was all right but when record was selected the drum and capstan failed to rotate although the record indicator lit and the tape laced up briefly before unlacing again. We found that the record/play-switching voltage double-diode block D626 was open-circuit on the record side. By coincidence we found a similar faulty device recently in the sound section of an older Hitachi machine. In both cases a couple of good old 1N4148 diodes wired in back-to-back proved to be a suitable replacement.

Hitachi VT65

Intermittent failure to record: After soak testing it for a week we found that the amplitude of the record f.m. signal input to the head-switching chip was occasionally low. Our problems were made much worse by the fact that even the slightest movement would restore the amplitude of the f.m. signal. The cause of the trouble was eventually tracked down to IC202, which is a hybrid thick-film chip.

Intermittent failure to record in colour: Initial tests showed that the tape-in light was permanently illuminated. Further tests revealed a bad tape-end sensor. The cause of the 'intermittent colour' was an extremely dirty video head drum. This was cleaned – the customer declined the quote for a replacement.

Hitachi VT8000

Wouldn't turn on: The businesslike 'zizz', as the drum motor runs up to speed in this family of machines, is characteristic and unmistakable. How we would have liked to hear it from this one! It wouldn't turn on in response to its operate switch, although had we known this it did work in the timer mode. This is starting to sound like one of those test cases! Anyway, the trouble was traced to a leaky 5.1 V zener diode (ZD905) in the switch-on network between the key and pin 41 of the microcomputer chip IC901.

Refused to unload and eject tape: This machine would load up and play. It then refused to unload and eject the tape, although the left-hand reel was driven normally to retract it. If the tape was unloaded manually and a fast wind command was given the machine would respond initially but the loading arms would also move forwards slightly. This disturbed the loading switch thus preventing acceptance of further commands until the loading arms were wound back again manually. The microcomputer's voltages were all correct and the right instructions reached the loading motor control chip IC905. Replacing this TA4194A device did the trick.

Would shut down after 5 seconds: All functions worked in this old machine, but only for about 5 seconds. It would then shut down. The cause was a stretched take-up spool carrier to counter belt – so the counter and the take-up spool rotation pulses stopped. A new belt solved the problem.

Intermittent fast forward: After carrying out a mechanical service there was still intermittent fast-forward operation. The cause was traced to the rearmost microswitch on the deck. It had a loose fitting case, which resulted in premature operation. A replacement cured the trouble.

Hitachi VT9500

Intermittent sound on E-E and playback: On removing the covers I homed in on the relays, but this time they were blameless. The fault was due to a dry-joint on C429.

No sound or vision in E-E mode: This was traced to a faulty TA4349 chip (IC909).

Front panel keys didn't work: The owner of this machine had been operating it for some time via remote control only – for the very good reason that the front panel control keys didn't work. The only one that did anything was the play button: when pressed, the record and play lights lit and the machine went into the record mode. Checks revealed that the HD38750A53 function control chip IC2003 was receiving the correct supply voltage and that its oscillator section was running. Fitting a new chip cured the problem.

Complete failure: This is a veteran machine. The unusual symptom was complete failure – no clock display, no indication lights and no functions. Mains current was flowing into it, however, and the several rectifiers were doing their stuff. The system control/regulator panel along the back of the machine is no fun to deal with physically. We found that the trouble was due to one of the 10 V dropper diodes here – D904 was open-circuit.

Tape loaded then unloaded: Fast forward and rewind were OK but when play was selected the tape loaded to the heads then unloaded as there was no drum rotation. A check on the supply at pin 4 of the drum chip IC503 produced a reading of 0 V instead of 9 V. The cause of the fault was the STK5720 regulator chip IC901 on the syscon PCB.

Apart from clock no functions: After checking the supplies on the rectifier PCB and finding that they were as given in the manual we next checked the voltage at the syscon chip's supply pin 21. The reading here was 0 V instead of 10 V. This supply is obtained from the 12 V regulator chip IC903, the feed being via two ISS133 diodes, D904 and D905. When they were checked we found that D904 was open-circuit while D905 was leaky. After fitting new diodes the machine worked normally.

Hitachi VTF770

Shut down immediately at power-on: Checks showed that the 18 V supply was missing. The cause was a crack around one of the mains transformer's pins. Some fresh solder restored normal operation.

Shut down about 3 seconds after switch-on: The display came up, but no tape functions seemed to work. The cause of the problem was no loading motor drive because the BA6209 drive chip had failed. Unfortunately the failure of this chip had resulted in the mechanics going badly out of sync. The mode switch had to be replaced and the mechanics realigned.

Lifeless apart from clock display: It had a fully laced-up tape inside. When the power button was pressed the channel indicator came on but the machine shut down again 2 seconds later. Fuse F852 (1.6 A) in the supply to the 14 V bridge rectifier on the power supply PCB was open-circuit. As it hadn't blown, a replacement went in. This restored normal operation.

Failure of the 1.6 A delay fuse: A defective capstan motor is usually the cause of failure of this fuse in the power supply at intervals varying from days to weeks. There may be other symptoms, perhaps a screech or roar during loading, eject or fast tape transport.

Wouldn't eject tapes: This machine wouldn't eject tapes. Fortunately the cause was simply a slipping belt.

Shows 'CODE-1' in the display when the operate button is pressed: Try replacing C711 (33 μF, 6.3 V) and C771 (220 μF, 6.3 V). They are both on the front panel, beneath the display. I've also had to replace the EEPROM (IC702) when the machine wouldn't change channels after replacing the two capacitors just mentioned.

Hitachi VTF860

No-go or intermittent failure to perk up from cold: This is generally caused by the kick- start capacitor C6 (1 μF, 250 V) in the power supply having dried out or being partially open-circuit. The fault doesn't put in an appearance with normal use in the home – unless the machine is deprived of mains power for any reason. It crops up when the machine is on the bench for diagnosis of the cause of some other fault – or when the customer gets it back home again!

Would shut down when loading in the record mode: The tape guides would reach about half way towards the full loading position, after which the display would go out briefly and the loading motor would shut down. This would last for about a second. The machine would then start up again and revert to standby. I suspected a fault in the power supply, and my diagnosis turned out to be correct. Q7 (2SC1741) was the culprit – it was breaking down when warm. I replaced the 2SK1611 chopper transistor (f.e.t. type) as well.

Dead: The main causes are D11, D12 or D13, RF1 (a 68 Ω resistor) or PR2 (ICP-N38). They are all on the secondary side of the power supply.

Hitachi VTM622

Poor stills in the SP mode: This brand-new stock machine came from the shop with the complaint of poor stills in the SP mode. This is not actually a fault with these machines, but it does catch the unwary. On this model preference is given to the LP mode. Thus functions such as picture search and still are of poorer quality in the SP than in the LP mode.

Tape damage: This brand-new VCR had a healthy appetite for tapes. The damage was not creasing: the tape was being stretched so much that oxide had fallen off in places to reveal the clear backing. A check on the tape back tension sent the scale needle to over 60 grams. The tension pole positioning was correct, and after removing the tension band we found that the supply reel was still very stiff. When the reel had been removed a black circle mark was seen on the tape deck chassis. The spacing washer under the supply reel had never been fitted.

No servo pulses on record only: This machine was OK with pre-recorded tapes. The cure was simply to clean the audio/control head. I wonder if this is going to become a problem, as with the VT410 series?

Hitachi VTM722

Audio problem: The E-E audio was low and distorted while playback of a pre-recorded tape produced only a cyclic chirping sound. We found that the always 9 V supply to IC401 was low at only 4.9 V because zener diode ZD854 on the power board was short-circuit.

Dead, no clock, no loading motor: The P50116 microcontroller chip in this model deals with a wide range of functions. It's responsible for deck control, tuning, the timer clock functions etc. Deck control problems can easily lead one astray. A VTM722 came in recently with what looked like a power supply fault. The machine appeared to be completely dead, with no clock and no loading motor movement. As the power supply checked out all right we came to the conclusion that the microcontroller chip was faulty. It's on the top of the main panel, in close proximity to the front escutcheon. A replacement was therefore fitted, which is not easy as the print is very fine, but the fault remained as before. I then remembered a conversation with Jim from Hitachi some time ago. He told us that these machines can easily become confused if the deck is out of sync. The microcontroller chip has to deal with so many functions that a wrong signal from the mode control switch can produce total lock-up. This is in fact what had happened. With the loading motor and the mode switch removed I was able to resync the mechanism and could then reset the mode switch to position one (eject). Up came the clock and all functions worked correctly. The eject mode is quite easy to find when winding the mechanism by hand. After removing the loading block simply turn the main cam until the eject gear beneath the deck – it drives the carriage – turns when the capstan motor is rotated. This is the eject mode. Then set the mode control switch to position 1 and replace the loading assembly. It's important that the cam and the mode switch are correctly aligned. When the capstan motor rotates the eject gear, as described above, you will usually find that by turning the cam to one end then backing it off until it clicks into its first position it is in the correct eject position. Be sure always to replace the mode switch.

Wouldn't accept tape: The power-up and cassette lights were on all the time and the machine wouldn't accept a tape. The cause of the trouble was dry-joints at the end sensors.

Fuse had blown in power supply: Fuse F852 in the power supply had blown. Checks showed that there was a dead short across the 12 V rail. Many items were disconnected before we found that the r.f. booster was the culprit. A splash of solder was shorting out two pins.

Hitachi VTM822

Severe warbling on record only: The fault was more noticeable in the LP mode. Surprisingly a new pinch roller and capstan motor failed to cure the problem, whose cause was eventually traced to the reel clutch assembly underneath. This item has been responsible for a number of faults we've had – usually no play, chewed tapes, no rewind or no eject. On this occasion, however, one of its cogs was vibrating. The vibration was travelling along the drive belt, affecting the capstan. That's our theory, anyway, and we're sticking to it!

Rewind action poor: This was because the rewind gear actuating slide didn't travel far enough to engage with the gear correctly. It took us a while to discover that the mechastate switch was responsible for this. Although this would be an easy part to remove and replace, it's supplied only as part of the whole 'loading block assembly'. Good old Hitachi!

Wouldn't accept a tape: This machine wouldn't accept tapes, and when you could manage to load one the tape would be chewed. Checks around the microcontroller chip IC901 revealed evidence of liquid spillage. A clean-up here restored normal operation.

Intermittent mode: There were intermittent problems with this machine: it would stop playing or go into other modes – in fact it seemed to have a mind of its own. Resoldering the microcontroller chip seemed to put matters right, but the repair bounced. The cause of the trouble was that the connectors between the main PCB and the operating PCB were dry-jointed. Resoldering them put an end to the playing about.

Hitachi VTM830

Poor colour playback: It was sometimes difficult to see the fault symptom on this machine: the customer's complaint was of poor colour playback. It could best be seen with a dark picture, where coloured snow was present. Much of the colour processing is done on a subpanel. As luck would have it we had another of these machines in, for drum replacement. I therefore swapped over the subpanels, proving that this is where the cause of the fault lay. A check on the waveforms at R326 (270 Ω) proved to be useful as the waveform at the end connected to the 2 H delay line CP303 did a nose dive, being hardly visible. When we replaced CP303 we were rewarded with a very good picture.

Would randomly go into odd modes: This machine would go into some rather odd modes at random. We found that the microcontroller chip IC751 was dry-jointed. After resoldering all the connections and a long soak test all functions worked correctly.

Sound problem: This machine didn't record sound and there was no E-E sound whilst recording. When the record button was pressed there was a noise as if the sound was being strangled: sad, really! The sound feed to the modulator and the scart socket is from pin 30 of the LA7297 record/playback processor chip IC401, which also switches on the 9 V supply to the bias oscillator. The bias oscillator transistor Q401 had died.

Dead, fuse F852 blown: When a replacement fuse was fitted the machine accepted a tape but wouldn't lace up. The XRA6209 (BA6209) loading motor drive chip had failed. After fitting a replacement I found that the machine loaded very slowly. The cause was the loading motor, which was taking amps off load! A new loading block, which incorporates the motor and mode switch, cured the fault.

ITT

ITT VR3907
ITT VR3918

ITT VR3907

Shut down 3 seconds after play: This was a good one. The machine would wind and rewind perfectly but in play or record it would lace up, run for 3 seconds then shut down. The counter is electronic and counted during those 3 seconds. We thought this indicated that the take-up reel pulses were OK, but we were wrong. They were there but of low amplitude because Q0610 had gone low gain. The pulses were adequate for the counter but of insufficient amplitude to tell the microcomputer chip that the take-up was working. It took rather a long time for the ITT supplier to provide the spare part – we were told that this particular machine is a Samsung clone.

Machine wouldn't switch off: The channel indicator was lit, so was the power-on LED. If the machine was playing and the power button was pressed the tape would unload but the lights remained on and the E-E voltages remained. We found that a transistor, TR2, in the power supply was leaky. A TIP42 is a suitable replacement.

Hum bars on E-E picture: Check the $100\,\mu F$, $100\,V$ smoothing electrolytic capacitor C4 connected to the $33\,V$ line.

ITT VR3918

Smell from mains transformer: This machine came in dead shortly after the guarantee ran out. When we powered it the mains transformer rapidly heated up and produced a terrible smell. A replacement from a stock machine was fitted but when we switched on the mains fuse blew. After disconnecting the secondaries we tried again. This time there were no signs of distress. When the bridge rectifier D5001–4 was reconnected the fuse blew, but there were no measurable shorts. So we changed the STK7226 power supply regulator. This time at switch-on there was no fuse blowing but there was still no 13 V line as protection diode D5009 had failed. It seems strange that there's no fusible protection for the transformer's secondaries, as the transformer gets very hot under fault conditions before the mains fuse fails.

Cassette lift fault: I expected this machine to be a JVC clone but as soon as I removed the covers I saw that I was wrong (I found out later that it's of Sanyo manufacture). There was a cassette lift fault – the machine was reluctant to accept a cassette. It could take four attempts before a cassette would be taken in, and would only partly eject the cassette once it had been taken down. The cassette lift is operated by a gear off the capstan flywheel, so I wound a cassette in manually and pressed play. The machine threaded up but the capstan didn't turn. Luckily I was able to obtain a manual: then battle commenced! The capstan motor wasn't being turned on in the play mode, and on the rare occasions when the capstan did turn, to operate the lift, the syscon didn't seem to be able to read the lift's position, even though the limit switches were all OK and the signals were reaching the syscon chip. A study of the block diagram showed that a signal should go to the syscon chip from the capstan FG. It was missing. A new LC74128017 chip brought it back and restored normal operation.

Tape ejected when record selected: At last a simple one! The erase prevention switch had become disconnected from the frame of the deck. As it didn't open, it didn't tell the microcontroller chip that the tab had been removed. I refitted the switch with a tiny spot of glue so that it couldn't fall off again.

When play was selected the machine would unload: Rewind and fast forward were perfect, but when play was selected the machine would instantly unload. The drum didn't rotate, hence no pulses to the microcontroller chip. During play pin 14 of IC4001 was low at only 0.8 V instead of 2.5 V. Fitting a new LC7142–8017 chip (very expensive) restored the correct voltage and drum rotation.

JVC

JVC HR7200	JVC HRD580
JVC HR7300	JVC HRD610
JVC HR7700	JVC HRD640
JVC HRD110	JVC HRD660
JVC HRD120	JVC HRD700
JVC HRD140	JVC HRD720
JVC HRD150	JVC HRD750
JVC HRD170	JVC HRD820
JVC HRD171	JVC HRD830
JVC HRD180	JVC HRD860
JVC HRD225	JVC HRD880
JVC HRD230	JVC HRD910
JVC HRD320	JVC HRDX22
JVC HRD400	JVC HRFC100
JVC HRD455	JVC HRJ200
JVC HRD4700	JVC HRJ205
JVC HRD520	JVC HRS4700
JVC HRD530	JVC HRS5000
JVC HRD540	JVC HRS5800
JVC HRD560	

JVC HR7200

Capstan speed slow: This machine was fitted with the later PU21235A motor-drive amplifier panel. The problem was with the capstan speed – it was slow. We found that Q210's emitter-base junction was open-circuit while its collector-emitter path was leaky.

Aerial input socket problem: These fine machines go on and on. A weakness is the coaxial aerial input socket which is directly connected to the booster unit. The socket has a tendency to work loose, resulting in poor contact with the aerial plug or even breakage. If the socket is broken it can be replaced, using a beefy soldering iron, but if it's intact the contacts can be carefully realigned to ensure good contact with the aerial plug, after which a reinforcement ring from a scrap TV coaxial socket should be fitted. I wonder why JVC didn't fit such a ring in the first place?

No colour playback: This oldie had no colour playback with known good recordings and no drum lock or chroma with its own recordings – no prizes for spotting the connection. The fault could be instigated and cleared by touching anywhere on the bottom PCB. We eventually found the dry-joint on one leg of C347, which is connected to one leg of IC402. The leg had a very fine ring around it.

JVC HR7300

Capstan servo fault: The pull-in range of the phase discriminator control R10 was very poor and it couldn't be set for 6.2 V at TP203. The waveform at TP5 was wrong, the positive part of the sampling pulse being much too large. A lot of time was spent checking around IC3, but no reason for the fault could be found. Over lunch I studied the block diagram and discovered that one end of C38 should be earthed in playback – it wasn't. Also pin 15 of IC4 should be at 8.6 V and wasn't. IC4 (IR2403) was faulty.

No E-E signals and no tuning signal: The symptoms sometimes found with this range of machines are no E-E signals and no tuning signal although the r.f.-through signal is OK and the deck mechanics all work. In these circumstances check for a dry-joint at the always 9 V regulator transistor Q101.

Drum would stop after a few minutes: We still see a large number of these excellent and on the whole reliable machines. With this particular one the drum would stop after a few minutes and wouldn't restart. An initial check on the supplies (we all do that, don't we?) proved to be a good move as the 12.5 V rail read 15 V. It's derived from Q1 on the power supply board. This was OK but further back the sensing transistor Q3 and its emitter zener diode D5 were faulty.

Black flashing line on recordings: The cause was traced to the 12.5 V supply being too high: we found that the 1 kΩ set-up potentiometer R5 was open-circuit. A replacement enabled the 12.5 V supply to be set up correctly, restoring normal operation.

JVC HR7700

Mechanical overhaul problems: I quoted for a mechanical overhaul – belts, idlers etc. plus loading rollers, audio/control head and video heads. Much to my surprise the customer accepted the rather high price – the machine is a good one for features but the customer had also just bought an FV14. When work started I soon hit problems. While changing the loading rollers I found that the small PCB for the loading photo-interrupter was cracked, so a replacement had to be ordered. Then when the machine was put on soak a further fault developed – there was no sound or vision in the E-E mode, just a black raster. The V-V switching worked all right and the root cause was absence of the E-E 12 V supply. This follows quite a complicated path, originating in the horrendous system control (mechacon) circuit. Here it's derived from the 12 V rail by the microcomputer chip via transistor X15 (2SA1020) which had a 900 Ω leak between its base and emitter.

Wouldn't record any picture at all: This machine played back OK but would only record noise. Someone else had recently fitted a new head. Naturally we assumed that it was OK and spent a lot of time making checks and setting up the circuits. When we replaced the head with one obtained from Ferguson normal results were obtained. The head we took out worked all right in a 3V22.

Tape did not lace in play: Rewind and fast forward were OK but when play was selected the machine loaded the tape half way then stopped – the drum also stopped. To unlace the tape you had to use the on/off button. All the sensors and switches were tried, then all the panels, but still no success. We finally tried a new drum assembly, after which the machine worked perfectly. The rotor base unit in this assembly has two magnets opposite each other at the bottom. One of them was missing, so there was no pick-up pulse.

JVC HRD110

Intermittent sound muting: The problem was intermittent sound muting, recording and E-E depending on the picture content. Dark scenes produced sound muting, bright scenes brought the sound back, while in between the sound fluttered and caused a disturbance to the picture's luminance level. The cause of the fault was traced to L7 on the tuner/i.f. panel. It should be tuned to 15.625 kHz but was found to be off frequency. Normal operation was restored when L7 was peaked, using an oscilloscope, but at resonance the core was fully tightened. An unnumbered tuning capacitor within L7's can was thought to be defective. Adding an externally mounted 330 pF ceramic capacitor remedied the situation.

Erratic servo lock playing back own tapes: This machine had a fault that was unusual to say the least: playback was OK but the machine's own recordings played back with erratic servo lock. If the machine was stood on end it made good recordings! There are two capstan FGs in this design, one in the motor and one on the belt-driven flywheel. The bracket that supports the bottom of the flywheel shaft had become somewhat bent. As a result the FG printed coil had moved away from the capstan reducing the FG output at TP402 to 0.2–1.1 V.

Capstan fault in play mode: Rewind and fast forward were normal but when play was selected the capstan ran flat out, giving the fast search symptom. Plug CN11 on the servo PCB was dry-jointed and loose. Resoldering didn't cure the problem: soldering the leads direct to the PCB did.

Shut down after lace-up attempt: A new loading belt was fitted but made no difference. The cause of the fault was traced to the after-load leaf switch unit, which is mounted close to the loading motor – one of the contacts had broken off.

JVC HRD120

Intermittent r.f. or converter fault: It was difficult to pinpoint whether the fault was in the r.f. amplifier or the converter – it was very intermittent. We decided to remove the aerial booster unit, solder all suspect joints and refit it, but the fault was still present. Taking the same course of action with the converter unit put matters right. Prior to doing this we'd checked the supply rails and found them to be correct. When we'd done all this we found that playback was faulty, the symptoms looking like the effect of misadjusted tape guides. The cause turned out to be different, however: the back-tension brake band was broken at one end. Replacement plus back tension resetting was required.

Syscon chip failure: A good rule of thumb with microcomputer-based mechacon and syscon circuits is that the main microcomputer chips themselves seldom fail. Before replacing them check the relevant d.c. lines, the clock pulses etc., and remember that the various buffer chips are more prone to failure than the microcomputer chips. We've recently had two HRD120s that proved to be exceptions to the rule. In the first the machine operated normally for about 10 minutes before jumping into the timer mode, after which the machine became totally non-functional. Liberal squirts of freezer didn't have any effect. We found that, in the fault condition, the input to the CPU chip from the timer switch at pin 35 wasn't activated. The conditions at the rest of the CPU's pins appeared to be correct, with the trains of pulses on data line pins 27, 28, 29 and 30 showing some activity on the scope. After removing half a tube of Japanese Evostick from the chip's 52 pins and fitting a replacement the machine worked normally. The second machine couldn't be switched on by the operate switch. Again the conditions at the CPU's pins all appeared to be correct, with the level at pin 36 changing when the operate switch was selected, but this time there was a distinct lack of activity on the four data lines that connect to the input/output expander IC202. A new CPU chip was again the answer.

Noisy capstan motor: We've found that the capstan motors in quite a number of these machines have become noisy. This normally has no effect on the quality of the picture and sound, and when the owner is confronted with the price of a replacement motor he's usually prepared to live with the noise. If a particularly bad motor is run in the play mode for a few hours, however, check that it still has sufficient torque to perform tape unloading properly – a loop of tape can be left outside the cassette and this will be damaged when the cassette is ejected.

Drifted off tune on BBC-1 only: My first thought was to put BBC-1 on another button and see if this drifted too. It did. The tuning voltage is obtained from a voltage doubler circuit, which according to the circuit should produce some 45 V at TP6. Ferguson told me that the voltage here should be a bit higher than this but if it gets very high, say 70 V, all the components in the voltage doubler circuit should be changed – D14, D15, C21, C22, C23, C24, C25 and R13. I did this and the machine was OK for about four weeks. Then it came back again. This time I decided to change the HZT33–02 33 V stabilizer D10. The machine seems to have worked all right ever since, but I would still recommend that anyone experiencing tuning faults with these machines should look at that doubler circuit. It had gone haywire in this particular machine and no particular component could be blamed.

Sound would pop and crackle: The record and playback picture quality were good but in both modes the sound would 'pop and crackle'. In the E-E mode there was no sound. The 2SD636 transistor Q5 should have

9 V at its emitter in this mode, but the reading was 0 V. A replacement transistor restored normal E-E sound but record and playback remained distorted. Scope checks on the audio waveform while a tape was being played back showed that a normal signal entered the LA7042 chip IC1 at pin 2 but the output at pin 3, after passing through a preamplifier, had a d.c. level that fluctuated wildly. Pin 3 is connected to the 150 Ω playback level preset control R6, which turned out to be faulty – in fact it fell to bits when it was gently removed during the fault diagnosis stage. A replacement restored normal sound.

JVC HRD140

Clock display failed after an hour: The cause was the TL066CP reset chip which changed the state of its output when warm, holding the clock chip in the reset condition.

Head drum twitching: The customer complained that the tape wouldn't play. When we tried we found that the tape would lace up and then immediately unlace. The head drum wasn't going round but could be seen to be twitching as if it was trying to start. Voltage checks revealed that the control from the servo was missing (no voltage at pin 2 of CN403). Tracing back we found that the 5 V zener diode D408 was short-circuit, but replacing this didn't get the drum running – we also had to replace the AN6671K head motor drive amplifier chip. This required complete removal of the mechanism to gain access to the PCB.

Tapes damaged in backwards (search) mode: The deck used in this machine is similar to the one in a whole range of JVC-based machines – the HRD140, 3V43 etc. Sometimes tapes are damaged in the backwards search (review) mode. This is due to the tape buckling and bubbling between the capstan and pinch roller. The cause is an out-of-vertical guide arm, the one between the capstan and the TLL reel. It's better to replace this than to bend it – part number PQ41384A.

No power indication/drum rotation: Here's a nice simple one for a change. In the event of no power indication and no drum rotation check circuit protector CP4 for being open-circuit.

Servos did not run at right speed: The servos did anything but run at the right speed. We found that the cause was lack of the FSc signal to the servo section. The cure was to replace the chroma subpanel, part number PU22046A.

JVC HRD150

Clonking noise in record and playback: The owner of this machine must have had super-sensitive hearing – or a shelf or trolley that acted as a sounding board! He complained of a barely perceptible clonking noise in the record and playback modes. In a very quiet part of the workshop we could hear it: the noise was coming from the area of the supply spool turntable. We found that the supply reel clutch pinion was very slightly eccentric. This was proved by watching and listening, while we spun the supply reel by hand with the back-tension band slackened. A replacement clutch assembly eliminated the trouble.

No response when 'operate' key pressed: There was no response when the 'operate' key was depressed although the deck display was present. The main power supply section was OK but the microcomputer chip IC601 had no 5 V supply. This comes from the unswitched stabilizer transistor Q602 which in this case had zero voltage at its base and emitter. We expected the fault to be in zener diode D604 but in fact the parallel 22 nF capacitor C605 was short-circuit.

Play symbol always on: The play symbol, a dotted triangle, was lit up the whole time the machine was switched on, whether or not the play mode was engaged. It was caused by leakage between pins 4 and 5 of the fluorescent display PCB. Someone must have managed to spill liquid through the cassette loading slot! Thorough scrubbing with surgical spirit removed the conductive deposit in this very high-impedance circuit.

Fails to go into record or playback: This machine would sometimes fail to go into the record or playback mode, switching itself off in the loaded state. The cause of this is usually hardened grease and a worn belt on the loading block, but a replacement block failed to cure the fault. An optosensor that's fitted to the underside of the deck was the cause of the trouble.

JVC HRD170

Remote control failure: The remote control handset was first brought in on its own, with the complaint that it didn't work. It lit up our magic mirror all right, so we asked for the VCR itself. This presented a problem: the owners were quite unable to manage without it . . . A loan machine got round that. The infrared receiver and preamplifier worked correctly, and strong pulses were reaching pin 26 of the microcomputer chip IC601. Ceramic filter CF601 was all right so the chip was suspect. A replacement – type M50731–610SP, at a net trade cost of £16.50! – solved the problem.

Channel change fault: If you encounter one of these machines with a channel change fault, be wary! We've now had two in which the channel could not be changed using either the front-panel keys or the remote control unit. Since neither the tuning nor the channel digit display changes you might reasonably suspect the clock-display and key-decoder microcomputer chip on the front panel. Not a bit of it! With both the machines we had in, the bug responsible was IC2 on the tuner/i.f. panel. It's type M50440–391SP.

Intermittent function fault: This machine had been to another dealer with the now familiar 'intermittent function' fault. In return for a large sum of money a new set of carriage end-sensors had been fitted. This had failed to cure the trouble of course and the machine's owner, being unable to obtain either satisfaction or a refund, brought the machine to us. As usual we found that the small earthing screw on the motor subpanel was loose. After putting this right we connected the machine to the mains supply and switched on. It immediately went into the fast-forward mode. When a cassette was loaded the machine acted normally. We noticed, however, that when the tape was ejected the loading motor at the side of the carriage continued to run for a further 5 seconds. Also the machine went into fast forward every time it was disconnected then reconnected to the mains supply. A quick look at the end sensors showed that the 2-pin plug on the left-hand sensor had been left off. Refitting this restored normal operation.

Tape-in indicator would work with no tape: When this machine was connected to the mains supply the clock flashed as usual but the tape-in indicator came on although a tape hadn't been inserted. When the operate button was depressed the indicator remained out and the machine went into the rewind mode for 2 seconds after which it shut down. An STK5481 chip is used in the power supply. This seemed a good place to start and a replacement restored normal operation.

Loss of capstan servo control: The problem with this machine was loss of capstan servo control – noise bars ran through the picture. The control track pulses were OK at pin 20 of the servo chip IC2, and someone had already changed this i.c. So further investigation was required. We eventually found that C25 (4.7 μF, 25 V) was the cause of the trouble. It tested OK but a replacement cured the fault.

JVC HRD171

No drum rotation: Fast forward and rewind were all right but when playback was selected the tape laced up then, within a few seconds, unlaced because there was no drum rotation. After wasting a lot of time we found that the voltage at pin 20 (drum start/stop) of the VC2025 chip IC1 didn't go high when play was selected. Pin 20 was internally shorted to chassis. A complete stator/MDA unit cured the problem.

Cyclical tracking bar in play: This machine worked in all modes except play, when a cyclical tracking bar would travel from the bottom to the top of the screen with a slur on the sound as the bar passed. A check on the control pulse at pin 6 of IC2 (M51796P) showed that a nice 5.2 V peak-to-peak square wave was present here. It should be passed to pin 20 of the V2023A servo chip IC2 via a 10 kΩ resistor but was missing at this point. A voltage check here produced a reading of 5.2 V instead of 3.4 V: pin 20 had shorted internally to the 5 V line. A new chip cured the fault.

No functions: The complaint with this machine was no functions. It took some time before we realized that the four circuit protectors in the power supply were going open-circuit intermittently – sometimes you would get a voltage reading, sometimes not.

JVC HRD180

No display, no record: We've had some mains power interruptions here lately. This was the reason why two identical machines came to be sitting side by side on the bench, with identical complaints – no display at all, and no record. When tested both machines worked perfectly in all respects. Two more cases of microcomputer lock-out! The micro-computer chips were obviously reset when the machines were powered up and switched on in the workshop.

Failure of STK5481 power supply i.c.: We've had several cases of failure of the large STK5481 power supply chip used in this and allied models. The switched 5 V output fails while circuit protector CP3 remains intact. You'll find that the symptoms are no 'on' LED indication, no drum rotation when playback or record is selected, and shutdown after a few seconds in the fast-forward or rewind modes. None of the replacement chips has failed to date, so maybe it was batch problems or perhaps the design of the chip has improved.

Poor quality playback picture: Here's a case where the fault symptom and the defective component were not clearly related, the sort of thing that causes us considerable hardship. The machine was a JVC HRD180 with which Mastercare was having problems. I was asked to help out as a JVC authorized service centre. More often than not the playback picture was very poor. The fault was easy to see – the poor quality playback picture had lots of very small herringbone squiggles all over it and a grainy look, with black tadpole-like spots trailing across on peak whites. The first thought that crossed my mind was the obvious one that the outputs from the heads were low. Our local Mastercare branch is very good about supplying engineer's notes, however. They showed that the heads and the HA11870NT preamplifier chip had been replaced, also that there had been some problem with the power supply as a regulator had been changed twice along with some circuit protectors. To be on the safe side I spent some time just cross-checking, in case some point had been missed. This included verification of the preamplifier operation, the LP/SP switching and the adjustment of the head tuning circuits. The latter were left temporarily misaligned as it was impossible to determine their correct alignment at this stage. The head drum seemed to lack power, and gave the impression that it was running on the slow side, but a locked picture was obtained. At one point early one morning a good quality picture came up and then deteriorated before I could check anything. Some recordings were made as a check. They played back reasonably well on another machine. So it was difficult to know exactly what were the fault symptoms and what were red herrings. Scepticism about everything seemed to be the only logical approach to adopt. Time was spent checking the operation of the luminance playback circuitry. This included changing the PU22282A chip IC101. All to no avail, except that the upper and lower edges of the f.m. waveform at pin 22 of IC101 were seen to be ragged and not smooth as with an HRD170 used for comparison (it's the single-speed version, and happened to be hanging around). I next hijacked the HRD170's preamplifier and tried it in the faulty machine. This move was inconclusive but suggested that the preamplifier was all right. I was still suspicious about the f.m. carrier, however: it seemed to contain excessive h.f. content, causing beat patterning. So if the preamplifier was OK the problem was possibly due to the filter circuit between the preamplifier and IC101. This was checked, but no evidence was found to suggest that there was a faulty component here. The only likely item left was the lower drum unit with the rotary transformer and motor. This was changed, but the symptoms persisted.

I next tried some panel jockeying to try to isolate the cause of the fault. The preamplifier was again swapped, this time for an HRD180 one. The servo panel was swapped and the power supply rails were again checked for level and ripple. I also checked the r.f. unit, but as the testing was on the video output this was a move of desperation. The HRD180 was then put aside for a while for further thought – actually the further the better . . . After a week or so the situation was reappraised. So far I'd eliminated the video heads, the lower drum and rotary transformer, the preamplifier, the luminance processing, the power supplies and the servo, also the tuner and r.f. modulator. A helpful suggestion from JVC was that the only thing left was the cabinet. Thanks, Kevin! Eventually one quiet Saturday morning the HRD180 was put back on the bench. The fault had not gone away. There was still a poor picture and the f.m. waveform had fuzzy edges. That bothered me: you know the feeling in the back of your mind, a nagging 'there's the clue'. Then it dawned! Clearly the cause of the problem, if it was not the f.m. carrier filters, was that the signal from the video heads was not of the correct frequency. This in turn meant that the drum speed was incorrect. As the servo and drive amplifier, along with the power supplies, had been checked and cleared, what else was there that could affect the drum motor? Only the stop/start signal from the syscon microcomputer chip, but as the drum stopped and started correctly this must be operational. Pin 7 of the M50965–612SP system control microcomputer chip IC601 should be high for drum 'off' and low for drum 'on'. It sat at 3 V and had a sawtooth waveform on it. So that was it, the sawtooth was modulating the speed of the drum motor and the f.m. signal coming off the tape! Once the microcomputer chip had been replaced clear pictures were obtained. Guides, switching points and the head tuning circuits were then set up and the repair completed. Poor picture quality due to a faulty syscon microcomputer chip, would you believe it? I'm sure that the trade price of a new machine was less than my invoice, despite the full time not being charged.

Wouldn't record new video signals: Sound was recorded and the previous video was erased, but the new video information was missing (if the full erase head was disconnected temporarily, the previous video was left). A check on the pre-rec board showed that the /REC line didn't go low. The cause was a dry-joint at the ribbon cable link (CN2) between the mechacon board and the video board.

Failure of front-panel keys: If the fault with one of these machines is permanent or intermittent failure of some of the front-panel keys (station selection etc.) to operate, look for dry-joints at CN1, where it's soldered to timer PWB no. 15, before getting involved with the front panel.

JVC HRD225

Would not unthread correctly: This machine functioned correctly apart from the fact that when unthreading the supply spool wasn't driven in order to pull in the tape as the loading arms retract. Now although this model has a reel idler to drive the spools it doesn't have a reel motor. Instead the idler is driven by a pulley, which in turn is driven by a belt from the capstan. Whilst unthreading the capstan motor wasn't being driven and we found that in this mode the drive transistor Q206 was without base bias, although the 5.7 V bias was present in the play and fast wind modes. A study of the circuit diagram showed that in play the capstan drive comes from the servo chip, in fast wind it comes from the CPU chip IC201 and during unthreading it's switched by the expander chip IC202 (pin 38). This pin was found to be permanently low, all the other ports relating to capstan motor operation being correct. So, having encountered a number of faulty expander chips in this model, I replaced the i.c. This made no difference! Further checks revealed that R272 (10 kΩ), a bias resistor in the drive transistor's base circuit, was open-circuit.

'Sound deteriorates when hot': We found that when the machine had been running for an hour or two the E-E sound became distorted and threatened to disappear altogether. The playback sound remained OK. A scope check showed that the audio signal coming from the receiver section was 'strangled' at TP4 on the tuner/timer panel because the inter-station mute circuit was coming into operation. This is based on the action of the 15.625 kHz tuned circuit T5. Careful adjustment of T5 to peak up the line-rate waveform at the collector of Q11 overcame the problem.

Picture looked as if the line hold was off frequency: You could see that the head drum was revolving too fast. After a few seconds the tape stopped and unloaded. Our first check was on the drum FG pulses at pin 4 of IC404. The waveform here and also at pins 6 and 11 was correct. Moving on to IC406 we found that the voltages at pins 15 and 13 varied while there was zero voltage at pin 14. The 12 V zener diode D412 was short-circuit.

Won't accept a tape: Check whether the supply spool rotates for a short time. If it does, check the l.t. supply to IC204. Should this be missing circuit protector CP1 (F15) is probably open-circuit. If it doesn't, check the voltage at pin 37 of IC201: this should be in the low state without a tape in the cassette housing. If it isn't low, check the up/down detector switches by replacement.

JVC HRD230

No eject, squealing noise from loading belt, then shutdown: This seems to be a fairly common fault. The symptoms, from switch-on, are no eject and a squealing noise from the loading belt, rapidly followed by machine shutdown. If the machine is in play when the trouble occurs, the tape is unlaced then the squeal comes, followed by shutdown. These things stem from a loose screw that secures the deck terminal PCB under the deck. If it's not tight, earth continuity to the mode switch and the take-up FG is lost – intermittently.

Intermittent field roll: The fault occurred when any tape that the machine had recorded was played back. It was OK in the E-E mode. The f.m. waveform was checked first. This cleared the drum input and output guides of any blame. Attention was then turned to the video processing chip IC101. The output at pin 24 was correct. It's fed back to pin 5 where the field sync pulses were found to be crushed. The culprit was the coupling capacitor C134 (2.2 μF, 50 V).

Machine operation 'haywire': Another easy job I thought – the customer complained that the machine's operation was haywire and my tests confirmed this. I replaced the STK5841 chip and the machine seemed to be all right, but on soak test it went haywire again. This time attention was directed to the mechanics, which were found to be dry and stiff. After removing the mechanism baseplate and cleaning and lubricating the mechanism the machine worked well.

Intermittent play and record: There's a well-known weak spot in certain JVC VCRs. It's cured by fitting a shake-proof washer to the under-deck PCB fixing screw to ensure a good electrical earth connection. This is not relevant to the HRD230, but we had one whose play or record would be suddenly interrupted at random times, reverting to the stop mode. It was because the reel pulses disappeared. To obtain a reliable earth connection for the optocoupler we had to fit a tiny shake-proof washer to the reel sensor PCB assembly's fixing screw.

Large hum bar: In the E-E and playback modes a single, large hum bar travelled from the bottom to the top of the display on the monitor's screen. A check on the switched 5 V line showed that a distorted 500 mV, 50 Hz square wave was present. We initially suspected C14, which decouples the switched 5 V output at pin 3 of the STK5481 chip IC1. It was OK, however, the cause of the fault being the chip itself.

No capstan rotation: The machine would cut out in play or record as soon as it had finished lacing up. The M54644BL drive chip IC604 was found to be faulty, a new one restoring activity. But the wow and flutter were atrocious. No wonder since the motor was extremely tight. A new one prevented another drive chip biting the dust.

Request for new drum to be fitted: The customer asked us to fit a new drum in this four-head machine. He'd cleaned the heads himself, then come to the conclusion that they were faulty. What he'd actually done was to break all four head tips. When we'd replaced the upper drum the original fault was apparent. Every time that play was selected, with a known good tape, a portion of the f.m. playback carrier was missing, giving the impression that the heads were clogged. At one point there was so little f.m. signal that the tape might as well not have been wrapped around the drum. Thus a mechanical fault was ruled out. In this model the drum PC and FG signals are combined and leave the drum together at pin 3 of connector CN3. A scope check here showed that the PG pulses were missing. After some careful testing we removed the PG pick-up and found that it was open-circuit. Unfortunately parts for the drum are not available individually. But we found an identical pick-up in a scrap HRD170, although the motor is different. Fitting it cured the fault.

Intermittent failure to play or record: The tape would lace up, but the drum wouldn't rotate. Voltage checks showed that the motor 12 V supply and the motor drive and motor run voltages were all present. When I removed and dismantled the lower drum I found that the ICP on the drum motor's PCB was dry-jointed. Resoldering provided a complete cure.

Would intermittently stop: This machine would intermittently stop while in the play or record mode. On test it appeared that the capstan motor was stalling. As there was no stiffness in the motor I came to the conclusion that it had a dead spot. A new motor cured the trouble.

Channels slightly off tune: But when fine tuning was carried out and the channel was stored in memory it reverted to the original mistuned condition. The culprit was the MN1220 EPROM chip IC101, which is conveniently plugged in piggyback fashion at the rear of the front control panel.

Bent verticals on playback picture: The playback picture was marred by bent verticals which extended from the bottom to the top. Clearly the head drum was hunting, but why? I've come across various causes of this in the past, including a defective drum motor, dried up electrolytics in the power supply, and drag because of loss of nickel plating on the upper drum. This time the cause was very simple. A squeal came from the earthing brush on top of the head drum. When the slightest pressure was applied to it the squeal and the hunting stopped. Cleaning the brush and applying light lubrication cured the fault.

JVC HRD320

Three buttons inoperative: The problem with this brand new machine, straight out of the box, was that three of its buttons were inoperative set–, set+ and Ch. set. On investigation we found that D11 on the timer/display board had been fitted back to front. It's part of the key-scan matrix. The diode was undamaged and fitting it the correct way round restored normal operation of all the buttons. The same symptoms would arise with other makes and models fitted with this type of timer/display board.

Very intermittent playback picture loss: Days would pass without the fault showing. Sometimes it would tease us for a few minutes at first switch-on from cold. We tried replacing the luminance chip IC101 but this didn't cure the problem. The symptom finally stayed for long enough to enable us to do some scope tests. These proved that low-pass filter LPF102 was the culprit. These filters have many internal joints, which probably explains why they are so often the cause of exasperating intermittent signal problems.

Would accept a tape very slowly, then ejects it and switches off: Sometimes the machine would try lace-up/fast forward/rewind etc., accompanied by unhealthy noises (crunching and groaning) from the mechanism, then shut down. Checks on the STK5481 regulator chip, always a good place to start, showed that pin 3 was at 1 V instead of 5 V and pin 4 at 0.7 V instead of 12 V. Replacing the regulator restored normal operation.

Clock reset: This would happen with annoying irregularity. Otherwise the machine was perfectly OK. Once in a while the display said 'video' and it locked up. I decided to phone JVC (after the usual fruitless search that occurs when a machine develops a mind of its own) and was told that the back-up capacitor C3 (0.1 µF, 5.5 V) had given up the ghost. A replacement restored normal operation – it's mounted on the display panel. My thanks to JVC for help with this one. The Ferguson FV21R seems to be very similar, so the fault might also be experienced with this model.

JVC HRD400

No E-E or recorded sound: If you come across an instance – possibly intermittent – of no E-E sound and no recorded sound, check for a hairline crack in the print at the rear of the main PCB (03) adjacent to the left-hand unsoldered lug of the r.f. modulator can. It seems that the modulator 'rocks' when the aerial plugs are inserted or withdrawn, stressing the PCB.

Intermittent record and playback: This was the result of dry-joints within the luminance module. Although it bristles with surface-mounted components, it's possible, with care, to provide a cure by resoldering.

No mechanical functions: There was no play, fast forward, rewind or eject, the machine going to standby after a few seconds. We found that the reel brakes were jammed on hard and the idler was jammed on the brake mechanism. The clutch spring, which is used along with the 'windmill' to release the reel brakes, was broken and when the mechanism had been stripped down I found that the main cam was also damaged. The slide encoder, the main cam and the clutch spring were replaced and after realignment of the mechanism everything worked correctly.

JVC HRD455

Periodic tracking bars: This machine actually wore Saba livery. The complaint was of periodic tracking bars, the tracking knob having no effect. We found that when play was selected the machine seemed to be in fast search for 2 seconds, then stabilized. Replacement chips in the servo section had no effect. When an off-air recording was tried we noticed that the channel number wouldn't change and the memory couldn't be used. The drum and capstan speeds were erratic and there was slight hum, the picture flicking about. A check on the switched 12 V line produced a reading of 19.5 V. All these problems were being caused by Q2 in the power supply: it was leaky all ways.

Machine dead: This because of a dry-joint at CN1 in the power supply.

'Won't accept a tape': The cause was failure of one or both of the cassette detect microswitches on the cassette housing. We replaced them both.

JVC HRD4700

S-VHS instability: The complaint was that it wouldn't record via the S-VHS sockets while the S-VHS output signals were inherently unstable. We found that it recorded and played back very well in the S-VHS mode. The playback instability occurred with titles that had been added by the amateur film maker who owned the VCR. We told him that the black-level clamping in his titling machine wasn't up to much – if indeed it had such a thing.

Intermittently noisy picture: It looked like head clogging, but this wasn't the cause of the fault. The lower drum assembly had to be replaced because of a problem with the ribbon cable that carries the r.f. signals to and from the heads.

JVC HRD520

Failure to complete tape loading: As soon as the guides had gone fully home the loading belt at the top right-hand side of the deck would slip and squeal loudly for some seconds until the machine shut down in standby. In this half-loading deck design the pinch roller is lowered into place at a late point in the loading cycle, when a peg on the underside of the pinch roller pressure lever drops into a groove on the control cam. This one was getting stuck in the tight double-bend there. We cured the problem by easing the profile of the inner groove and lubricating the groove and peg with white Molykote grease.

No playback sound: We found that the sound system was muted because the /EE control line was low at 2.6 V – it should have been at over 10 V during playback. The source of this control line is pin 32 of the microcomputer chip IC601. Leakage inside this chip was pulling the line down – we proved this by disconnecting the pin, whereupon the line rose to 10.4 V. Replacing IC601 cured the trouble but the curious thing was that the chrominance, luminance and other sections of the VCR still functioned in the playback mode despite the /EE line being at 2.6 V.

Mechanical malfunctions: For mechanical malfunctions such as the tape being ejected while still laced up try changing the control cam, part number PQ32413. The latest type is made of grey plastic instead of white.

Tracking lines on pre-recorded tapes: On inspection you could see that the tape wasn't riding along the knife-edge on the lower drum. As the rotary guide locking screws were loose the guides were adjusting themselves as the tape played. All that was necessary was to reset the guides and tighten the screws. Had the phantom fiddler passed this way?

Counter failed in fast transport: The half-loading mechanism in this machine enables the counter and index functions to work in the fast-forward and rewind modes when the tape isn't wrapped around the drum. This machine's owner insisted that it sometimes failed to count in the fast tape transport modes. We found that the counting worked perfectly if fast forward or rewind was entered from stop after play, but if fast forward or rewind was selected immediately after tape insertion the guide pole failed to pull out a tape loop and there was no count. The mode switch was responsible. Since we had none in stock we dismantled and cleaned the original one, which had tiny black spots on its stator contact bars.

Loading-mechanism or mode switch fault: If you get one of these machines with this sort of fault, e.g. the pinch roller drive peg hits the end of its cam, check that the slit washer (item 40 in the mechanical parts list) is present and correct. If it falls off, the sliding plate assembly's teeth can jump across those of the control cam.

Intermittent tape looping: This machine had been to several service departments, always with the same complaint. After a great deal of time we traced the cause of the fault to the mode switch, which is available from CPC under part number TNPV60622–1-1.

Muted sound with lines on picture: Check the f.m. waveform. This will show that the guide poles are misaligned. Under these conditions the VCR may record and play back its own tapes all right but it won't play back pre-recorded tapes.

Pictures marred by mistracking bars: This machine produced off-tape pictures that were marred by multiple thin mistracking bars spaced progressively closer towards the top of the screen. The cause was failure of the entry guide to go fully home – its base jammed in the plastic guide rails that had become distorted or warped. Some careful paring with a file or a very sharp knife gets things moving again.

Mistracking: Had I been more thorough this one wouldn't have bounced on me. The mistracking was because the exit guide locking grub screw had been loose. The guide had rotated when the owner used a cleaning tape. I reset the guide and locked it. A few weeks later it was back with a similar problem, this time because the same thing had happened to the input guide.

Picture rolls in forward search: Check whether the brass retainer for one of the slant pole guide blocks – usually the supply side one – has come out. It may have been forced out of the guide block because of tape reclaim failure when unloading: some customers then pull the cassette out so hard that the tape breaks! The mode switch is well worth checking. It may be that the tape loops around the guides etc. A component tester is great for checking mode switches. Check all combinations of the switch pins: if you see any raggedness, throw the switch in the bin. The trace will be in either one position or the other, with no in-betweens. Try the tester with a new one for experience.

Loop of tape on eject: The well-known cure for this and various other mechanical problems (in this and similar models) is to replace the mode switch – we change the main cam as well. One machine we repaired in this way bounced straight back with the same fault – and a very irate customer waving a damaged Snow White tape! We found that there was some toffee-like substance on the felt strip of the back-tension band and the corresponding periphery of the supply spool turntable. As a result the turntable 'picked up' the band now and again during eject and jammed, preventing tape take-up during the unthreading process.

Damaged tapes on eject: I replaced the mode switch as usual, but the tape was still a little loose and sometimes caught on eject. The cure was to replace the capstan belt and clean the capstan brake.

Tape ejection when half loaded: The cause is usually the mode select switch. It can also be responsible for no play or no reverse search.

Only rewinds 15 per cent and would then stop: Further attempts with the rewind button had no effect. On test we found that the start-sensor photodiode voltage, which is normally 5 V in darkness, fell to 3.8 V as rewinding progressed. At this point the machine shut down. The tapes were the cause of the problem. Four TDK ones, which were semi-translucent. When we shone torchlight through them we could see more light than with four other brands we tried. Hopefully this is a one-off batch problem. Has anyone else encountered this problem with TDK VHS tapes?

JVC HRD530

Machine would stop in the middle of play or record on rare occasions:
Luckily we were watching the deck when it had a spasm and saw that the
take-up reel stopped. We subsequently found that when the fault
occurred the voltage applied to the reel motor dropped but the current
through it increased. In fact the motor was going short-circuit
intermittently and had to be replaced. To be on the safe side we also
replaced the drive chip in case it had been damaged by the increased
current flows – more than an amp.

Hum bar in the E-E mode: It's not the easiest power supply to work on,
but scope and meter tests revealed that the 40 V supply was low at 30 V,
with a lovely ripple. Even better, the offending capacitor C5, a 47 µF,
63 V electrolytic, was sitting above the mains transformer winding, next
to C6 which was also suffering from heat stress. This is a 100 uF, 63 V
electrolytic. Two new electrolytics solved this one.

Loading fault: You sometimes find that the half-loading arm is not able
to extract tape from the cassette. The problem is caused by static. As a
result of this, the tape sticks to the lid of the cassette. The cure is to
replace the housing lid guide, part number PRD43315.

JVC HRD540

Playback in reverse direction: This is a rare fault with any model, but
several of these JVC machines have done exactly this. The sound is
backwards: the picture is present with some mistracking bands on it,
and there's colour. The play mode is entered but the tape runs in the
reverse direction. The cause of the problem is the PU61003 capstan
motor: due to an internal fault in its drive chip the forward/reverse line
is shorted. In all cases so far a new motor has provided a cure.

**Machine would play, rewind and fast forward for a few seconds then
stop:** This machine would play, rewind and wind fast forwards for a few
seconds then stop, although the tape counter was working. Pressing
pause cured the stopping. All was revealed when we consulted the
service manual. The supply and take-up reels both have a rotation
sensor. One supplies a pulse, the other only a 5 V d.c. voltage. A call to
our suppliers revealed that this part is in demand and was out of stock.
Has it suddenly become a common fault?

Cassette loading motor would run with no cassette inserted: The
customer told us that a cassette had got stuck in this machine and that

he'd removed it himself. When we plugged the machine in we found that the cassette loading motor would start to run with no cassette inserted. Then, after a few seconds, the machine would switch to standby. If a cassette was inserted and the machine was switched on the cassette would load down then immediately be ejected: the housing motor would continue to run. We decided to load a tape manually and thread it up. The result was some operation but the capstan motor didn't rotate and there was thus no reel drive. The machine sat happily in pause, however. Attention was turned to the capstan motor, which was found to be without its 12 V supply. Tracing the source of this back we came to an open-circuit circuit protector (CP5, type ICPN38) on the main PCB. A replacement was fitted and the machine was given a good soak test. This proved that there was no underlying cause for the failure of CP5.

No E-E sound: The cause was failure of the 4.7 Ω safety resistor R47 in the l.t. feed to the audio circuit.

Intermittent E-E and playback sound: Application of freezer anywhere around Q5 and Q6 on the tuner/i.f. PCB instigated the fault, as did any attempt to measure the voltages around Q6 (FMS2) which proved to be the culprit. This tiny, five-legged surface-mounted device is not listed by any of our usual suppliers, but a replacement was obtained from JVC without difficulty.

Deck jammed: It's very common for the deck to jam, leaving the loading motor whirring away. The cause of this is the half-loading arm/gear, JVC part number PQ43570B: the teeth wear and the gear then jumps a tooth or two, jamming the deck. Sometimes the first tooth breaks off the master cam, JVC part number PQ20822–2-7. Both items are available from Willow Vale. It's best to replace them both regardless. When you've removed the old half-loading arm, don't discard it immediately – it can be useful for realigning the deck. You'll see that there's a hole just above the cassette lamp: if you push the half-loading arm into this hole it will hold the slider plate beneath in exactly the right position ready for fitting the master cam. Don't forget to align the audio/control head, which has to be removed to fit the half-loading arm.

Fault with tape guide: With this model and those that use a similar tape deck you may encounter an intermittent fault condition in which the entry tape guide stops short of its locating V block. When this happens there's gross mistracking and, sometimes, tape damage. The cause may well be that the fastening pin ('stopper 2') is not pushed fully home into the pole base assembly on the deck's underside. Thus the loading pusher arm (32 in the JVC parts diagram) 'flops' on the shoulder of the pole base.

Slow functions: When the rewind or fast-forward button was pressed the tape loaded up fully and went into the selected mode, but at the visual search speed. Replacing the end sensors cured the problem.

Severe tracking error: About four tracking bars were present towards the top of the screen. On investigation we found that the left-hand guide pole didn't engage with the end stop. A thorough check for a foreign body, i.e. something that might have caused the problem by blocking its path, was carried out. As we couldn't find anything we came to the conclusion that pole was just a bit too loose, snagging before it entered the end stop. The cause of this was stopper 2 (item 17). Because it was a poor fit, it had worked loose. A replacement (part number PQ43525) cured the problem. An improved pole kit is available to deal with severe cases, part numbers PTU96102E (supply) and PTU96103E (take-up).

Cause of tracking problems: This can usually be traced to defects with the pole bases as mentioned above. You may find that the rotary guides are loose, the result being that they adjust themselves as the tape passes. The brass insert can become dislodged, so that the pole base does not go fully into the V block. Finally, the inclined guide can become loose, so that its alignment changes – or it can even fall out!

Apparently dead with no clock display: Checks showed that the unswitched 12 V and 5 V supplies were present but not the switched 12 V and 5 V supplies, and there was no power-on signal to the main micro IC601. If the power-on line was taken low manually, the switched supplies appeared but there was still no go. The microcontroller chips seemed to be getting all their supplies, but there were no strobe pulses from micro IC1 to the clock display. I took a chance and ordered a μPD725216ACW-A35 from JVC. Fitting it restored normal operation.

JVC HRD560

Machine wouldn't tune: We replaced the tuner, the tuning memory chip, then the tuner/timer chip for key scan control, all to no avail. Finally we replaced the system control chip. That did it! Another of these machines came in dead. We replaced circuit protector CP1 in the power supply then, on final test, noticed that the capstan motor was inclined to act erratically – in fact at one time it went into reverse. A new motor had to be fitted.

Fast forward/rewind very slow: An E180 tape took over 30 minutes to fast wind! Throughout this time it remained fully laced with the drum rotating. The cause of the fault was failure of the supply reel spool rotation sensing optocoupler, circuit reference PS2. In addition it's a good idea to change the slider mode switch and ensure that the new type of main cam is fitted – this can be identified by its black colour.

No drum or capstan operation: The cause was traced to CP1 in the power supply being open-circuit.

No play: On test the drum didn't rotate and the machine shut off. A look at the circuit diagram led me to CP401, which protects the 13 V motor supply. It was open-circuit, a replacement restoring the supply to the drum, which then rotated at the correct speed. This wasn't the end of the matter, however. The capstan seemed to be running as though there were no control pulses, which turned out to be the case – the control amplifier had failed. It's incorporated in the HD49733NT servo chip IC401. Replacing this finally restored normal operation.

Snowy bars at the top of picture or part of picture missing: Check the loading arms. They usually become loose, or become disengaged altogether. You can tighten them or reset the arm with a little glue to hold it, but for a good repair it's best to replace them.

JVC HRD580

Recordings marred by horizontal black flashes across screen: This is the sort of interference you get from a latchety aerial plug or an intermittent tuner. The effect could be seen on the E-to-E pictures. Some heating and freezing on i.f. panel 07 revealed that one end of R38, the demodulated video feed, was dry-jointed.

Made a 'knocking noise': We found that the drive belt bush at the back of the capstan motor had split. A replacement capstan motor had to be fitted – this is becoming quite a common fault.

Sound fault: Operation in the E-E and record modes was OK, but playback produced only a loud hum. Checks around the BA7765 sound chip showed that there was a problem with the switching lines – Q6's collector didn't go low for playback. The command comes from pin 34 of the main microcontroller chip IC601, where we found that the voltage rose to only 9 V instead of 12 V for playback. IC601 (JPC2002B-263) was faulty.

Wrong type of mode switch fitted: We seem to be getting quite a few of these machines in which a previous engineer has replaced the mode switch but fitted the wrong type. The symptoms are that the tape laces up and the machine then plays for a few seconds before shutting down. The part number for the mode switch, which has a black body, is PU60973. Don't fit the red type.

Wide line across picture: The line was about 1.5 inches wide and looked like a very wide noise bar – the sort of thing you get with poor back tension or a badly worn lower drum. The cause turned out to be a faulty 3.3 μF, 50 V chip capacitor (C6) on the drum motor PCB.

JVC HRD610

Works well in play until pause or search selected: The picture then had lines about every 3 mm across it. Wet finger checks soon established that the cause of the fault was IC301 – it's a small subpanel. When a replacement was fitted all was well again.

Playback picture had five thin tracking lines across picture: Also the sound was muted. The cause was that the exit guide was loose. Slight adjustment of the guide's height, followed by tightening the grub screw, corrected the fault. It's worth noting that a defective mechacon chip can produce almost identical symptoms but in this case the audio track will be erased whilst you are looking at the symptom.

Would record only one timed event: Manual recordings were fine, as were single-event timer recordings. But if more than one event was programmed in the mechanism would jam when the second event occurred. The machine would then switch off, and the second set of information was lost. The cause of the trouble turned out to be the mode switch: when the deck tried to start from the fully loaded position it locked up. A new mode switch put matters right.

Picture rolling: When we examined the video f.m. waveform we saw that there was a gap in the envelope: the head-switching point was out. It's set up using the presetter remote control unit from JVC.

When a tape was inserted it would start to lace up, then stop and the machine would switch off: I tried this several times: each time the point at which the machine stopped varied. When I removed the mode switch I found that it was starting to break up. A replacement put matters right.

JVC HRD640

The wrong mode switch: Don't get caught out like I did with this machine! It came in with a faulty mode switch. We fitted a replacement but the machine bounced back with all these symptoms: momentary formation of a tape loop while unlacing; failure to come out of the pause mode, followed by shutdown; and intermittent tape spillage during eject. What we'd fitted was slide switch PU60973 which looks right and fits perfectly but is intended for an earlier model. The correct part number for the HRD640's mode switch is PU61247.

Would not play after pause or reverse search operation: This machine would then unlace and enter the stop mode. The cause turned out to be the mode control switch.

Dead machine with 'Set Clock *' in display: This means the child lock is set. To clear it use the remote control handset to send a power on command – the customer did send you the remote control unit, didn't he?

JVC HRD660

Deck normal, no front panel display: The deck functions were normal but there was no front panel display. Absence of the $-30\,\text{V}$ supply at pin 2 of CN1 was the cause. We found that the safety resistor R2 ($47\,\Omega$) had gone open-circuit. A long soak test proved that there was no contributory cause for the failure of R2.

Playback in LP mode poor: Tape playback in the SP mode was good but the pause, search and LP modes were poor. I found that the LP heads were not being switched on because one end of R19, a chip component on the head amplifier PCB, had never been soldered.

Pulley ring on capstan motor split: The result being either cyclic interference to playback accompanied by a ticking noise or complete failure if the pulley and belt actually come off. The pulley is a toothed gear. JVC can supply a replacement, part number PTU96031–678C, to eliminate the need to replace the capstan motor.

No play: The tape would be pulled out of the cassette in the normal way, but would not be positioned between the capstan and the pinch roller. As a result there was no forward tape motion. The pin that was responsible for this problem had parted from its plastic holder and was nowhere to be seen (the deck reference number is 47). It's reminiscent of the infamous limiter post used in some Matsui machines, and could also become a stock fault.

JVC HRD700

Dead with a faulty STK5481 power supply chip: It's becoming increasingly common to find one of these machines dead with a faulty STK5481 power supply chip. You may or may not find one or more of the CPs that protect the supply lines open-circuit.

Fluorescent display won't light: Before delving into the rather inaccessible electronics on the front panel assembly it's worth having a look at R5 on the 01 power transformer panel. It may have gone open-circuit, deleting the −30 V supply. As is common with these safety resistors, there may be no external cause of its failure.

Tape looping and crushing: This machine suffered from a rare intermittent fault: about once a week the spool motor would fail to rewind the tape into the cassette when entering the stop mode. The result was tape looping and crushing. A replacement mode switch solved the problem – the original one seemed to be putting hash and noise into the microcomputer control chip whenever the loading motor was on the move.

Field roll in playback mode: It's becoming quite common to find that the entry and exit guides in these machines are starting to work loose, so I set up the tape path. Then, while playing back a tape, I noticed a severe hum bar on the picture. It seemed to be intermittent. Checks showed that there was ripple on the switched 5 V supply when the fault was present. A replacement STK5481 power regulator chip restored normal operation.

JVC HRD720

Dead machine, fully loaded tape inside: There was no clock display, no nothing, although there were outputs from the power supply. We didn't have the manual, but did find one for the Ferguson FV45X which seems to have the same power supply. Armed with this we soon found that the unswitched 12 V supply was missing because circuit protector CP2 (N20) was open-circuit. A meter check showed that the maximum current being drawn was 400 mA. A replacement CP cured the fault.

If tape rewound to beginning it would then be ejected: However, if a partially used tape was inserted the machine would work. Suspicion fell on the BOT sensor, but scope checks in this area showed that the sensor's output was influenced by the EOT sensor and vice versa. Checks around the microcontroller chip led us to a subpanel where we found that D611 had been fitted the wrong way round. Refitting it correctly restored normal operation.

No play/record, cut-off in rewind: In addition the drum didn't rotate when a tape was inserted. Voltage checks showed that the motor 13 V supply was missing at plug CN401. The cause was CP401 (ICP-FI5) which was open-circuit. A replacement restored normal operation.

JVC HRD750

Diagonal blue lines on red backgrounds in the picture: This complaint was received with some scepticism in the workshop. Sure enough, on test the reproduction of reds was flawless, even with the Madonna tape that accompanied the machine for the purpose of demonstration. We had to get the TV set, a JVC one at that, before we could sort this one out. The cause of the trouble was interaction between the TV set and the VCR, and was cured by physically separating them by a foot or two. The effects produced by this sort of radiation interference can vary tremendously with different combinations of VCR and TV receiver.

Intermittent failure of circuit protector: The trouble with this machine was intermittent failure of circuit protector CP802 in the motor 12 V line. When failure had occurred the machine would perform no deck functions, switching itself off after 8 seconds. The cause was an expensive one: an intermittent short-circuit within the capstan motor.

Dead, apart from a hissing noise: This machine was dead apart from a hissing noise. On investigation we found that there was arcing between the 350 V line and the chopper transistor's heatsink. Cutting the track and fitting an insulated wire link as a replacement restored normal operation.

Kept switching from hi-fi to normal: It was almost always in the normal mode, when a pre-recorded tape was being played. The cause was a slightly misaligned exit guide. I'm getting a few of these machines with loose entry/exit guides.

Display would dim and then go out: When this machine was switched on from cold it appeared to work, but after a few minutes the display would dim and then go out. A check showed that the −30 V supply dropped to −10 V. The cause was soon traced to IC3 on the tuner board.

Playback pictures flickered: This was because there were no PG pulses from the head drum. We had to replace the lower drum assembly.

Failure to function, intermittently: This can be caused by sparking between the chopper chip's heatsink tab and the adjacent PC land which goes off to R1. It can be accompanied by failure of mains fuse F1. I dealt with this by cutting away and scraping off some of the PCB foil to the side of the heatsink tab.

Total loss of action: Some of these excellent machines are suffering from dried-up electrolytics in the power supply due to age. If the problem is total loss of action, check C14 (1 μF, 50 V) and C13 (180 μF, 16 V) in the chopper circuit.

No deck functions: You may well find that circuit protector CP802 has failed. A replacement generally gets the machine going – until CP802 opens again. The root cause of the problem is the capstan motor.

Complete loss of action: There were no output voltages at all from the power supply module. For once the culprit was not the kick-start capacitor C14, although we replaced this as a matter of course. It was the STR10006 chopper chip.

JVC HRD820

Intermittent loading fault: It would also stop dead in play for no obvious reason. We suspected the syscon sensors but it was the microcontroller chip itself that was faulty.

Tape damage: When a dummy tape was inserted and play was selected the take-up spool carrier was seen to rotate in a jerky stop–go manner. Rewind and fast forward were sluggish and noisy. Suspecting a clutch problem, I removed the bottom cover. A small toothed pulley mounted on the capstan flywheel drives the clutch via a toothed belt. It had become loose, and a crack was evident down its side. Non-JVC account holders can obtain the pulley from Willow Vale – part number 87660PG.

Only half-loaded tapes: When a tape was inserted it would be loaded to the half-load position. If play was then selected the pinch roller would move down but the guides would stay where they were. On examining the underside of the mechanism I found that the pin and circlip which hold the plate assembly in position with the guide arm gears had come out. So I refitted it. Then, when play was selected, the machine jammed. What had happened was that the machine had received attention elsewhere. The previous engineer had glued up the brass part of the entry and exit guides – and managed to glue the guides to the deck! I had to strip the guides from the deck, remove all traces of superglue from the guides and the runners in the deck, then relubricate the runners and reassemble. This cured the problem. If the brass part parts company with the guide, remove the guide, use one drop of superglue and refit the brass part. Wipe any excess from the guide quickly. I normally remove the head drum assembly to refit the guides. Use the stoppers to attach them to the guide arm. Before returning the machine to the customer, check several times that the guides go to the play position fully.

JVC HRD830

No Nicam sound: The f.m. sound was normal, but when the machine switched to Nicam the result was no sound at all. Circuit protector CP1 N10 had failed due to an internal short in the M65109BSP micro-computer chip. Replacing both items restored the Nicam sound.

Capstan motor rattles/roars in play: This was a strange and unusual fault! The capstan motor would rattle and roar in the play mode, the playback picture showing that there was no capstan phase lock (noise bars cycled over the picture at a rate of about three a second). If the CTL pulses were removed – by playing a blank tape, lifting the tape from the ACE head or shorting out the CTL head winding – the motor would settle down. After a long search we found that C405 in the servo circuit was leaky – it read about 800 Ω.

No timer recordings: The machine went through the motions then, after loading up, ejected the cassette. Manual recording was OK. A replacement record inhibit switch cured the fault.

No or intermittent tuner signals: Check the 2SD1863 transistor Q13 on the tuner board. It's easy to find this transistor: the board around it becomes discoloured as the transistor overheats and becomes defective. A 2SD1207 seems to be a more manly transistor for the job.

Playback picture would jump: The sound would jump from hi-fi to linear. A check on the off-tape f.m. signal envelope showed that a slice was missing. To cure this the drum motor (available only as a lower drum assembly) had to be replaced.

Dead with no clock and no functions: This machine was dead with no clock, no functions and the chopper transformer buzzing. We found that zener diode D28 on the secondary side of the power supply was short-circuit while the mains rectifier's reservoir capacitor C12 on the primary side was open-circuit. Replacing these items restored normal operation.

JVC HRD860

No E-E sound, recorded sound buzzed: I checked the sound along to IC6 where it disappeared. Further checks showed that the supply here was low at only 2V. Following this back I found that the full voltage appeared at the connector to the panel. The plug was covered with a gluey substance, presumably to stop it moving. When this had been cleaned off the fault was no longer present. What amazes me is how the machine could work for months before the substance decided to foul up the sound completely. Funny business!

Fast-forward and rewind problem: This machine worked perfectly except for fast forward and rewind. When these modes were selected they would start but fast operation, which should commence after about 10 seconds, didn't take place. Fortunately I've had this problem many times before. So I changed the reel sensors. It's quite a common fault with these machines.

Alignment was out: According to the customer the fault occurred after he'd rewound a tape. Once the mechanism had been realigned it worked perfectly, but I replaced the reel and tape sensors to be on the safe side – they often cause problems. I've not seen the machine since, so presumably it has been cured.

Complete/intermittent power supply failure: If the problem with one of these VCRs is complete or intermittent power supply failure, in addition to replacing the start-up capacitor etc., check for a dry-joint at the emitter connection of the chopper transistor Q1. A section of the solder pad seems to detach itself from the rest of the print land. To be sure of the repair, use a tiny length of tinned copper wire to bypass the immediate printed land area.

Intermittent playback: The complaint with one of these machines was that playback was sometimes OK, sometimes there was only half a picture and sometimes none at all. The cure is easy: replace the 4.7 μF, 63 V capacitor on the head drum PCB. This is becoming quite a common fault with these machines.

JVC HRD880

Tape stuck, no functions: We found that a key scan port associated with IC1 was stuck high. Everything worked correctly when the chip had been replaced.

When cold would power-off in fast forward or rewind: The machine was fine when hot and played all right. When stop and fast forward were selected during play the machine unthreaded and then started in fast forward slowly for a few seconds, but at the point where the fast-forward system would normally accelerate the power unit shut down and all the displays went out. Checks showed that the 17 V motor power supply was switched to the capstan motor by the microchip and Q601 for 8 msec before the power supply shut down. The capstan motor current rose to approximately 250 mA, which was normal, so the capstan motor was OK. The trouble was that the power supply couldn't provide the current. A substitute power supply solved the problem, and we subsequently found that C19 (270 μF) was down to about 6 μF.

Sometimes failed to accept a tape: This machine would sometimes fail to accept a tape. The cause was a broken tooth on the lift gear. We had to replace the lift assembly as lift parts are not available separately.

Dead machine: There was no clock and no functions. Checks in the power supply showed that the 12 V supply was missing. CP2 was open-circuit.

JVC HRD910

Failed to play, damaged tapes: On test I found that the drum and capstan motors turned very slowly. Checks on the servo PCB showed that the motor 13 V and SWD 5 V supplies were OK. Further checks on this panel revealed that the fsc. input at pin 42 of IC1 was missing. It comes from the video area, where there was no signal at crystal X1. I next found that the SWD 5 V supply here was missing. Moving back, I discovered that CP803 was open-circuit. It's not shown on the circuit diagram, but you'll find it near the corner of the main PCB in the pinch roller area.

Tape stuck in machine: It couldn't be ejected nor could any other function be selected. This wasn't surprising, as the loading motor drive chip IC1 had a large hole in it, probably caused by the motor. Replacing IC1, the loading motor and the circuit protector restored normal operation.

Intermittent loss of colour in the LP mode: SP was OK. Our prime suspect was the main video processor chip on the video subpanel, but a replacement made no difference. By coincidence I found that moving the head amplifier brought the colour back. All that was required was to tighten the screws which secure the head amplifier's can to the deck chassis.

Unstable picture in top half, snow in bottom half of picture: This tip could save you a lot of heartache – as well as money! The symptom we had was an unstable picture in the top half of the screen and just snow in the bottom half. Scope checks showed that the output from one head was greatly reduced. Replacing the upper drum marginally improved the top half of the picture, but had no effect on the snow . . . Logically, the cause of the problem had to be the lower drum. But before we frightened the customer with the price of a new one we phoned JVC Technical. We were told that there's a 3.3 µF capacitor, which is not shown in the service manual, on the lower drum PCB. Replacing this cured the fault.

JVC HRDX22

No E-E and playback sound: The cause was traced to dry-joints at several of IC301's pins. A good solder-up is all that's required.

Jammed when it went into play: This centre-loading machine jammed when it went into play. I stripped it down and replaced the half-load gear and control cam, but it still jammed. The cause of the problem was the fact that the control cam's spindle was not seated properly in its plastic moulding. Pushing this down until there was a click provided a complete cure.

Faulty cassette lift operation: This one caused me to scratch my head for a few minutes before I realized that the phantom fiddler had been at work. The machine would accept a cassette, but something prevented it going fully down. I noticed that the pin was broken off the guide arm (exploded view reference 47) and was wondering whether the gears had jumped some teeth when I saw that the half-loading arm (reference 49) had two pins! Yes, someone had glued the broken pin on to the wrong arm.

JVC HRFC100

Tape counter didn't work in fast-forward or rewind modes: The real-time tape counter in this VHS/VHS-C compatible machine worked in the record and playback but not in the fast-forward and rewind modes. This was because the left-hand half-loading arm (item 25 in the exploded deck diagram in the manual) was bent, diverting the tape path past the control track head.

Chews up VHS-C tape: Towards the end of rewind a VHS-C tape (but not an ordinary VHS one) would be cruelly chewed. When small cassettes are being fast rewound the tape guides are extended a little from the cassette shell. The tape was riding up and over the upper collar of the entry guide because it was loose and able to vibrate and lean backwards. We cured the problem by pushing home the entry guide's retaining stopper on the underside of the deck. It's item 11 in the exploded deck diagram on pages 4–6 of the manual.

Wouldn't accept tapes: The guides would half load then the machine would cut off to standby. The cure was to replace the cont plate assembly. This controls the alignment of the mechanism part of the plate and was broken.

Power supply blew up: Occasionally you will find that the power supply in these dual VHS/VHS-C machines blows up. The usual cause is the chopper transistor Q1 going short-circuit. I replace Q1 with a BUT11AF and Q2 with a BC637. After that the power supply runs normally.

JVC HRJ200

No action and failure to eject tape: Check CP1 in the power supply. It's an N20 type, rated at 800 mA, and often fails. You will probably find that the current through the replacement is normal, at about 550 mA, but to prevent further failure earth the cassette cradle with a bracket, JVC part number PQ46086. It seems that static discharges can produce current surges through the protector.

Failure of the mode switch: This is quite common with JVC machines, but it's the first time we've had the fault with a 'mid-mount' model, however. The symptoms – both very shy and spasmodic – were failure to rewind and tape looping at eject.

Intermittent rewind and tape chewing: We noticed that when rewind was selected the VCR would sometimes load then unload slowly, leaving a loop of tape. We also found that this could happen in the play mode. Replacement of the mode select switch (part number PU60622–1-2) put matters right. We have had this fault several times now.

JVC HRJ205

Display would vanish when play selected: The customer said that when he inserted a tape and pushed play the machine would stop working and the display would 'go peculiar and disappear'. I'd seen this one before and went straight to the ICP fuse in the power supply. Sure enough it was open-circuit, a replacement restoring normal operation. This is becoming a common fault. When I phoned JVC Technical for advice I was told that the cause is being looked into. Until they come up with something, keep a good supply of these ICPs handy in your kit.

Rice-pattern effect on playback: This can be caused by excess grease on the drum discharge brushes. The head drum motor and drive are on the top of the drum assembly in these machines. To repair, remove the two screws that hold the drum drive board to the drum. Then loosen the grub screw that holds the bush to the shaft, noting its position carefully. Remove the upper drum and clean the grease off the brush assembly. Reverse this procedure to reassemble the unit. Remember to check/adjust the head-switching point.

No E-E tuner operation: There was just a blank raster. Unfortunately the construction of this unit makes fault diagnosis in this area very difficult. Replacing the tuner and i.f strip restored normal pictures and sound.

Appeared to require new head drum: The customer thought that this machine needed a new head drum. In fact the cause of the trouble was an incorrectly set head-switching point. This had possibly been misadjusted by the engineer who tried to cure the static problem!

Double image: Playback of a tape with vertical lines in the display showed this up. The fault was also intermittent. We found the cause to be dry-joints at the delay line. When this item was removed from the video processing board we found that there was a crack in the print to its earth pin. Remaking the print and resoldering cured the fault.

JVC HRS4700

No functions, CP2 was open-circuit: After replacing CP2 and running the machine on test the protector again blew. This time I monitored the capstan current in the play mode and found that it intermittently peaked. Replacing the capstan motor and CP2 cured the problem confirmed by a couple of days' soak test.

Faulty functions: What a brilliant fault report! Says it all doesn't it? Normal operation was restored by replacing the CAT chip. To non-JVC types, that's the memory i.c.

No functions: No power-up, no nothing. Voltage checks showed that the display chip's reset port was at 2 V, which is not a good thing. The fault persisted when this chip had been replaced. Doesn't it make you mad! Replacing the timer chip and, for good measure, the CAT chip got the machine going.

Poor picture on S-VHS recordings: Although this is a high-specification VCR, we've had a number of complaints about poor pictures with S-VHS recordings. A sample tape was brought along with one machine that came our way recently. The fact is that off-air recordings look worse in S-VHS than they do in standard VHS. Because the S-VHS bandwidth is wider than that of the received signal, the difference is filled with h.f. noise. With standard VHS this noise does not arise. Thus S-VHS playback of an off-air recording looks worse, because of the extra noise. The customer was given back his tape, with some high-grade camcorder recordings (of our dogs running around). These did justice to the machine's excellent S-VHS capabilities.

JVC HRS5000

Muted E-E and playback video: When we attempted to tune it the channel display would revert to number 1 after a few seconds. In all modes pin 13 of IC3 was high at 5 V due to a dry-joint. Putting this right restored the playback video but there were still no E-E signals. After tracing the picture mute line back to the tuner control PCB we found that pin 23 of the M50445–398SP tuning and channel change microcomputer chip IC1 remained high at all times. Replacing this chip restored everything to normal.

Overloaded video and crushed sync: The problem was overloaded video and crushed sync in the E-E mode. Its cause was traced to C14 on the signal processing panel being short-circuit. As a result the relevant pin of ICI was at 5 V instead of 1.4 V, upsetting the clamp detector stage.

Intermittent slow playback: This all-singing, all-dancing machine came in because of intermittent slow playback with some tapes. After watching the tape supplied by the customer we came to the conclusion that the tape transport system was running too fast. As lengthy efforts failed to instigate the fault we left the machine on the soak test bench. About two weeks later the fault finally showed up. The loading cycle hadn't been completed and as a result the pinch roller hadn't been pulled on to the capstan shaft. Thus the reel motor was driving the tape. A replacement mode switch cured the fault.

JVC HRS5800

Remote control OK, but no on-deck control: There was also no change when the audio/mix switch was operated. IC1 was at fault – one of its scanning ports was down.

No display, no action: If complete failure to operate (no display, no action) is the problem with one of these machines it's likely that the UNSW 12 V line is at zero voltage because the 22S2 fusible resistor R15 in the power supply is open-circuit. The resistor seems to fail for internal reasons – we've never found it to be overloaded.

Various faults: This machine came in with a list of faults: intermittent sound; picture not stable; and the left VU meter not working. There was no sample tape, and I had little to go on as the machine had come from another dealer. I checked the tape path and set it up. This cured most of the problems. I then braced myself for a complicated VU meter drive problem. There was relief when I discovered that it had been selected as a tracking indicator.

Loading difficulties: The cause was a broken spring in the idler/brake control. All suspect gears and cams were replaced to restore reliable operation.

Logik

```
LOGIK VR950
LOGIK VR955
LOGIK VR960
```

Logik VR950

Capstan servo hunting, tracking control had no effect: We traced the cause of the fault to the control, which was worn and open-circuit. Because of spares problems with these machines we used a standard horizontal $470\,k\Omega$ preset, fitted with a suitable spindle that matched up with the front panel. This arrangement worked well. We then found the cause of the control's failure – a badly worn head drum that produced a poor, unstable picture and gave rise to an overworked tracking control. All was well after fitting a new drum from MCES.

No playback chroma: This was traced to the coupling capacitor C3103 $(0.01\,\mu F)$ being leaky. While searching for the cause of this fault, someone had changed just about everything else in the colour section.

Intermittent loss of tuner signals: Investigation revealed that there were several dry-joints on the tuner's pins. Resoldering these restored normal working. We've since had two more of these machines with the same problem.

Servo lock lost with own recordings: Playback of pre-recorded tapes was normal but as there were no recorded control pulses servo lock was lost with the machine's own recordings. We found that the inverter transistor Q0214 on the main panel PC6 was faulty.

Machine wouldn't switch off: This was because Q3 (2SC815) was short-circuit collector-to-emitter the power supply was always on.

Hum bar on E-E, record and playback: This was the result of C6 $(2200\,\mu F)$ having fallen in value. It's in the power supply.

Poor playback: The poor, low-contrast pictures produced by this machine led us to suspect that the playback luminance amplifier Q0310 was faulty. We've had this transistor fail on other occasions. This time, however, the culprit was its collector resistor R0364 (3.3 kΩ) which had gone high in value.

Tape chewing: I thought I had an easy idler job, but when I ran the machine up it laced all right and started to play. Then the tape looped around the pinch wheel and stopped. In wind and rewind the tape moved, but very slowly. Further investigation showed that the operation of the reel motor was sluggish. Checks around the BA6209 driver chip IC0212 revealed that the supply at pin 8 was low at 9.2 V instead of 12 V. Also the chip was running warm. The 13 V supply was correct and regulator Q0221 was OK, but there was nearly 3 V across R0282. The circuit gives the value of this resistor as 3.3 Ω, but it read 12.6 Ω when checked. A replacement of the correct value cured the problem. Incidentally this machine is of Samsung manufacture.

Rewind fault: This machine accepted a cassette but when rewind was selected the tape was rewound for a few seconds after which the machine shut down with the standby LED blinking. When fast forward was selected the reel motor refused to turn – there was just a click, then the machine shut down again. There was also no reel motor rotation in the play mode, so the tape was looped. The fast-forward command comes from pin 22 of the syscon chip IC602. We found that the voltage here changed from 0 to 2 V when fast forward was selected. This voltage change should have appeared at pin 2 of the BA6209 motor drive chip IC0212. It didn't because of a hairline crack in the print near this pin. When this was linked across all functions worked but there was again shutdown after a few seconds. This was caused by the reel optocoupler, which was producing distorted pulses.

Wouldn't play or record: The arms didn't lace up because the pin had dropped out of the sector gear. As a result it didn't move when the main cam did. A new sector gear was installed and the pin, found loose in the mechanism, was fitted into it. The old gear was faulty – the hole for the pin had become enlarged.

Infrared sensor broken and loading arms flopping about: The owner said that she'd tried to remove a jammed cassette and had damaged it in the process. What in fact appeared to have happened was that the nylon gear sector – it's the fanshaped bit on the loading mechanism – had split where the steel pin is located, allowing the pin to slip out. Hence the looseness of the loading arms. A spot of superglue was all that was required to repair the infrared assembly. A new gear sector and pin – they are separate items – had to be ordered from Mastercare. Imagine our surprise when, a few days later, the postman delivered two packages from Mastercare, one a box containing the gear sector, the other a jiffy bag containing the pin! Anyway fitting the parts and removing a thick ring of oxide from the capstan restored normal operation.

Wouldn't play, no drum rotation: Also the 'operate' LED didn't light up. Although some of the power controlled circuits were working, the PC12 line was low. The cause of the trouble was D4 (1N4002) on the power supply PCB.

Worked OK then drum speed increased: This machine worked well for about a quarter of an hour. The drum would then speed up and nearly take off! The capstan motor would run flat out and the machine would shut down. As both motors were affected we decided to check the power supply. In the fault condition pin 6 of plug F02 (5 V output to the servo) was at 2.4 V. When this pin was removed the voltage returned to 5 V. As there were no shorts across this supply a 6 V battery was connected. This restored normal operation, so the 2SC1008 5 V regulator transistor Q5 was replaced. After a long soak test the machine was returned to its owner.

Tuning drift with 33 V line low or unstable: Replace C2 (47 μF, 63 V) in the power supply and the 33 V regulator IC901 which is on the PCB behind the clock.

After 60 minutes stops in play/record: This machine would stop in play or record usually after about an hour. Up to the shutdown point the head and the take-up spool rotated normally. Suspecting a reel pulse problem. I connected the scope to the collector of Q610. The waveform consisted of four healthy square wave pulses, but of only 2 V p-p instead of 5 V p-p amplitude, followed by very noisy and even lower-amplitude pulses. When the spool carrier was removed and the ten silver-plated reel pulse reflectors were examined they appeared to be clean. But I then noticed that there were three ring spacers on the shaft. So the spool carrier was sitting too high! Removing two of them and refitting the carrier produced ten healthy 5 V p-p square-waves. Why the shutdown delay? As the take-up spool fills with tape it rotates more slowly. Thus the 'poor' portion of reel pulses remained for longer. As a result the system control thought that there was no tape drive.

No record picture or sound: These machines have never been favourites of mine. The customer's complaint was that there was no record picture or sound. It transpired that someone had had a good twiddle. The f.m. record level, carrier and deviation controls had to be set up correctly. Good results were obtained after doing this.

E-E picture distorted and part of it was blanked out: The cause was hum on the preset tuning voltage supply, which is identified as PRST VTG. Replacing C4 (47 µF, 100 V) cured the fault.

Logik VR955

No erase bias: This was caused by an internal fault in the bias oscillator coil unit L0504. It's perhaps worth noting that the Samsung unit is much cheaper.

No sound in record mode and no erase: Another common fault caused by the bias oscillator, of course, which is contained in can L0504, adjacent to the plug from the head. Although it's no doubt easier to replace the complete assembly it can be repaired on the bench. The culprit is the entombed 2SC1318 transistor. I replace it with a BC337, carefully reassemble the module and pack in some heatsink compound. This always works.

No erasure: This problem can manifest itself as the old sound left on the tape and floating colour blobs on playback of the machine's own recordings. But you need look no further than L0504, which is a little oscillator module in a screening can. It's prone to staging a mini bonfire inside. Willow Vale can supply replacements under part number 79710CB.

Loss of test signal and tuner supply: The cause was traced to L105, a 33 µH choke on the bottom PCB. A replacement from a scrap machine put matters right.

Machine wouldn't record: Checks showed that there was no record 9 V supply because Q110 was open-circuit. This was in turn caused by bias coil L504 having gone short-circuit. Replacing these items, also an idler and the pinch roller, restored correct operation of the machine.

Logik VR960

Buzzing on sound due to faulty aerial socket: Strange, I thought, but on test this proved to be the case. When the aerial input socket on the modulator was pulled down the E-E sound disappeared, leaving a loud buzzing noise. A scope check on the audio input to the modulator showed that the signal was still present when the fault occurred. No dry-joints could be seen when the modulator was removed but after going over all the connections with a fine-tipped iron the fault had cleared.

Would unlace after a few seconds: Rewind and fast forward were OK. When play was selected, however, the machine laced up then, after a few seconds, unlaced and shut down. It wasn't the limiter post this time but the loading belt which was slipping. Normal operation was restored when a new belt had been fitted. We noticed that a slight crack was developing in the limiter post so this was replaced as well – we didn't want a 'same symptom as before' situation.

Almost dead: The display was alight, and standby was operational. But the machine wouldn't work. Circuit protector ICP201 was open-circuit.

No operation light and no functions: If you get this problem the STK5332 power regulator chip is probably faulty.

Matsui

MATSUI VP9301
MATSUI VP9401
MATSUI VX1000
MATSUI VX1000Y
MATSUI VX1100
MATSUI VX2000Y
MATSUI VX2500
MATSUI VX2700
MATSUI VX3000
MATSUI VX6600
MATSUI VX755
MATSUI VX800
MATSUI VX820
MATSUI VX880

Matsui VP9301

Would load a cassette and shut down: When I watched it, I found that the drum sped up just before the fault occurred. By careful manipulation of leads I narrowed the fault area to the heads. So I removed the stator PCB that sits on the top of the video heads and found a nice little crack in part of the print. Careful repair got the machine working perfectly again.

No playback sound: Tracing the signal path back I came to C5015, which was dry-jointed. Resoldering it restored the sound.

Poor i.f. stability and streaking: This can be cured by replacing the three $0.47\,\mu F$, $50\,V$ capacitors in the i.f. module. We use $1\,\mu F$ replacements.

Static interference on playback: This was caused by the drum assembly. Unfortunately the upper and lower drums cannot be separated for cleaning, so a complete new unit had to be fitted.

Serious wow and flutter problem with the sound: As I had a spare capstan motor in the workshop I fitted it. This cured the fault – sometimes you have a bit of luck!

Matsui VP9401

Plays with no tab: I've had two of these machines with the same symptoms but different faults. The symptoms were that the machine would accept a tape, play it if the tab is missing, but there's no display and the front panel buttons don't work. In one case the control chip IC601 was the cause of the fault. In the other case we found that a digital waveform was superimposed on the oscillator signal: because of a dry-joint at CN601's earth connector, the oscillator's earthing was open-circuit.

Dead machine: Once we'd opened it up we found that fuse F502 was open-circuit. A replacement got the machine working, but there were no mechanical functions and it went into the standby mode after 3 seconds. This is a mid-mount machine, so I took it all apart and removed the PCB from the casing. The loading motor chip IC1004 had obviously been getting very hot: on closer examination I noticed that there was a small eruption on its plastic encapsulation. A new i.c. cured the problem.

Wouldn't load: As there was no drive to the loading motor I pulled the machine apart to gain access to the loading drive chip. This was very hot and its casing was damaged. The machine worked when a replacement chip had been fitted, but the chip was still overheating. A new loading motor put that right. The machine then worked fine.

Failure to load and a burning smell: A previous engineer had diagnosed 'incorrect data from IC1001': he obviously hadn't noticed the large hole in the loading motor drive chip. Replacing this item cured the trouble.

Rewind problems: This machine wouldn't rewind tapes and would stop when rewind search was selected. The beginning of tape sensor Q1002 was found to be dry-jointed.

Tips when putting the deck back: A warning about this and similar models: when you put the deck back into position after replacing the mode switch, make sure that you do not crush the central LED tower as you can short the two unprotected leads together, causing strange mechanical symptoms. It's easy to do this, not so easy to find the cause of the resultant faults.

Odd mechanical functions: We've had this with several of these machines. The thing to do is to carry out a visual check on the little brake lever coupling that sits under the cassette housing. When its securing clip wears or breaks, it becomes loose. The result is half actions etc. Replace it if it has come adrift.

Tape spilling out in reverse search: With this deck the usual cause is the mode switch. A drop of bearing oil should also be applied to the capstan motor. This time, however, these measures didn't work. We found that the fast-forward/rewind clutch assembly was slipping, although the clutch itself wasn't faulty. In reverse search the back-tension band doubles up as a soft brake. This is where the cause of the trouble lay. There was too much braking pressure. A replacement lever sub-brake, part number 850P600311 (item 334 in the exploded view of the VP9301 deck), cured the problem. It's driven from the master cam. Because a small plastic leg had broken off, it rode up and didn't release the brakes properly. This problem could become as common as the limiter post failure in earlier Matsui machines. Shop around for a replacement.

Low sound: Check C3606 (10 μF, 16 V) by replacement.

Matsui VX1000

Severe sound wow: This would get worse towards the middle and end of a tape. We found that the reel brakes weren't being released during playback because the operating lever had come adrift. A broken plastic lug on the main chassis was the cause of this. Unfortunately the only cure is to replace the deck chassis.

'Half the old programme showed': We found that the tape failed to make proper contact with the full erase head because the back-tension arm didn't move to its correct position. The cure was to grease the operating lever that contacts the back-tension arm and check the back tension.

Intermittently chewed tapes and forward search was slow: When we put the machine on test it slowed down after a while. Application of freezer to the capstan motor got the machine going normally again, while warming the motor with a hairdryer shut it down altogether. A new capstan motor was the costly answer.

Matsui VX1000Y

Appeared to be poor drum servo lock or a head fault: I have had the following problem several times now. The symptoms can appear as poor drum servo lock, a head-switching type fault or as if one head has failed – the fault can also be intermittent. To ensure a permanent cure remove the FG pick-up on the drum, clean the two pins and the PCB connections thoroughly, then resolder. I would add that with a lot of the faults experienced with Matsui/Saisho VCRs it pays to look for bad soldering or poor connections. Doing this will probably cure many of the faults that come your way.

Couldn't be tuned in: We found that the tuner had no tuning voltage because R6045 was open-circuit. This resistor can also go high in value – the result is tuning drift.

Sound was very poor: A check showed that the sound from the phono sockets was good. The r.f. converter turned out to be the culprit, a replacement restoring good sound.

Wouldn't tune in signals: I found that the voltage at tuner pin BT was virtually zero wherever the unit was tuned. The voltage at the 33 V regulator was correct but R45, a tiny 33 kΩ resistor, was open-circuit.

Didn't record sound: There was no output from the bias oscillator. We found that there was liquid spillage in the vicinity of T5001 and the surrounding components. After removing them we cleaned the board thoroughly. Then T5001, I5002 and C5031 were replaced. This restored sound recording.

Matsui VX1100

No power: A check on the voltages around IC501 showed that the operate voltage was missing. This comes from IC601 on the timer board. A replacement, which took quite a while to obtain, restored normal operation.

Servo, sound erase faults: If you have any servo, sound erase etc. faults with these machines remove the mechanism and check the connections between it and the main panel. Resoldering the connections will clear many problems.

Bias oscillator intermittent: If you went straight into record after switching on it worked, but if you came out of record then went back after a few minutes it didn't. The problem was that C5017 wasn't being fully discharged, so that when the oscillator tried to start up again the change in voltage wasn't enough to get it to run. I replaced most of the components in the circuit to no avail then, in desperation, connected a 3.3 kΩ resistor across C5017. This provided a complete cure.

Intermittent E-E sound: Getting out my faithful old screwdriver, I did some highly technical fault tracing by tapping around the boards. This soon led me to a very sensitive scart panel, and on closer examination I noticed that C4513 was dry-jointed. Resoldering it cured the fault.

Intermittent mechanism faults: These are now common with this range of machines, e.g. failure to eject, not playing etc. If you get this problem you can clear it by cleaning the mode switch. For a permanent cure, however, replace the switch with the improved type.

No LP playback colour: Recording was normal. The cause of the fault was eventually traced to C4316. One of its legs was dry-jointed.

Matsui VX2000Y

No remote control operation: The handset worked all right with another machine so I connected a scope to the output from the IR receiver can. This showed a healthy waveform. I followed the signal along the print and found that it disappeared when it passed (or should have passed) through the hinge-type edge connector.

Machine appeared dead: The power supply was working. The culprit turned out to be TC01, the orange trimmer capacitor that sits near the microcontroller chip in part of the clock circuit. A replacement trimmer restored the machine to life.

Noisy fast wind: These machines have no fast wind buttons and you need the handset for servicing. This one came in without the handset. The faults were noisy fast winding and an intermittently wobbly playback picture. The fast wind groaning was caused by a noisy capstan motor. Dismantling it then cleaning and lubricating the bearing cured that – there was a huge amount of sticky mess in there. Back-tension arm oscillation was the cause of the wobbly playback. This was in turn caused by dirt on the tension band pad. Cleaning sufficed.

Intermittent rewind, snaps tape at the end of rewind: The cause of the trouble was a dry-joint at D01, the sensor LED on the deck PCB.

Matsui VX2500

Coloured blobs on picture and the sound of a previous recording could be heard: I thought that this would be a nice, simple repair. I made a test recording and sure enough there was no erasure. This usually means a dodgy connection at the full erase head. But there was nothing wrong here. So more detailed checks were required. The voltage at the base of the bias oscillator transistor Q5002 was found to be incorrect, the result being that it was cut off. I then found that the 5.6 kΩ bias resistor R5001 was open-circuit. A replacement restored normal operation: not so bad after all!

Power supply failed to come up: Checks showed that Q07 was open-circuit.

Intermittent shutdown: This would occur after several hours' use in the play or record mode. The cause turned out to be intermittent loss of the FG pulses from the drum motor because C18, which couples the pulses to the servo chip, was dry-jointed. The drum would suddenly run at full speed: as the head-switching square wave was then so far off frequency, the microcontroller chip would switch the machine to standby.

No picture: When attempts to clean the video heads didn't help we found that there was no drum servo lock. This was put right by replacing IC2001 (OEC6014B). With the machine running on its side the drum assembly made a noise. We found that the collar under the dome-shaped flywheel was coming loose. If this collar is not in the correct position relative to the video heads the flywheel, which contains the magnet for the PG head to detect, will also be incorrectly positioned. As a result only a part of the picture will be seen on the screen, the rest of the display consisting of noise.

Chewed tapes: The pinch roller was OK, but while checking the take-up I noticed that the limiter arm didn't move – the spring had slipped off. I put it back, secured it and gave the machine a good clean. It worked well.

No remote control functions: I found that the handset was transmitting, while labels showed that the machine had recently been to another repair centre! A scope check showed that data appeared to be reaching the timer microcontroller chip. When I took another look at the handset I noticed a reset button beneath the battery cover. After actuating this and fitting new batteries everything worked correctly.

Sound problem: The customer complained that there was no sound with his recordings. In fact the previous sound was present, which is a characteristic of bias oscillator failure. What we actually found, however, was that the amplitude of the oscillator's output was only about a quarter of what it should have been – 50 V or so peak-to-peak is normal. The 2SB698 transistor Q02 was short-circuit.

Playback and E-E pictures very poor: As this is one of the later versions with a scart socket, I tried a scart connection to the TV set. The picture was then OK. A replacement r.f. converter cured the fault.

Machine wouldn't record: On test it soon became apparent that the cause of the fault lay in the receiver section, where it was possible to select stations but the signals were badly broken up. As the problem looked like poor demodulation, we decided to replace the LA7577 chip IC01. This restored normal operation.

Matsui VX2700

No control track pulses: There were no control track pulses because the relevant section of the audio/control head was open-circuit. Replacement cured the problem. This is becoming a very common fault. The audio section of the head can also go open-circuit, the symptom then being no sound.

Wouldn't load/unload: This machine wouldn't load or unload: the load motor drive chip IC1003 had failed.

Wouldn't eject: This machine wouldn't give the customer his tape back. The cause of the fault was in the power supply, where the switched 5 V feed was missing. A new STK5342 chip (IC501) restored the 5 V supply and gave us back the tape.

Intermittently reverts to standby: This would happen after long periods of varying duration. Resoldering a number of dry-joints around the microcontroller chip IC1101 cured the trouble.

'Tracking problems': What appeared to be tracking problems was in fact a badly broken-up playback picture. When we examined the operation of the mechanism in the playback mode it was apparent that someone had been fiddling with the back-tension arm. As a result, it jammed. After readjusting it and checking for correct tension, we were able to see the original fault symptom – the capstan was running at the wrong speed. If the back tension, and hence the load on the capstan motor, was reduced manually the tape speed altered noticeably. A new DD motor assembly was required.

Matsui VX3000

Loss of playback picture: A fault you sometimes get with these machines is loss of the playback picture, resembling head wear or failure. As often as not, however, the cause is poor head amplifier earthing, which is achieved via the screening can. The cure is to solder a short length of copper braid from the top of the screening can to the copper static discharge strip on the head drum.

Loss of tuning overnight: On the bench, however, no channels could be tuned in. R6045 (33 kΩ) was open-circuit.

Tape went in and came straight out: Checks on the switching at the lift and on the end sensors were fruitless. A new OEC0017B system control chip had to be fitted.

Capstan motor loses torque: If the capstan motor appears to have lost its torque, operating slowly with some tapes, try cleaning and lubricating the bearing. This may enable you to avoid having to replace the motor.

Wouldn't accept deck commands: This machine would power up and accept a tape. But it wouldn't accept any deck commands and would then shut down with a short capstan motor run. These machines are designed to load the tape around the drum on accepting a cassette. The one we had failed at this initial point. Checks in the power supply showed that the PC12 V supply was missing at pin 2 of CP501. We had no circuit diagram, but the supply appeared to be derived directly from pin 6 of the STK5342 chip. So we replaced this chip, and were rather disappointed to find that the situation was the same as before. But we felt that the cause of the fault couldn't be far away. This was so: R508 (10 kΩ), which is connected between pins 4 and 6 of the STK5342 chip, was open-circuit. After replacing this resistor we found that the working voltages at pins 4 and 6 are 24.5 V and 13.5 V respectively. To prevent unnecessary callbacks, we stripped out and cleaned the mode switch assembly and replaced all the belts.

Matsui VX6600

No E-E or playback picture: We found that the cause was L01 in the on-screen display circuit. It had gone open-circuit.

Wouldn't load a cassette: When we checked it we found that the BA6247 loading motor drive chip was getting hot. Was it the motor or the chip? We disconnected one lead to the motor and fed it from an external power supply. As it was in order a new chip was obtained and fitted, restoring normal operation.

No record picture, or overloaded picture: This machine played back all right, but in the E-E mode there was only a blue screen. On channel change a very ragged and overloaded picture appeared for half a second, then the blue screen returned. Checks in the vision i.f. strip revealed that C17, a 0.1 μF, 50 V tantalum electrolytic, was open-circuit. It decouples pin 4 of the vision/sound i.f. chip IC6001. Failure of this capacitor can result in recordings showing i.f. pulling and a.g.c. overloading.

Wouldn't store channels: This machine would tune in but not store channels. The cause of the fault was the little μPD6525C memory chip on the operation/display PCB. Take care when ordering: it's advisable to quote the Matsui part number or obtain it from Partmaster.

Matsui VX755

Wouldn't play tapes and would leave a loop of tape in machine when the cassette was ejected: The complaint with this machine was that it wouldn't play tapes and left a loop of tape inside. We removed the cover and inserted a cassette. When play was selected the machine laced up but there was neither drum nor capstan rotation. Thus after a couple of seconds the machine unlaced, without taking the tape back into the cassette. The cause of the trouble was the OEC9005 chip IC2001: fitting a replacement restored normal operation.

Intermittent stopping during play: Replace the take-up reel sensor.

Would pause for a quarter of a second: This machine would run all right for a couple of hours. It would then appear to go into the pause mode for about a quarter of a second, run normally for a few seconds then pause again. When we looked in at the top we saw that the capstan motor was pausing, with the tape not stopping long enough to switch off the machine. Checks at the motor drive and supply pins showed no reason for the pauses, so we assumed that the motor was duff. In fact all that was required was a spot of oil in the motor's upper bush.

Matsui VX800

Intermittent failure to eject: Intermittent failure to eject a tape as a result of carriage overshoot is a common fault with these machines. Carry out the following modification to overcome this problem. Remove the blue lead from the cassette loading motor and replace it with a BY127 diode (cathode to the motor), with a $27\,\Omega$, $0.25\,W$ resistor in parallel with the diode.

Slow functions and tape chewing in play mode: The symptoms were slow rewind and fast forward, also tape chewing in the play mode. We initially suspected the clutch/idler assembly, but while running through the mechanical operations without a tape in we noticed that in the play mode the reel motor didn't rotate. A check showed that there was only $0.5\,V$ across its terminals. The culprit turned out be the 2SD1246 transistor Q2022, which was short-circuit base-to-emitter. It forms a low-resistance chassis return for the motor, which instead was relying on a couple of low-value resistors that are in parallel with the transistor.

No drum rotation: Fast forward and rewind worked but there was no drum rotation. Also there was no E-E picture – just a blank, noise-free raster. The cause was F2001 of course. It's an N20 ($800\,mA$) ICP. To avoid having to remove the front PCB etc. you can replace it from above by lifting out the power supply. After replacing F2001 everything worked all right until play was selected. The tape then loaded but as soon as it touched the head drum the tape stuck to the drum and the motor stalled, blowing F2001 again. This was due to sticky 'gunge' on the drum. A good clean and polish plus a new ICP put matters right. Normal current through the ICP is $260\,mA$ in the E-E mode, $360\,mA$ in the play mode.

Fuse blows intermittently: While the machine sat on the bench it behaved normally, but as soon as it was moved the 2.5 AT F502 fuse blew. We found that the bottom PCB had warped beneath the power supply. As a result the fuse blew every time the PCB came into contact with the metal bottom case.

Mechanical loading problem: This machine's mechanics failed to complete the loading cycle in either play or record. On inspection we found that the cam gear was bone dry. A replacement together with some grease restored normal operation.

Matsui VX820

No operation and no clock: We checked that all outputs from the transformer were present then followed the wiring down to the power supply chip on the main panel. There was a burn mark under transistor Q02, as if it had been overheating. It appeared to be a 12 V regulator. Fitting a replacement put matters right.

Loading fault: This machine would wind and rewind but when play was selected the loading started then jammed prior to engagement of the pinch roller. We first suspected that the loading belt was slipping. But when we turned the machine upside down to remove the base cover a small metal pin about 5 mm long dropped out. On investigation we found that this came from a plastic arm just to the front of the pinch roller. After replacing it, using a dab of Araldite, the loading worked perfectly.

Dead: A dead machine with not even any clock operation should lead to a check on the 12 V regulator transistor Q02. You'll often find that it's short-circuit base-to-emitter. You may also find that the 13 V zener diode in its emitter circuit is faulty. Q02 is associated with IC01, the main power supply chip at the rear centre of the main base panel. During normal use a lot of heat is generated here, which no doubt contributes to the failure of Q02. We use a BD131 as a replacement.

No functions: A faulty mode switch proved to be the cause of no functions with a Matsui VX820. To replace the switch the carriage must first be removed. The switch can then be taken out by releasing the retaining screw and unsoldering the three leads that are attached to it. Reassembly is the reverse of this procedure. Take care to align the two slots on the switch.

Damaged by power surge: According to the customer this machine had been afflicted by a power surge or something similar. All the unswitched supplies from the multi-regulator were present but there were no switching signals for the others. This was caused by loss of the 12 V supply as the 2SD1207 regulator transistor Q2 was short-circuit base-to-emitter.

Wouldn't accept a cassette: This was a NICAM (Nasty Intruder Caused Absolute Mayhem) job. The original complaint was that the machine wouldn't accept a cassette. Its owner had accepted the kind repair offer of a friend at the local Electricity Board. When she retrieved it some time later it was completely dead. A replacement 12 V regulator transistor (Q2502) restored some life, but the machine still refused to accept cassettes. If a tape was wound in manually the start and fast-wind functions worked, but with no end-sensing operation. The left-hand PT361 sensor was open-circuit (the original fault?), but a replacement made no difference. The right-hand sensor is decoupled by a 10 nF capacitor (C1012), which had been carefully replaced by a $4.7\,\Omega$ resistor! The correct component restored loading and end-sensing, but with no playback picture. Meterman had removed the head amplifier module and refitted it with the PCB edge connector misaligned. After correcting this we had a working machine, the owner had a large bill, and I suspect that her SWEB friend was about to receive a shock.

Machine dead: Voltage checks showed that the STK5332 power supply module was faulty, so a new one was fitted. The machine then powered up, but there was no drum or capstan rotation. Voltage checks around the servo chip IC2001 were inconclusive, but scope checks showed that there was a distinct lack of activity in this area. Replacing this i.c. restored normal operation. Presumably the faulty STK5332 had destroyed the servo chip. This machine also appears as the Hinari VXL35 and the Orion VHML.

No power up, only a twitch: When the mains supply was connected to this machine the motors gave a twitch and the 'operate' LED came on for half a second. Checks showed that the 12 V rail was low at only 3 V. It's provided by the series regulator transistor Q02 whose base is biased by R06 (1.5 kΩ) and the 13 V zener diode D07. The zener diode proved to be leaky.

Matsui VX880

Low E-E and playback luminance: The chrominance and sound were not affected. PF01, a coil which couples the signal to the r.f. modulator, was faulty. When it was dismantled we found that it was full of green 'gunge'. The machine has a history of spillages and the results are beginning to show up.

Would accept tape, immediately reject it, then turn off: We noticed that during the attempt to load the tape the cassette-in symbol flickered on, then off as the tape was ejected. The after-load leaf switch mounted on the carriage was OK. We then made a careful study of the action of the mechanism during the cassette loading operation. This showed that the leaf switch is activated by a lever driven by the side-mounted rotary cam, which is in turn driven directly by the loading motor. The lever pressed the leaf switch home all right but immediately returned back a small way so that the switch went open-circuit again. During the loading process the motor spun normally, but as soon as the cassette reached the bottom the motor reversed by about three revolutions. A meter connected across its terminals showed no reverse voltage when this occurred, thank goodness. A new loading motor cured the problem.

Intermittent poor lift operation: This machine was sometimes dead, and appeared to overshoot the end stop during the eject cycle. Both leaf switches were tested and found to be OK. Fortunately another similar machine came in – we tried its lift and got the same results. This led us to the mode control switch which, when checked with a component tester and scope, produced some very ragged on-to-off changeovers. A replacement put matters right.

Refused to carry out functions or give the tape back: One of these machines would accept a tape and its display showed the functions selected, but it wouldn't carry out any of the functions and refused to give the tape back. We found that circuit protector ICP201 in the 18 V supply was open-circuit. A replacement plus resoldering of Q02's connections restored normal operation.

Mitsubishi

MITSUBISHI HS304
MITSUBISHI HS306
MITSUBISHI HS330
MITSUBISHI HS337
MITSUBISHI HS347
MITSUBISHI HS349
MITSUBISHI HSB12 etc.
MITSUBISHI HSB20
MITSUBISHI HSB27
MITSUBISHI HSB30
MITSUBISHI HSB32
MITSUBISHI HSM57

Mitsubishi HS304

No chroma in playback: On removing the top cover I could see a lovely break in the print, bang in the middle of the main panel. Repairing this restored the colour. It's a very long stretch of print that eventually connects the anode of D2A1 to C6D1.

Would ignore function control: This machine would ignore any tape operation function after a couple of hours' use, e.g. if the machine was in the playback mode you couldn't stop a tape without switching the machine off. If rewind was selected and the tape was allowed to rewind fully you would then find that the machine wouldn't eject. The cause of the problem was a poor leaf switch contact, FL-SW-2, which is the inner of the two leaf switches on the right-hand side of the cassette loading housing. It's the one that makes when a cassette is pushed in and is held closed by the loading cam, going open again only when the tape is fully ejected.

E-E and playback sound intermittent: This machine was being checked after coming back off rental. The customer had pointed out to the engineer who collected it that the sound was intermittent in E-E and playback. While thinking about getting to the heads to clean them I spotted the cause of the trouble – the r.f. modulator's audio pin had never been soldered.

Herringbone pattern on picture: This old faithful has been through our workshop a few times over the years. Its latest problem almost led us to tell the customer that it was by now past its sell by date. The playback picture was sometimes obliterated by herringbone patterning. The severity of this patterning was reduced when a hand was brought near the head drum or surrounding area: it would almost disappear when any metal part was touched. Having had similar problems with other makes of VCR, where earthing (common) links between the chassis and PCBs made a difference, we set about checking the earthing around the head drum. Sure enough the drum connector PCB has an earthing land which relies on the fixing screw making connection with the deck. Fitting a star washer and tightening the screw produced a vast improvement but didn't cure the problem completely. A friendly Mitsubishi engineer suggested that we scrape away the protective 'goo' around the head amplifier chip, as with time it causes leakage between the pins. Doing this finally put matters right.

Mitsubishi HS306

Damage to function controls: In this model most of the function controls are mounted vertically on the main board and are thus likely to be damaged as a result of heavy-handed use. I recently had to change the play and rewind controls: they worked all right but due to heavy use they had holes impressed in them and the customer controls wouldn't operate them. The movement of the switches and the plastic bracket that retains them is such that heavy use flexes them back from the vertical and increasingly heavy pressure seems to be necessary to get them to operate. Once the damage has started it's inevitable that it will continue. Some people will always use unnecessary force and I think it was a mistake to mount the controls in this way.

Very intermittent tuning drift: This machine would run all right for days with the top cover off. Heat and freezer had no effect. So we left a meter connected to the stabilized 30 V line and covered the machine with a blanket. When the fault appeared the voltage dropped. Careful inspection of the 30 V regulator on the main panel showed that there was some brown, foreign matter at the connections to D913 and D914. The problem didn't recur when this had been removed.

Intermittent stop: Check IC5A4 – it's mounted on the metal bracket along the front.

Poor sound with intermittent failure to record: This was cured by replacing the REC bias preset VR3A1.

Problem with loading arms: When play was selected the left-hand moving guide didn't go into the V block fully unless you gave it a push with a pencil. I suspected wear in the plastic gear cams, but a new pair made no difference. Finally, to cut a long story short, changing the cast-alloy shuttle block itself (part number 32 in the exploded view) restored normal operation.

Mitsubishi HS330

Wouldn't switch from E-E mode: In playback the tape laced up but the sound and vision wouldn't switch from the E-E mode. We soon found that the switched 9 V line was missing as the 2.5 A fuse F901 in the mains transformer's 14 V a.c. winding had gone open-circuit. It had died rather than blown, and a replacement held. A long soak test produced no further problems.

No playback colour: We scoped the video output when a known good tape was tried. This showed that the colour burst was present, so the next step was to check the alignment of the colour playback circuit. The voltage-controlled crystal oscillator was found to be running at 4.4328 MHz instead of 4.4336 MHz. Adjusting this as laid down in the manual restored the playback colour.

Sound would vibrate after warm-up: Having tested the machine for ages and heard no 'vibration' I questioned the customer to find out whether she meant wow and flutter, which is not uncommon with this model. Not so. It seemed to be a buzz. So I had a poke around and had success – a buzz appeared on the playback sound. Its cause was traced to a dirty connection between the copper-coloured spring metal that earths the top of the cassette housing and the regulator heatsink. A clean and retension cured the problem.

Mitsubishi HS337

Indicator light on operate button wouldn't come on: However, when the operate button was pressed the clock display changed from clock to counter as usual, reverting to clock after approximately 5 seconds. We found that the microcomputer's power-on pin went low and returned to its off setting after 5 seconds. All the unswitched supply lines were correct: the problem was to check the switched lines in less than 5 seconds. This problem was dealt with by removing the control transistor Q9A3. Once this was done all the switched lines with the exception of the 5 V line came up. The problem was due to a defective joint at the emitter of Q9A2.

No operation, caused by dry-joints: We've had two or three cases of 'no go' with these models – sometimes intermittent. The symptoms have been as follows. On pressing the 'power' button the channel indicator has come up momentarily on 1, but the 'on' LED has not lit at all. The problem is caused by failure of the SW5 V line as a result of dry-joints between Q9A2 and the control PCB. Note that the circuit diagram in the manual is correct but on the board itself and the board layout diagram in the manual Q9A2's base and emitter connections are transposed. Other joints in the area are worth checking. The 'on' command from the syscon microcomputer chip establishes the TU-12 V line which in turn brings on the SW5 V supply.

Intermittent standby: This machine would intermittently go to standby. The cause of the problem was dry-joints on transistor Q942 in the 5 V switched supply on the bottom PCB.

No rewind, fast forward or record: Check the record inhibit switch. It may be misaligned, broken or damaged as the result of use of a C format cassette.

Picture slowly 'evaporated': The sound was OK. There was no E-E or playback output although there was a signal at the scart connector. So we knew that the signal was there but wasn't getting to the modulator. The cause of the fault was C2E5 (100 μF, 25 V). Being close to a power transistor it had slowly and surely baked dry!

Mitsubishi HS347

Chewed tape due to crystal failure!: From time to time on this model crystal X6A0, the chroma reference crystal on the YC panel, would go open-circuit, deleting the colour and with it most of the urge of the drum and capstan motors – the crystal is also used as a servo reference. When the crystal stopped the head drum slowed, the capstan motor pulsed and, since the capstan motor powers the reel drive, a loop of crushed tape was left hanging from the front of the ejected cassette.

Poor rewind: The customer also complained that the machine didn't always carry out a timer recording. At first I couldn't see a connection, but the faults did have a common cause. The reel idler was so worn that it could hardly wind forward either. Thus if a tape was inserted with the start leader showing the idler tried to more it along but couldn't. When play or a timer recording was then tried it wasn't carried out. A new gum idler unit was required.

Intermittent timer failure: We found that this fault would occur when the tape was at the beginning after a rewind. After full-speed rewind the tape leader was left showing and the low-speed forward take-up didn't take the tape back in. Thus when the timer recording started the tape leader showed at the take-up end sensor and the unit stopped. Although the take-up torque and the fast-forward/rewind torque were OK the low-speed take-up torque was very low. A new reel idler cured the problem.

No functions and no display: There were no functions and no display. Checks showed that the standby 5 V supply was missing. The cause of this was traced to Q9A0, which was short-circuit base-to-emitter. A replacement restored normal operation.

Mitsubishi HS349

Intercarrier buzz on all channels in the E-E mode: The buzz was being recorded. Component checks in the sound i.f. circuit revealed that C307 was open-circuit. It's one of those capacitors that look like a resistor.

No picture, only lines: When a test tape was tried it was obvious that the drum speed was wrong. We found that D4A4 in the drum servo circuit was broken, a new 1N4148 putting matters right.

Rolling picture or top half blanked on playback: When the machine was dismantled the picture returned to normal. Tapping the servo board made the fault come and go and a check on the drum flip-flop waveform showed that when the fault was present it was not a true 1:1 square wave. At this point the machine decided to work correctly and no amount of tapping and banging would provoke the fault. We inspected the servo board under a powerful magnifier but couldn't find any dry-joints. So to do the utmost to prevent a comeback we resoldered all the components and wire links connected with generation of the drum flip-flop signal, from the drum pick-up pulse on CA1 through Q4B0 and Q4B1 to pin 24 of 14A0 and then from pin 11 of 14A0 to plug CE.

Machine kept stopping in play: The reel tacho signal at the collector of Q5AO was intermittent because the reel sensor PCB's earthing screws were making poor contact.

Mitsubishi HSB12 etc.

Accepts tape, laces up but no play: The following problem can occur with most Mitsubishi VCRs in the HSB 11/21/31 and HSB 12/27/32 series and later. The symptoms are that the machine will accept a tape and lace up but won't play, won't wind properly, damages tapes and the tape counter doesn't work. Insert a cassette and watch the action of the guides during lace-up. Then eject the tape and observe the at-rest position of arm TU-G (C-033) which is sometimes referred to as the half-load arm. You'll see that it has been prevented from going all the way back by arm TENS-REG-T (C-031) which has moved too far to its left. It should not move behind arm TU-G but should stay to the right of it. The consequence of this is that when a cassette is loaded arm TU-G is outside the cassette and as a result can't carry a loop of tape towards the capstan and pinch roller during lace-up. Thus when play is selected the tape doesn't move and during wind the counter cannot operate as the tape isn't held against the audio/control head. As with most machines which incorporate a real-time counter, indexing and jog/shuttle/reverse play functions there are several guides and tension arms that are not found in more basic models. In this design arm TU-G has to carry the tape through the gap between the capstan and the pinch roller during the lace-up so that it passes across the face of the audio/control head during both forward and reverse play. Arm TENS-REG-T operates as a reverse back-tension arm during reverse play and search. It also has, at its pivot end, a small brake arm that bears on the take-up reel turntable. If this brake's friction pad becomes dislodged, arm TENS-REG-T can move an extra five degrees or so to the left. This in turn means that when arm TU-G moves back during the unlacing process it's obstructed by arm TENS-REG-T and comes to rest outside the front edge of the cassette, producing the symptoms mentioned above when the next cassette is inserted. If the missing brake pad is still inside the machine it can be refitted using a touch of Evostick. Otherwise, replace the complete arm assembly, part number 591B551010. Note that there are two tension springs at the pivot end of the arm, omit one of them at your peril!

Would cut out after a few seconds: This was because the arm guide and arm tension lever didn't take up the tape. The arm tension post lever has a felt pad that rests against the take-up reel: the pad was missing. A replacement lever, part number 591B551010, cured the problem.

Tuning problem: There was a strange fault with this machine. It could be tuned in when in the preset tuning mode but in the normal mode channels couldn't be obtained. I suspected the tuning or EAROM chip, but before ordering replacements I decided to check on the supplies. This was a worthwhile move: the $-30\,V$ supply at pin 2 of the EAROM chip was low at about $-10\,V$. Moving back to the power supply I found that the fusible resistor R904 was open-circuit. Its replacement, using the correct type, restored normal operation.

Failure to wind/rewind tape: This can be caused by the machine trying to operate through the reel-drive clutch, slide-bar B having failed to latch. We've never had the fault stay long enough to be able to confirm a diagnosis, but have found that replacing the latch magnet coil (part number 299P124010) and its driver transistor Q5B5 has, along with a check on their interconnections, provided a cure.

No play, no counting in fast modes: A look at the deck showed that the tape wasn't being loaded correctly: the half-loading arm was jammed by the soft-brake arm on the take-up side, and as a result didn't take the tape around the audio/control head – hence no counter. The pad on the end of the soft-brake arm had fallen off, leaving the arm in the wrong place all the time. As the arm was OK and the pad was lying by the take-up reel disc we decided to stick it back on. The machine then worked perfectly.

Played at the wrong speed: The complaint with this machine was that it played at the wrong speed and produced Mickey Mouse sound. Sure enough, that's exactly what it did. When we removed the top we saw that the loading arm was sticking and in the wrong position. This was because the brake pad lining had come adrift from its lever. A replacement brake lever cured the fault.

Mitsubishi HSB20

Nasty buzz on E-E sound: There was a nasty buzz on the E-E sound with this new machine. We fed the output from a colour-bar generator into the machine and found that the buzz disappeared when the generator's chroma signal was switched off. Attention was therefore turned to the 6 MHz filter circuit CF151. By making comparisons with a good machine we found that although the output waveforms at pin 18 of IC101 were similar they were different at the input to CF151. After replacing various components in this area to no avail I was getting somewhat puzzled. L153, which is connected between CF151 and chassis, had been measured but as a last resort I decided to swap it over with the coil from the good machine. This cleared the fault. Both coils were identically marked and gave exactly the same resistance reading, so I can only assume that the faulty one had a couple of shorted turns or perhaps a crack in its core.

No E-E picture and sound: Playback and the tape functions were OK but there was no E-E picture or sound. Channel selection was shown on the display and the tuning indicator searched normally for a station to lock to, but as none was found the search continued in vain, the monitor's screen remaining blank. A check on the voltages around IC101 in the i.f./a.g.c. can showed that pin 2 was at a much lower voltage than its normal 5 V. So the a.g.c. wasn't working and the signal was cut off. The cause was the 0.1 μF, 35 V tantalum capacitor C105 which was leaky.

Intermittent failure to accept tape: This machine had been to another dealer who, despite replacing the loading belts, had failed to cure the fault – intermittent failure to accept a tape. The cause of the trouble was the cassette-in switches on the front of the loading unit. A replacement, part number 439C021010, cured the fault.

Runs slow, refuses commands, eats tape: This one gave us a hard time. It was reputed to run slow, refuse to accept commands, eat tapes and in general exhibit some pretty expensive sounding symptoms. After many hours of soak testing the capstan motor seized up. So we fitted a replacement, gave the machine a soak test for several hours then returned it to the customer. Two days later it came back with the same complaints. After several more hours of soak testing the fault put in an appearance. It transpired that the microcontroller chip was applying a gradually increasing braking voltage to the capstan drive chip. A new micro was the answer.

Mitsubishi HSB27

Tapes loaded too fast: We've had the following fault on several of these machines. When the tape is inserted it's loaded very fast, i.e. snatched out of one's hand. Then when play is pressed the machine acts as though it's in the search mode. In addition the fast-forward and rewind modes operate a lot faster than normally. The cause of the trouble is no capstan FG. Replace the capstan motor and PCB assembly. We've also had this fault with the more recent Model HSMX1B.

Machine cuts out after 45 minutes: We've had this fault on three occasions: the machine cuts out approximately 45 minutes after the beginning of a tape. The cause of the fault is low-amplitude take-up reel pulses. They can be checked and compared with the supply reel pulses at the collectors of Q5A4 and Q5B4. The cure is to replace the photo-interruptor. It's possible to replace this item without dismantling the bottom PCB.

Dead machine: As it was a Mitsubishi I had a quick look round for open-circuit safety resistors before I got too involved and sure enough R5K3 on the lower board was open-circuit. A short to chassis could be measured from one end of the resistor, however. I traced this to the deck where a spring was shorting one end of the latch magnet to the chassis: the reel disc unit had sprung apart and as a result the spring had fallen out.

High speed cue mode did not work: This machine worked correctly in all modes except the higher times-nine speed cue mode. We noticed that in this mode the capstan motor was stopping and starting. The cause of the trouble was a worn lower drum assembly – this was making the tape drag.

Worn lower drum assembly: This can cause many problems such as poor cue and review, picture jumping, poor tracking and no picture. The diagnosis can be confirmed by monitoring the f.m. waveform envelope, which will usually be impossible to set correctly. The fault can give trouble in the SP and LP modes.

Intermittent rewind/fast forward when hot: The other functions were OK. A new mode switch didn't cure the fault, but a new loading motor assembly did.

Mitsubishi HSB30

No playback picture: We traced the playback f.m. into and out of IC2A1 and then to transistor Q2B2 where it disappeared. Replacing Q2B2 cleared the fault – it was open-circuit.

Counter didn't work correctly: In fact the digital readout didn't change at all. At the same time a noise bar moved through the picture, indicating absence of the CTL pulses. Replacement of the audio/control head cured both faults. This is rather intriguing: why complain about something as trivial as the counter when the playback picture quality is so poor? As a footnote I received another VCR with the same complaint/symptoms later the same day. This one was a brand-new Akai, but this time all that was necessary was to clean the control head.

Intermittent drum speed: This machine had been playing around for some time. A couple of calls had been made but nothing amiss had been found. Eventually we saw what was happening. The drum motor would slow to the point where line lock was lost on the TV set. When the wires that connect the drum motor to the servo panel were moved normality was restored. We removed the drum motor and found that the connector plug hadn't been pushed home fully. Refitting it restored normal operation.

A blue, muted screen in E-E mode: This signified loss of the signal. Sure enough nothing discernible emerged from the M51496P i.f. chip IC101. The voltages around this chip were reasonable except for those at pins 1 and 2, where the expected 4.9 V was much reduced because C104 (0.22 μF, 50 V) was leaky. It's of the much maligned tantalum variety.

Mitsubishi HSB32

Machine played OK, but there was no fast forward/rewind: This was because the brakes stayed on. They should be held off by an electromagnetically controlled plastic lever. The magnet was energized but the lever didn't latch on the brake cam correctly. As the lever is plastic and the brake cam metal it seemed reasonable to assume that the plastic item would wear first, so it was replaced. This made no difference. A replacement brake cam put things right. It's called brake cam C, part number 591b554010.

Dead: There was no 5 V output from the power supply, at pin 3 of connector PZ. The reference voltage was present at pin 5 of PZ. IC901 (LA6324) was faulty.

Stuck in pause: On inspection I found that the tape wasn't being loaded up to the capstan shaft because the half-load arm didn't return to its eject position. It fouled on the tension lever. The reason for this was that the tension lever's brake pad had come off, allowing the lever to move too far to the left. A replacement tension arm cured the fault.

Mitsubishi HSM57

Pattern at bottom of picture: A fault you can get with these machines is a half to three-quarter inch pattern at the bottom of the picture, worse with pre-recorded tapes. Mitsubishi has come up with an answer. The cause of the fault is the fact that both video heads are in contact with the tape at the same time. As a result there's crosstalk between the heads, hence the patterning. Mitsubishi will, if requested, supply a new, specially selected head drum (upper and lower). This doesn't cure the problem completely but does make a big improvement.

Fast forward/rewind brake stays on: Clean all the graphite grease off the plate beneath, using solvent cleaner. Do not regrease it.

Noise on Nicam: Sure enough when stereo was selected in the E-E mode there was a loud rushing noise that swamped the audio signal. Mono EE sound, also hi-fi record and playback sound, were OK. A problem in the area of the digital-to-analogue converters seemed likely – otherwise I would have expected the internal muting to have been in operation. Checks around the TD6710AN chip soon bore fruit: the 16.93 MHz resonator X7A2 wasn't oscillating. A replacement restored the Nicam sound to normal.

NEC

NEC N9077

Fast forward/rewind very slow and the machine shuts down: We had had to clean the heads on this new machine several times, which seemed odd. So we brought it in for a check over. The customer's tapes were new and of good quality, and apparently he never used hired or rented ones. After several days testing the fault showed up. Fast forward and rewind became very slow, and in the playback and record modes the machine shut down because the take-up reel had stopped, the tape being laced tight enough across the upper drum eventually to stall it. Attention was turned to the reel braking mechanism where the cause of the problem was found to be an intermittent brake solenoid. It energizes to take off the reel brakes and occasionally didn't do so. Fitting a replacement put matters right. It was also necessary to obtain and fit a kit of three transistors and one diode. This is available from SEME. Part numbers are 35543418 (2SC1741A), 355D1931 (2SC2785), 35S62518 (2SD1227) and 36001026 (IS133).

'Can't tune out radar interference': Radar at Squires Gate is a major problem in this area, making channels 35/36 almost unusable in some parts of the Fylde. In this case, however, radar was not the cause of the problem. The white dashes across the screen were longer than the familiar radar blips and didn't have the characteristic repetition frequency. In fact the problem was caused by tape dropout as the dropout compensation circuit wasn't doing its job. A look at the circuit suggested that VR204 controlled this function, but a couple of experimental tweaks had no noticeable effect on the fault. There's no circuit description in the manual, and the function of IC12U2, which precedes VR204, is not shown on the circuit diagram. It's identified in the parts list as a 'PAL 1H CCD', i.e. a one-line delay CCD chip. Pin 12 of the LA7323 luminance processor chip IC1201 feeds pin 6 of IC1202. IC1202's output, at pin 4, goes to VR204 via an emitter-follower (Q209) and a filter (FL202). VR204 feeds back to pin 10 of the LA7323 chip via another emitter-follower (Q208) and a coupling capacitor (C208). The waveform at TP207/8 didn't look much like that shown in the manual and didn't vary when VR204 was adjusted. As Q208/9 and C208 checked OK I decided to order and fit a replacement line-delay chip. This was a mistake – it was the LA7323 chip that was the cause of the problem.

Tape stuck in machine: Checks showed that the supplies from IC1 (PQ12R04) were missing. A replacement was obtained from SEME under part number. 37101407.

Nokia

NOKIA VR3615
NOKIA VR3716
NOKIA VR3722/42
NOKIA VR3761
NOKIA VR3783
NOKIA VR3784

Nokia VR3615

Dead, no clock display and functions: Checks showed that Q803 was short-circuit, removing Q801's gate bias. Replacing Q803 restored normal operation.

Power supply buzzed and ran hot: We found that the voltages on the secondary side of the power supply were too high, and the overvoltage protection diodes were conductive. The cause of the problem was D804, which was short-circuit.

Intermittent rewind and fast forward: Also when eject was operated the machine would sometimes leave the tape hanging out. A replacement mode state assembly cured the problem.

Hum bars, E-E picture too bright: Playback was normal. I found dry-joints at various components in the i.f. unit, i.e. Q102, Q101, Z103, Z102 and L104. After resoldering these the machine performed correctly.

Picture intermittently too bright: On checking the video waveform at the input to the modulator I found that it was producing over-modulation. The cause of the fault was traced to a damaged solder pad at the bias feed resistor R189, which is connected to the video buffer transistor Q181. Normal operation was restored by repairing a small piece of print.

Wouldn't come out of standby and wouldn't accept tapes: At switch-on the drum and capstan started but the loading motor didn't shuffle. The power supply outputs were all present and correct, but when a check was made at pin 8 (Vref) of the loading motor drive chip IC602 the reading was very low. It should be around 8 V. Zener diode D607 was found to be virtually short-circuit, a replacement restoring normal operation.

Wouldn't tune in any channels: When the tuner's VT input was checked during search a ripple was seen to be present – and the search wouldn't go below 12 V. A check on the PWM output from the microcontroller chip showed that this was noisy, irregular and erratic. I then noticed the presence of discoloured manufacturer's flux in this area. After treating the area with PCB cleaner the machine worked normally.

Bands on colour on playback: This fault happened only with pre-recorded tapes with colour guard etc. Replacing IC301 cured the fault.

Intermittent failure to rewind or leaving tape out on eject: When the fault showed up I noticed that the reel gear didn't come over far enough to touch the reels. Thus no take-up on eject or rewind. A replacement reel gear assembly cured the fault.

Intermittent loss of playback or E-E picture: We found that the on/off 5 V output from the power supply occasionally disappeared. The cause was dry-joints at Q854.

Wouldn't rewind: Close inspection of the 'relay plate' revealed that the pin below the main cam was bent. Straightening it restored the rewind function.

Nokia VR3716

Sound marred by wow: This happened after a while. The cause was a faulty main clutch.

No display, drum motor running and the capstan driving the supply reel: The cause of the problem was dry-joints at oscillator crystal X702.

Sometimes cut off in record: The cause was excessive noise on the key-scan lines. To reduce this, fit two 330 pF capacitors on the function PCB, at plug OA, between pins 4 and 6 and 3 and 6. Also check plug/socket OA/AO for bad connections.

Nokia VR3722/42

No functions, would not accept a tape or power up: The clock display showed four small zeros. A check showed that the back-up 5 V supply was low. D7001 was found to be open-circuit while the back-up capacitor C7001 was gradually discharging and not being recharged via D7001. Replacing D7001 cured the problem. You get the same fault when C7001 is leaky.

Loops of tape when cassette is ejected: There are two common causes. First a stiff capstan motor. As a result there's no take-up on eject. The cure is to replace the capstan motor. The second cause is a faulty mode switch. This switch can also be responsible for erratic functions, failure to accept tapes and failure to eject them.

No stored channels and wouldn't tune any in: I found that the pulse-width modulation at pin 52 of IC301 was of low amplitude. C6003 was short-circuit and had damaged IC301. Everything was all right when these two items had been replaced.

Nokia VR3761

Switch to standby at switch-on: Normal operation was restored when we replaced the 2SC4484S transistor Q5402 on the PWA board. It was short-circuit all ways round.

Wouldn't come out of standby: When the switched power supplies were checked we found that there was no switched 5 V output. Q5402 (2SC4484S) was found to be short-circuit base-to-emitter, a replacement restoring normal operation.

Remained in standby, pulsed on/off: This machine wouldn't come out of standby properly and would occasionally pulse on and off. Checks while the voltages on the secondary side of the supply were pulsing on and off showed that noise was present on the main data lines associated with IC301. Replacing IC301 cured the problem.

Sound problem: The customer said that the sound was faulty when he gave a tape to a friend to play. On checking we discovered that the linear sound was the old sound track while the hi-fi sound was the new sound, i.e. the machine didn't erase the linear sound when it made a recording. The cause was dry-joints at C2019 and CN202 on the linear audio PCB.

Nokia VR3783

Sound did not match picture on other machines: The complaint with this machine was that the sound was OK with its own recordings but when the recordings were played back via a friend's machine (not hi-fi) the sound didn't match the picture. We found that the linear sound on these recordings was from an old sound track. There was no erasure because the bias oscillator transformer T2001 was short-circuit. A replacement cured the problem.

No clock display and no functions: A check at CN511 in the power supply showed that the output voltages were all OK. So we moved on to the secondary rails on the main PCB, where PR541 was found to be open-circuit. Replacing this item restored the always 5 V supply and normal operation of the machine.

Would not stop in rewind: This machine wouldn't slow down and stop in rewind, breaking tapes. Refitting the LED tower plug to the sensor PCB cured the fault.

Loud tone during playback and intermittent tone recorded: The cure was to refit the full erase head to the audio/control head PCB.

Nokia VR3784

No E-E or playback signals at scart connector: A check on the waveforms at the scart PCB showed that the VD OUT AV signal was not present at pin 9 of CN861. On tracing back I found that the print at one side of C1602 had lifted.

No functions, stayed in standby: All the outputs from the power supply were present, but the secondary switched supplies were missing. This was because the power-up output from IC301 remained low. Replacing IC301 cured the problem.

Faulty hi-fi sound: The left channel output was very low – in fact virtually missing. Checks carried out around the AN3961NFBP-A hi-fi processing chip IC231 showed that the input side was OK but the left channel output was very low. Replacing this chip cured the fault.

Orion

Orion D1094

Various intermittent faults: We've had various faults, usually intermittent, with this model. Symptoms have been no audio playback or record, no control pulses, no erase etc. Check for dry-joints at the vertical PCB to the rear of the deck. It connects the deck to the main PCB.

Stopping short of full tape eject: This note applies to this VCR, to the Tatung Models DVR634VN, DVR832V, TVR734VN, TVR932V, and probably others – the problem is with the Orion deck. Symptoms are intermittently stopping short of full cassette eject or when the cassette is half way in, and intermittent deck functions like load and play. The cause is a dirty or tarnished mode switch. You can clean it, but replacement is better.

Would not record: Then off to another town, to two different VCRs with very similar faults. The first was an Orion D1094 whose mode switch had recently been replaced to prevent the usual jiggery-pokery you get with these machines. Since the replacement, however, the machine wouldn't record – it would immediately go into the play mode, whether the tape inserted was with or without its safety tab. It transpired that when the mechanism had been removed to replace the mode switch the lever that operates the record safety switch had not been positioned correctly. While on the subject of these machines, here's another point to note. If, after replacing the mechanism, you have no sound, a playback picture that looks as if the heads need cleaning, or another obscure fault, check that the plugs and sockets which connect the mechanism to the main board are seated correctly. I've had faults caused by plug and socket problems with new machines. The final coincidental call was to a Philips VR2574, the one with the JVC deck. It had recently been returned from the workshop and had identical symptoms to the Orion machine. When it had been in the workshop its reel sensor had been replaced. To do this, you have to remove a PCB on the underside of the mechanism. When he'd replaced it the engineer had not resoldered the record safety switch tags. I could go on about such coincidences, and am sure that other readers will have had similar experiences.

Osaki

Osaki VCR33

Tape chewing due to faulty reel idler: It didn't take long to find the cause of the fault in this machine – tape chewing due to a faulty reel idler. But the fact that the Panasonic mechanism it uses is similar to that in the NV370 may be of interest since the appropriate VUD kit or individual components are easy to obtain. Don't use the Panasonic pinch roller though – a Sharp unit from Willow Vale will do. The rest of the machine is not of Panasonic origin. The whole lot looks very similar to the Susumu XR1 which was marketed by Clydesdale.

Intermittent tape chewing: This fault report illustrates what can happen when incorrect parts are fitted and is similar to the one above. Intermittent tape chewing and going into the fault mode was the complaint. Now this model is a GoldStar clone and uses a deck that bears an uncanny resemblance to the Panasonic D mechanism. On inspection we found that the gear which is part of the plate assembly A10 had split. So we ordered one and fitted it. The machine then appeared to work correctly – until the bottom cover was fitted, when the fault returned. To cut a long story short, we discovered that the loading motor from a Panasonic deck had been fitted. Its pulley fouled the bottom cover, which was minus some of its screws, thereby stalling the motor. All was well when the correct GoldStar loading motor was obtained and fitted.

Channel-dependent cogging and pulling from cold: This would clear after an hour or so, and was cured by replacing the a.g.c. reservoir capacitor. The offending item is C704 ($1\,\mu F$, $50\,V$).

Panasonic

PANASONIC G DECK	PANASONIC NVFS90
PANASONIC K DECK	PANASONIC NVG10
PANASONIC NV180	PANASONIC NVG12
PANASONIC NV2000	PANASONIC NVG21
PANASONIC NV333	PANASONIC NVG25
PANASONIC NV366	PANASONIC NVG40
PANASONIC NV370	PANASONIC NVG45
PANASONIC NV430	PANASONIC NVG7
PANASONIC NV688	PANASONIC NVHD100
PANASONIC NV7000	PANASONIC NVHD90
PANASONIC NV7200	PANASONIC NVJ30
PANASONIC NV730	PANASONIC NVJ35
PANASONIC NV788	PANASONIC NVJ35B
PANASONIC NV830	PANASONIC NVJ40
PANASONIC NV8600	PANASONIC NVJ42
PANASONIC NV870	PANASONIC NVJ45
(D1 DECK)	PANASONIC NVJ47
PANASONIC NVD80	PANASONIC NVL20
PANASONIC NVF55	PANASONIC NVL25
PANASONIC NVF55B	PANASONIC NVL28
PANASONIC NVF65	PANASONIC NVSD200
PANASONIC NVF65B	(K DECK)
PANASONIC NVF70	PANASONIC NVSD25
PANASONIC NVF75	PANASONIC NVSD30
PANASONIC NVFS1	PANASONIC NVSD40
PANASONIC NVFS100	PANASONIC NVV8000

Panasonic G DECK

High back tension: We've had several of the of G deck machines (Models NV-G40/G45/G48) with high back tension. In each case the error has been corrected by replacing the back-tension arm spring with a new one from the Panasonic spares department. The new springs are a different colour, duller and not as silvery in appearance, and are minutely shorter. Only a small number of machines in this range seem to be affected. Note that the back-tension specification has been altered. Measure it with a tentelometer at the beginning of a 3-hour cassette: the reading should be between 22.5 and 27.5 g. This applies only to the G mechanism.

Tape wouldn't lace up properly: We recently had in a couple of Grundig machines, Models VS500 and VS540, that use this deck. As the tape wouldn't lace up to the capstan the pinch roller touched it with no tape in between. The P5 unit wouldn't come into position fully, failing to sit in position in the pinch cam rift. The P5 pull-out sector gear which controls this was slightly cracked at the corner. A replacement restored normal operation. We've also had several Panasonic machines with this fault.

Intermittent raucous squeal during threading or unthreading: This mechanism often seems to throw up new faults – new to us, anyway! The trouble this time was a very intermittent raucous squeal at the completion of tape threading or during unthreading. It came from the brake pad that operates on the capstan flywheel. Clean the pad or replace the arm.

Carriage jammed: If the carriage has jammed but when you remove it you find that the rest of the mechanism is correctly aligned the cause of the problem is a worn right-side plate and its corresponding connection gear. Inspection will show that the gear teeth are severely worn. Both items must be replaced or the machine will bounce. This has become such a common problem that I now replace these items automatically.

Panasonic K DECK

Damaging tapes: When we opened the machine up and watched the loading we saw that the loading arm pulled the tape up to the control head but didn't lock into place fully. On careful examination it could be seen that the arm was slightly bent over. Realigning it provided a complete cure. It wouldn't surprise us if this became a regular problem with the deck, especially if users try to retrieve tapes from semi-loaded machines as some do. The construction of this arm (P5) is not rigid.

Crunched tape under pinch roller: The K deck is less troublesome and easier to deal with than its predecessor the G deck. This one crunched the tape under the descending pinch roller, however, because arm P5 did not move far enough fast enough. The main lever (a plastic moulding, part number VXL2307) was found to be cracked in the region of the P5 arm's driving notch.

Faulty tape: As with many modern mechanisms that are otherwise reliable, the K deck can suffer when a faulty tape or the brakes within a cassette jam and the customer tries to remove it himself. You realign the deck and return the machine to the customer in good working condition. It then jams again for no apparent reason. In this event it's worth checking the following items:

The side plate (right-hand side). It can cause problems, but check that the spring which gives tension to a lever has not fallen off (difficult to describe, but easy to see if it has come off).

The main shaft drive arm (VXP1339) that drives the cassette lift. It tends to spring outwards and slip. If you look down the right-hand side of the carriage you will see that it is driving the lever down on its edge, not the whole cam.

The loading motor drive cog. It can be replaced without upsetting the alignment of the mechanism. You may find that the small, underside cog has some stripped teeth. The part supplied by Panasonic (VDG0868) seems to have been made stronger.

The P5 arm unit (subloading arm VXL2306) can get bent when the tape is ejected. It can sometimes be bent back, but it's well worth keeping some in stock.

For odd operation, check the tape sensors and IR sender. To do this you have to lift the deck off the mother board then lift the board out. This is a quick operation with later models that have fewer screws. If care is taken, the mode switch (VSS0365) is easy to replace and align. The loading motor's supply comes via a plug which is connected to the mother board. It can produce odd problems and is worth replacement if you are in any doubt. Some odd faults I've had have been as follows. A broken take-up arm unit (VXZ0313). The metal arm on the loading post had fallen out, jamming the mechanism. The pressure roller has a nylon peg that engages with a drive cam: it has been known to snap off. Those who have not worked on this deck before should note that the cassette lift loading drive relies on the lift assembly to tension the drive cog. The operation of this is not very easy to see. You'll find that a service manual for the K deck video tape is a great help.

Panasonic NV180

Wouldn't play: These portable VCRs have had a longer production run than most, having been kept going to support Panasonic's excellent range of semi-professional cameras. This one came to us with the complaint that it wouldn't play. On investigation the capstan appeared to be seized solid. When the machine had been dismantled we found that the rotor of the direct-drive capstan motor was rubbing against the stator coils. Normally such contact is prevented by means of a springy clip in the bearing assembly – it clips into a waist in the motor shaft. What had happened was that the upper bearing (deck topside) had become dislodged, maybe due to an impact. Pushing this bearing home and relocating its protective cap did the trick.

Cassette jammed: This is a 12 V machine and was used in a coach. It was brought to us because a cassette was jammed in. To release, you turn the machine upside down and remove the bottom cover at the front. At the front of the machine, near the colour auto switch, there's a release hole through the PCB to a metal plate on the cassette deck. Press this with a screwdriver and the deck will release. See pages 2–3 of the service manual. When we'd done this we discovered that the machine was dead due to a Wickman fuse (Fl0J). You'll find it in the middle of the servo board at the back, near TP1003. It's in the 5 V line and is not shown in early circuit diagrams, so you can be puzzled if you don't know it's there.

Could not be moved: This portable didn't like being portable! Any attempt to move it would be like pressing the stop button, but leave it alone and it would operate perfectly. When the cabinet was removed the fault disappeared. Every plug, connector and lead was checked and tightened, but the fault returned upon reassembly. With a certain amount of language not normally heard in the workshop I removed the casings again. After some time the cause of the fault was traced to the 5-pin edit socket connections on the front panel: two of them would short together when the unit was tilted. I cured the problem by fitting sleeving over the exposed solder connections.

Panasonic NV2000

Combination of faults: This one came in with no comments under 'fault' on the job card – always a bad sign! After switching it on the first fault noted was that the machine didn't switch from E-E to V-V. This was because the EXCEPT REC 9 V supply was missing, soon traced to Q6043 (2SA719) which had a 4.7 kΩ collector-emitter leak. Next the drum speed took about 40 seconds to settle down. Adjustment of the drum free-run control took care of this. I then decided to refix the two main panels to the metal frame, which was hanging around them loosely as someone had removed all the screws. I'd previously noticed that three out of seven chassis fixing screws were missing and bolts were fitted where there should have been screws and vice versa. Next I found that there was no colour in the playback mode – the chroma reference set trimmer had been 'graunched'. A replacement was fitted and set up within minutes, but the a.f.c. also had to be realigned – a twiddler had been at large ... We're not through yet: there was no sound in E-E or playback because the REC 9 V supply was permanently present, due to Q6041 in the syscon being short-circuit. When all this had been put right and the whole thing had been reassembled correctly the machine worked well. The correct screws were fitted!

No clock or tuner, capstan slow etc.: This machine had a mixed bag of symptoms: no clock display, the tuner not working, the capstan running slow and erratically, and hum bars on the monitor screen. All were cleared at one stroke when C1009 (1000 µF, 35 V) in the power supply section was replaced.

No capstan lock in playback mode: This old top loader came to me with a note on the card saying that there was no capstan lock in the playback mode: there was no mention of the hum that obliterated the sound and vision in the E-E mode! A check on the power supply revealed that D1011 (10E1) and its reservoir capacitor C1009 (1000 µF, not 100 µF as the sheet says) were both leaky – the electrolytic was getting decidedly warm. The effects of this were hum on the 12 V supply to the r.f. converter and the demodulator circuits and on the 18 V supply to the capstan servo circuit. This fault highlights the need to check all symptoms before you plunge in.

Capstan ran slowly for the first 10 minutes: The customer told us that the fault had developed gradually over the last few months. When we hooked up the machine in the workshop we selected the test signal to tune it to our test set and noticed that there was a sizeable hum bar, which ran down the picture. Investigation of the power supply showed that the modulator's feed is provided by Q1008 which supplies a regulated 12 V output derived from an 18 V line. The 18 V supply's reservoir capacitor was found to be almost open-circuit. When it was replaced the hum bar had gone and the capstan speed was correct from cold.

No E-E signals, channel LEDs out: Q1006 was open-circuit base-to-emitter.

Panasonic NV333

Poor recording and sound breakthrough: The customer brought this machine in a few weeks ago, complaining about poor recording and sound coming through from other channels. After soak testing the machine for several hours no faults showed up so we returned it – suspect, faulty customer . . . A few days later it came back, this time with a tape to show the fault – non-erasure of the previous sound track, with the chroma signal remaining. The problem was traced to dry-joints on the bias oscillator transistor Q4014.

No playback colour: This machine actually wore Blaupunkt livery. It was wanted back in a hurry! Although there was no proper colour there were signs of unlocked colour flickering about occasionally. The VCO and reference oscillator frequencies were both correct so to save time I decided to change the AN6371 and AN6363 colour signal processing chips. This didn't cure the fault, and with four or five camcorders and some cameras wanted urgently life didn't look too good. With this model it's possible to check the a.f.c. by comparing the line sync pulses at pin 3 of IC8002 with those at TP8006. I found that there was no lock in playback although the a.f.c. system was locked solid in record. There was a difference in the level of the sync pulses at pin 3 of IC8002 between playback and record, but this is normal. The only other discrepancy I discovered was that pin 9 of the chip was at 2 V instead of 5.55 V in playback. This led back via a switch (Q3011) to the preamplifier and dropout detector chip IC3002. The voltage at pin 15 was low at about 2 V instead of 4.8V with dropout pulses – something to do with advancing the a.f.c. loop in the event of a dropout. Anyway, I found that if pin 9 of IC8002 was linked to the 5 V rail the machine played back in colour without need to replace the preamplifier chip. Very naughty, but the machine was old, the video heads in poor condition and the customer didn't want any more expense. After all, the customer is always right (if he pays for it).

Wouldn't switch off: Even if the timer button was pressed this machine wouldn't switch off. We found that the 2.5 A fuse on the mains transformer PCB was open-circuit – it's in the 14 V a.c. winding which provides the 9 V and 5 V lines to the syscon. The switching is carried out within IC6002 (M53216P), and when this chip's supply pin was lifted the fuse remained intact. A new chip restored normal operation.

Tape riding up and spooling in review: Until recently the cause of the tape riding up and spooling into the machine in the review mode has always been a worn upper drum. Recently, however, we've had some machines where the same end result has been caused by the lower drum becoming too smooth. Sometimes a good rub with Brasso will provide a temporary cure, staving off replacement for a while. This is worthwhile as the cost of a new direct drive unit is considerable.

No playback picture: When I read the job card I thought it would be a simple head cleaning job, but I was wrong. No picture in fact meant a blank raster. After a lot of mucking about I found that D3012 (MA165) was short-circuit, damping the output from IC3003.

Tuner didn't work: This was because there was no 30 V supply to the tuning potentiometers. The 30 V stabilizer IC7001 was short-circuit.

No clock display: This fault gave us a few headaches. The machine, an older version, had no clock display. Probably IC7501 on the front panel we thought. However, as soon as we removed the screws that hold the front and main PCBs and swung these out the clock display lit up, went out, lit up and went out again. A dry-joint we concluded, wrongly. We finally got to work with our trusty component tester and found that diode D1015 (MA165), marked D15 on the print, was leaky although not short-circuit. This diode is connected to the emitter of transistor Q1002 in the power supply section, feeding a regulated 17 V supply to the clock PCB. A 1N4148 is a suitable replacement.

No playback sound: There was a buzz when the audio head connections were touched, but it wasn't very loud. Signal injection (with the same finger) showed that the playback switch/amplifier transistor T4006 was failing to come on, due to insufficient base current. R4015 had risen in value from 33 kΩ to 330 kΩ, which threw us until we double checked the ohmmeter range.

Intermittent failure to eject and other intermittent mode failures: When I checked the machine it behaved like a video possessed. On application of power, sometimes the record LED or the pause LED would light, play was intermittent, and at other times the machine would return to stop after a few seconds. Occasionally the cassette housing would eject 5 seconds after the button was pressed! My first thoughts were that perhaps the microcomputer control chip IC6001 was faulty or that maybe the mode switch was defective. With this machine, however, I've found that the microcomputer chip is usually innocent when there's a syscon fault. Changing it made no difference, neither did removing, cleaning, adjusting and replacing the mode switch. Detailed checks were then made in the syscon circuit. As a result I discovered that transistor Q6008 had an intermittent base-emitter open-circuit. Q6008 is driven by Q6009: they are employed by IC6001 to pulse scan its mode sensor input.

Flutter on sound and other symptoms: There were several problems with this machine. First, there was considerable flutter on sound. This was quickly remedied by replacing the capstan motor. All belts, the idler and the reel clutch were then changed. When the machine was tried it went into a state of confusion: the record LED was on all the time, the machine wouldn't switch off and the clock and timer couldn't be set. After a bit of checking we found that the 2.5 A fuse associated with the mains transformer's 17 V a.c. winding was open-circuit due to over-loading on the 5 V rail. The cause of this was internal shorts in the MN1405VKK microcomputer control chip.

No tuning voltage: This was traced to R7019 on the panel mounted vertically at the rear of the deck. Its value is 10Ω and there appeared to be no contributory cause to its failure.

Machine would stop in fast modes: After running for just a split second in the rewind or fast-forward mode this machine would stop. Playback was OK. We thought that the fault was a mechanical one but eventually found that D1003 in the power supply was open-circuit, reducing the relevant voltage to about half. We were surprised by this: since playback was perfect one would have thought that the power supply was OK.

Intermittent shutdown during play: We ran several tapes through before the fault showed up. There was excessive tape tension – even with the back-tension lever pulled off the tape tension was still high. A check revealed that the supply spool was stiff on its shaft. Stripping this down and greasing it provided a cure.

No record f.m.: This machine had a fault I've not come across before: there was no record f.m., but playback was OK. Another company had told the customer that the heads were faulty. True, they weren't very clever. Reverse search was poor but normal playback was fine. I started by checking the various record supplies, of which there are several. This revealed that the 'except rec high' line was at 5 V in the record mode when it should have been low. Following through the circuit I came to transistor Q3020: it was OK but the $220\,\mu F$, 6.3 V capacitor (C3094) connected across its emitter and collector was short-circuit. Fitting a replacement restored normal recording.

Capstan servo wasn't locked: The symptom was a noise bar that floated through the playback picture. This is typical of a no CTL pulse fault, but checks on the CTL section of the circuit showed that it was working correctly. Further checks revealed the fact that the 9 V supply to the tracking control was missing. This is the 'except Rec 9 V' supply. It comes from the syscon panel where Q6003 was found to be open-circuit.

Faulty audio: We found that the VCR also refused to work mechanically, the audio fault was in the E-to-E mode: approximately 1 second of audio was heard about every 5 seconds. The audio mute line was high most of the time but dipped low to allow sound to be heard intermittently. The audio mute command comes from the syscon chip IC6001. A check at its reset pin 27 showed that a continuous line of pulses was present here. The reason for this was traced to D6038 which was leaky.

Dead: This was because the STR1096 power supply chip didn't produce a, 6 V output at pin 5. There was a normal 15 V input at pin 1 and a 9 V output at pin 4, thus switching problems were ruled out. A new STR1096 provided a complete cure. As the customer wanted the machine serviced we fitted one of the excellent full refurbishment kits available from SEME and others. Fit only original Panasonic parts, clean, grease and lubricate as per the manual and no problems will arise.

Severe tracking bars: They couldn't be removed by adjustment of the tracking control – although the control was effective with some tapes. After wasting time cleaning the tape path and adjusting the tape guides we found that the tracking shifter control R2035 was at one end of its travel. Adjusting it with the tracking control at its centre 'click' position provided compatibility with all tapes.

Would accept a tape but wouldn't accept any instructions: Failure of the tape to even shuffle on insertion led me to the μPC358 dual operational amplifier IC6006 in the reel motor voltage regulator. A replacement restored normal operation.

Snowy r.f.–r.f. signal, no E-E output and no reception: The cause of all this turned out to be a dry-joint at pin 1 of plug P7003 on the TV demodulator PCB.

No E-E and no channel indications: The playback sound and picture were normal. One end of the 3.9 Ω resistor R7020 in the power supply was found to be dry-jointed.

Thick hum bar appeared when play or record was pressed: The cause was failure of one of the diodes in the bridge rectifier circuit that provides the 15 V motor supply.

Panasonic NV366

Drum motor appeared to have dead spot: We found that the cable connector on the motor was partially off owing to a tight run of cables. Rerouting the cables and fitting properly cured the trouble.

Channel 1 LED flashing, no channel change functions: This can very often arise after replacing the three gold memory back-up capacitors. If you remove these by twisting them off the board (as you should if they are leaking) you can break R7693 (22 kΩ) in half, producing the new fault.

Take-up spool rotated without a cassette being loaded: With a cassette inserted all tape modes worked but there was no auto-stop when the tape had been fully rewound. The cause of all this was dry-joints at plug/socket PL6010 on the main PCB. Because of its position at the rear edge of the board it is subjected to physical stress when the board is hinged down, there being very little slack in the wiring to the socket.

No rewind, and only 2 V at motor plug: The cause of the fault was traced to Q6022, a replacement putting matters right.

Panasonic NV370

No playback colour: The a.p.c. loop was not locking as the reference frequency was way off. The PCB module component was replaced.

Hum bar on the picture in all modes: As it was a single hum bar the ripple was at 50 Hz. This eliminated the main 12 V supply which is derived from a full-wave bridge rectifier. We found that C1102 (2200 μF, 25 V) on the power transformer panel was open-circuit, putting large dents on the regulated 12 V line from Q1101.

Low-gain TV pictures: We assumed that the problem was failure of the amplifier in the r.f. converter, but a replacement made no difference. A more careful examination revealed that there was a large hum bar when the test signal was switched on. The 12 V supply to the converter was found to be low and on moving back to the power supply we discovered that the 18 V supply reservoir capacitor C1102 (2200 μF) was leaking. Replacing this item put matters right.

Totally dead: This machine had no clock or channel displays and no deck functions. Voltage checks on the right side power supply confirmed that this was operating correctly. The next checks were made in the power supply section of the main panel, where we discovered that the 5 V rail was missing because R1001 (0.39 Ω) was open-circuit. This resistor feeds the collector of the 5 V regulator transistor.

'No functions': At switch-on an eerie 'heartbeat' noise came from within and continued until the machine switched itself off a few moments later: it was caused by the capstan motor shunting back and forth. Meanwhile eject wouldn't work. This effect is usually due to a missing 5 V rail, Q501 on the head drum assembly being open-circuit. The 5 V line was intact this time, but the 12 V line was missing. We quickly traced the cause of this to an open-circuit safety resistor (R1101) in the unregulated 12 V supply. It appeared to have failed for its own internal reasons.

Displayed a half black/half white screen in the E-E mode: The cause of the fault was traced to C1102 (2200 μF, 25 V) being open-circuit.

Would only load or eject a tape: The complaint was that this machine wouldn't do anything except load and eject a tape. The digital tape counter showed a flashing 'd', indicating the dew condition. There was a high on the dew input pin (23) of the microcomputer chip IC6001, and voltages that would give this result were present around the operational amplifier chip IC6004 that acts as a comparator. The reference signal at the inverting input (pin 6) was 0.375 V while 0.83 V was present at the non-inverting input (pin 5). The dew sensor is connected from pin 5 to chassis and is the lower half of a potential divider. We found that there was 0.83 V at both sensor connections as the earthing screw on the small connection PCB, to which the sensor is connected, was loose. When the screw was tightened the voltage at pin 5 of IC6004 dropped to 0.007 V and the machine worked normally.

Hum on E-E vision: The E-E picture was half white and half black, i.e. there was hum on vision. C1102 (2200 μF) in the power supply was open-circuit. The playback picture was only slightly affected.

Rolling etc. on some tapes: The picture was OK with some tapes but with others there was rolling and a small white bar was visible at the bottom of the screen. Playback of the machine's own recordings was OK except for the white bar. After a lot of deck realignment we established that the fault was being caused by the drum assembly. The rotor base at the bottom had shifted slightly out of position, so that the pole switching was incorrect. Slight adjustment by loosening the hex nut in the base and realigning it put matters right. All was now OK except for cue. When this was selected the machine ran in the cue mode but when the button was released it still ran in the cue mode although the cue sign disappeared from the display. The AN3822 capstan chip was faulty, pin 17 being virtually open-circuit when a check was made between here and chassis. A common problem with these machines is a dew indication. Touching the dew sensor on the deck with the tip of a hot soldering iron for about 20–30 seconds usually provides a lasting cure. If the dew indication is erratic, however, the sensor is at fault.

VCR loop-through not working: This machine worked fine as long as the TV set was on the AV channel, but when you switched to a TV channel the VCR cut off the loop-through. The cause was loss of the regulated 12 V supply to the r.f. amplifier. D1106, a 13 V zener diode in the regulator circuit on the power supply panel, was short-circuit.

Fine horizontal lines varied with tuning: The E-E picture was marred by fine horizontal lines that varied with the tuning. A.g.c. decoupler C702 on the tuner/i.f. panel was open-circuit.

Lines across picture: Tracking bars that the control couldn't remove were present. The cause of the trouble was insufficient back tension because the pad had dropped off the brake band. Watch out for this with all Panasonic machines that use the metallic band.

Panasonic NV430

Capstan ran at high speed: This one was being serviced after coming back off rental. When I switched it on, the capstan ran at high speed and the loading ring arms continually laced and unlaced. When the bottom cover was removed the mode switch retaining screw was seen to be loose while the switch itself had moved. Refitting it correctly had no effect, however, so a new one was fitted. This improved matters, but full order was not restored until the main cam gear and the brake actuating arms were realigned. All that was then required was a new VDV0152 loading belt.

Appeared to be dirty video heads: Cleaning made no improvement, however. When the scope was connected to the head preamplifier module to check for head wear the fault cleared itself. A check on the inside of the module revealed dry-jointed connectors and earthing connections. Resoldering the faulty joints cured the fault. It's worth checking that the module fixing screws are tight and thus providing good earthing.

Optical tape-end detection failed: The problem with this machine was that the optical tape-end detection wasn't working. As detection didn't occur at either end of the tape the infrared emitter circuit was the most likely suspect. On these machines the infrared LED is pulsed on and off by the system control chip IC6001 via the 2SD636 emitter-follower Q6006. Meter checks showed that the LED, transistor and two assorted resistors all read correctly. But no light reached the end sensors. Scope checks then revealed that while 5 V peak-to-peak pulses were arriving at the base of Q6006 the pulses at its emitter were of only 1 V amplitude. Replacing this transistor restored normal operation, but it read OK on the meter's diode check when tested out of circuit.

Wouldn't come out of reverse search: If the machine won't come back out of the reverse search mode until stop is selected check the AN3821K capstan motor drive chip IC2002 on the servo panel.

Some functions did not work: The rewind, stop and eject keys worked but the others didn't. All functions worked when the remote control unit was used. The cause of the fault was that R6554 (2.7 kΩ) on the operation PCB was open-circuit.

No colour at top and bottom: The E-E picture was OK. After initially suspecting a fault in the chroma circuitry we turned our attention to the electrolytics in the power supply. Replacing C1001– C1004 cured the fault.

Tape loops: If the tape loops after going from play to stop then eject, i.e. not after fast forward or rewind, the mode switch requires cleaning or replacement.

Didn't produce a clear E-E picture: There were wide black bands and bad distortion on the E-E picture. Not surprisingly, we found that one of the capacitors in the power supply had died of old age. The culprit was C1002 (47 μF) in the 45 V supply.

Dead, but clock flashed briefly: This rather complicated two-speed machine was dead, although the clock flashed on for a second. Almost always the cause of this fault is the two 100 μF, 63 V electrolytics in the power supply. Check the condition of the 3300 μF and 47 μF, 63 V electrolytics if there are signs of hum on the picture.

Panasonic NV688

Wouldn't switch from LP operation: Sometimes this machine wouldn't switch from LP operation in the record mode, while in playback it would select LP by itself with an SP recording and what can only be described as LLP with an LP recording. After a lot of tapping, prodding and flexing of boards we found that the supply to the SL/LP discriminator IC2005 was varying and at times non-existent. There was a dry-joint on a wire jumper that's part of the supply.

Incorrect symbol on circuit diagram: Here's an example of confusion caused by an incorrect symbol on the circuit diagram. The problem was no playback, recording or signals – just a buzz and noise bar in the E-E mode. We soon found that the regulated 5 V rail was low at only 2 V. No excessive load was apparent and transistors Q1003/4 and zener diode D1006 in the regulator circuit all checked out correctly when cold resistance tests were carried out. We decided to check the 2SD1275 series regulator transistor by replacement. It's listed as a Darlington device, although the circuit shows it as a straight npn transistor – which it seemed to be from our meter check. In fact the Darlington bit had shorted out. A replacement solved the mystery, with no thanks to Panasonic's circuit diagram.

Machine dead, some LEDs on: This ageing LP VCR was pretty dead: there was no clock display and the machine didn't work, although the LP and power LEDs were permanently on. The cause of the problem was that the regulated 6 V supply, from which the 5 V supply is also developed, was missing. There was voltage at the emitter of the regulator transistor Q1201 and the correct base bias was present, but there was no output at its collector which was dry-jointed. Q1201 is miles away from the power supply area, being mounted on the back of the DD unit.

Panasonic NV7000

Wouldn't play/record at end of tape: This machine, a well-worn old soldier, wouldn't play or record a 3-hour tape if the supply reel was full or nearly full. Instead it would shut down and unthread shortly after completion of tape threading. What was happening was that the supply reel was virtually unbraked, and the inertia of a full reel of tape would unwind a few centimetres of tape during threading: normal forward tape motion didn't take up this slack before the lack of supply reel rotation triggered the syscon to produce an emergency stop. On investigation we found that the felt lining had parted company with the back-tension regulator band, and the soft brake had softened to disappearing point . . .

Poor picture: On test the playback showed that the CTL pulses, with both its own recordings and pre-recorded tapes, were missing. All checks pointed to the audio/control head but this proved to be innocent when fitting a replacement left the symptoms as before. We then changed the i.c. that contains the CTL amplifier, but again there was no difference. Replacing C2039 and C2040 provided the cure.

Intermittently smeary: It was as if the picture was out of focus. Flexing or patting the bottom PCB, one half of which carries the chrominance and luminance circuitry, would instigate the fault. Sometimes the machine would behave itself for a day or two, then we were back to square one. The problem was that there was no time to scope any signals when the fault appeared. So I went on tapping the board until I came to the luminance playback level preset which seemed to be the cause of the fault. After changing it I gave the machine a soak test then returned it to its owner. Five days later it was back with the same fault. To cut a long story short I eventually found that filter FL3002 had a dry-joint at one end.

No picture: Sound was OK but there was no picture. In fact there was a blank white raster on playback. As the E-E picture and sound were OK we suspected either a blanking or recording fault. It turned out that the record-on switch transistor Q3013 was short-circuit, a replacement restoring the picture.

Snowy playback pictures: If it looks as if the heads are worn out, check the continuity of the rotary transformer's windings before changing the drum. I've had three machines with this fault and in each case one winding was found to be open-circuit.

No clock, stuck on channel 1: There was a very strange fault with this old machine. The symptoms were no clock display and stuck on channel 1. Everything returned to normal when the machine warmed up. It didn't take us long to find out that the 6 V regulator chip IC1501 in the power supply was sensitive to freezer. Although it's a 6 V regulator the output was found to be 5 V. Cooling it down made the clock go off but there was no change in its output voltage. A scope connected to the output also confirmed that there was no difference between the fault and working states. In fact we could find no reason for the clock going off when IC1501 was cooled down, but replacing it cured the fault.

Panasonic NV7200

Chrominance high, luminance low on record: Playback was fine. On removing the bottom cover I noticed a very small amount of spillage on the board just below the white clip control. This was cleaned off with 1–1–1 trichloroethane (RS Solvent Cleaner). The waveforms were wrong not far into the circuit, but were not drastically out. So I decided to see if the circuit could be set up as per the manual. After carrying out the dark and white clip and the luminance and chroma record current adjustments the machine performed well. One thing to note is on the white/dark clip adjustment: you should adjust the lower peak of the waveform for 175 per cent not 150 per cent as on the oscillogram.

Failure to record: For failure to record, check the voltage at pin 9 of the f.m. modulator IC3001. If 9 V is present check Q3003.

No channels, no E-E and no capstan: The symptoms were no channels, no E-E signals and no capstan operation. We found that the 45 V supply at pin 5 of P1003 and pin 4 of P1002 was missing. Checking back revealed that R1019 (120 Ω) was open-circuit.

Enters still-frame mode in cue: During the space of a week we had two of these machines with the same fault. The symptom is that at the beginning or end of a cue or review session the machine appears to enter the still-frame mode for a few seconds then reverts to stop. The mode switch can sometimes be responsible for this. In both of these machines, however, the cure lay in replacing both loading belts. Slippage of these belts is more often associated with failure to complete tape loading.

Cuts out after 2 seconds in play: This ex-rental machine had been put aside by a junior engineer. The complaint was that it cut out after a couple of seconds in play. At first glance the symptoms on the card suggested that the large loading belt was worn, but on test the machine laced and ran before cutting out. It also cut out in the rewind and fast-forward modes. A loss of reel pulses then? Yes indeed. I've had the magnetic base of the supply table fail in these machines and while beginning to check for this I found the cause of the problem – the spool-securing circlip was weak and the spool had risen above the brake. Thus the base was too far away from the sensor to be able to produce good pulses. Those not familiar with these machines should note that the pulses come from the supply instead of, more usually, the take-up reel.

Clock went out when record was selected: Also the channel number went back to one. We found that the machine did the same if play was selected. It did this as soon as the loading motor came under pressure and the drum started, so we were looking at a supply problem. Checks showed that the unregulated 13 V supply dipped to 8 V when the problem arose. Disconnecting the heaviest loads in turn proved their innocence until we tried the DD unit, when the fault cleared. But fitting a replacement didn't help. This confirmed my original suspicion that it was a supply problem. The reservoir capacitor C1005 had been changed and fitting another one didn't help. A new set of bridge rectifier diodes solved the problem.

Loud knock from mechanism when lacing up or in cue: This did not happen in the review mode. It suggested noisy loading motor bearings, but in fact the cause was a very worn capstan bearing.

No tape functions and won't eject: There was a tape in the machine but there was no response from the function selectors and their indicators although the power-on LED and the channel indicator were as normal. Our past experience has often been that this problem is caused by either dirty contacts or a broken cassette-in leaf switch. Not this time, however. With a cassette in there should be 12 V at each tag of the switch (contacts closed). There was no voltage at either tag, so we had a 12 V feed problem. When we traced back to pin 2 of plug/socket P6014 on the system control panel we found that the 12 V supply was missing here as well. L6001 was open-circuit.

No drive: This machine had been serviced by one of our apprentices – he'd fitted the parts in the VUD kit, changed the heads and one or two other items, but there was now no drive. The absence of a familiar sound when ejecting was noted: the solenoids weren't being energized, so the brakes were on. R6083 (2.2 Ω fusible) which feeds the unregulated 20 V supply to the solenoid drive circuits was open-circuit because the 2SA768 transistor Q6027 was short-circuit – it had run hot and melted its solder. The reason for this failure was the presence of an extraneous conductor that had fallen on to the syscon board from the mechanism – a circlip that was missing from the brake band under the back-tension post.

Would unlace after lacing up: This quite ancient machine would unlace about a second after lacing up in either the play or the record mode. After we'd checked various bits and pieces we came to suspect the supply Hall-effect sensor. I was about to find a scrap machine to rob it of its sensor when a colleague had the bright idea of checking with a magnet. His idea was to take the supply spool off, select play and move the magnet back and forth to the sensor. Why? Because he'd had exactly the same fault with an Hitachi machine and had found that the magnet on the underside of the spool rather than the sensor was the cause of the trouble. When the magnet was used the machine played happily. So we 'borrowed' a supply spool from a scrap machine.

Operation of capstan motor unstable: As the operation of the capstan motor was not stable the pictures in both the playback and record modes were unstable. When I opened the machine up I found that all the chips in the servo department had been replaced. As a first step I checked the supplies and found that the voltage on the 45 V line read low at 35 V, falling to 25 V. C1010 and C1013 had dried out, replacements restoring normal operation.

Panasonic NV730

No functions, display out: When the on/off button was pressed all this machine did was beep and light the 'on' LED. The fluorescent display was out due to loss of the −45 V supply and hence the −30 V supply to the display. There was also no power-on signal at the main regulator chip IC5001. This was because the 45 V supply was missing. Both these supplies are derived from a common winding on the mains transformer via a 4.7 Ω safety resistor (R1002) which had gone open-circuit.

Works for a short while then stops: We put the machine in play and soak tested it. After half an hour the fault showed: the machine unthreaded and switched off. When it was tried again in play the machine threaded up then the capstan motor oscillated, after which it unthreaded and switched off. Voltage checks proved that the AN3822 capstan motor drive chip IC2004 was faulty.

Display problem: The on, off and the two week bars in the display were permanently on. The cause was simple – the metallized plastic screening card, fitted at the front right of the machine to cover some of the ribbon cables that connect the display PCB, had been pushed too far forward. It was touching one of the exposed legs which stick out of the top of the fluorescent display tube, earthing it.

Tape tangling: The machine had been in the workshop about six months ago, when the reel idler had been replaced for the same fault. We gave it a test run, which lasted for several days before the fault appeared. This revealed that there was no voltage drive to the reel motor in any mode. A look at the circuit diagram suggested to us that Q1504, a regulator transistor that supplies the voltage to the motor switching circuitry, could be responsible for the trouble. On investigation we found that all three connections were dry-jointed. It's mounted on the right-angled heatsink at the rear right-hand corner.

Weakness of worm-wheel unit: The weakness of the worm-wheel unit, part number VXP0575, fitted to the front-loading mechanism in this machine has been mentioned before. A possibly puzzling symptom associated with this and other front-loading assembly faults is that the left-hand side supply spool rotates backwards for several seconds before auto-shutdown of the machine.

No picture: Two machines came in with no picture in the E-E or play modes. In both cases I first blamed the modulator, then had to start serious fault-finding. I'd no circuit diagrams for the first machine, a Sharp VC7300, but managed to track down the fault to the HA11703 chip. The second machine was a Panasonic NV730 where the chip that carries out the same functions, i.e. head signal amplifier and E-E/video switching, is an AN6337S. It's on the folded luminance-2 panel. Unfortunately this chip seems to be unavailable – and the board is hideously expensive. Worse still my meter probe slipped whilst I was monitoring the power supply lines. This damaged the tuner and the BN5115 on the demodulator panel, the result being low gain. I was able to replace these items with parts from a scrap NV366, with some modifications. Note that the manual may not correspond with the actual demodulator PCB or the aerial booster/modulator unit.

Intermittent stopping in the play mode: The machine had come to us as another local firm had unsuccessfully tried to sort it out. It was obvious that previous work had been done on the deck – the guides had been cleaned and the take-up idler roughened up. After a long soak test the fault showed up as no take-up, with tape spillage into the machine before the system control came to the rescue. Q1504 at the rear right of the machine (from the front) was found to have dry-joints.

Switch-off after 5 seconds: At switch-on very little happened with this machine except that it switched itself off again after about 5 seconds. As with many VCRs this one's syscon pulses the deck motors at switch-on and looks for a response from the sensors as a condition of maintaining the power lines. In this case the motors didn't pulse because the 'motor unreg' line was missing. There were dry-joints on the legs of transistor Q1501 on the power supply subpanel.

In play mode the 'record' display lit at random: The machine played and recorded normally. Our first clue was that the clock was incorrect. Replacing the clock microcomputer chip IC7501 put that right but the original fault remained. It was caused by the clock crystal X7504 being open-circuit.

Critical tracking in LP mode: The SP mode was OK. A check showed that one head's output was low in the LP mode. Fitting a new head made no difference, so we quoted for a new lower drum assembly and replacement video heads (they come together when you purchase the lower drum) and a service. Surprisingly, in view of the high cost, the customer gave us the go ahead. Correct operation was restored when the work had been completed.

VCR suddenly stops and crumples tapes: These machines are now old but since we've had a couple in recently both with the same nasty intermittent fault the following note may be useful to others. The symptom is that the machine suddenly stops during record or playback: when the cassette is ejected you find that there's a crumpled loop of tape hanging from it. The cause of the fault is dry-joints of the connections to the reel-motor voltage-regulator transistor Q1504 which is mounted on an L-shaped heatsink in the rear right-hand corner of the machine.

Slipping mode control belt: When this machine was switched on the capstan and reel idler would shuffle to and fro a few times then stop. The machine wouldn't accept a cassette. There were no channel indicators or power-on display, although the clock flashed. The cause of the fault was the $0.12\,\Omega$ resistor R1101, which was open-circuit. It's in the 20 V feed from the power supply.

Panasonic NV788

Complete loss of playback and E-E sound: The fault with this machine was complete loss of both the playback and E-E sound. There was no switched, regulated 12 V supply to the audio panel. Tracing it to source via the video panel finally brought us to the syscon board, where regulator transistor Q6062 had 12 V at its base but nothing at its emitter – the base-emitter junction was open-circuit. We replaced this 2SC1847 transistor with a BD137. Restoration of power to the audio section also brought up the backlight in the LCD counter display. Had we known about this we'd maybe have found the culprit sooner.

'Still' mode fault: The complaint was of a poor picture. On test, however, the picture was all right except in the still mode, when almost half the picture was lost in noise. Having seen this problem with other machines I checked the playback tension, suspecting that this was low (10 g/cm instead of 30 g/cm). The cassette carriage assembly was removed, a cassette was put in it and another one was held in the machine. A check on the back tension then proved that it was correct. I tried again with the cassette carriage refitted and spotted the cause of the problem – the back-tension post arm, which runs against the tape, moved too far to the left and rested against part of the carriage assembly. With the assembly removed the arm moved far enough to give the correct tension. The cure was to move the fixed end of the brake band to pull the arm further to the right when running then adjust the tension spring for correct back tension. This problem could very easily arise if the brake band is replaced and the position of the back tension arm, when playing a tape, isn't checked before refitting the carriage.

Poor picture, worse in pause: A field engineer had visited the customer's house and cleaned the heads. This had improved the playback picture but didn't help very much in the still/slow-motion modes, so the machine was brought into the workshop for further investigation. We found that the top third of the picture was noisy in the still mode, the rest of the picture looking all right. The back tension was checked and adjusted without producing any improvement and new heads made no difference. A slight improvement was obtained by adjusting the inlet guide. We then looked at the f.m. output waveform at TP3512 – see Figure 7 below. In the playback mode the normal L and R heads are used and the output was correct. In the still/slow modes, however, a fifth head L' is switched into circuit, replacing head R. As shown in Figure 7 it was the output from this head that was causing the problem. Since heads R and L' share the same playback amplifier the only cause of the difference could be in the switching which is done by relay RY3501. We didn't have a replacement in stock so the relay was removed, its cover was opened and the contacts were squirted. When the relay was refitted the fault had gone and the machine gave perfect

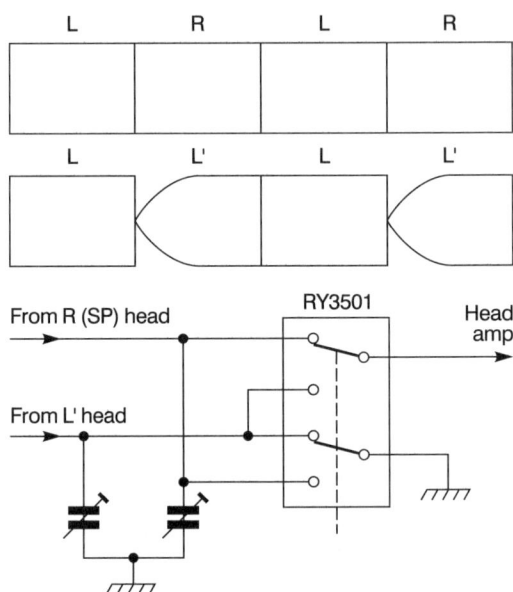

Figure 7 *F.M. output waveforms and head-switching relay*

stills and noise-free slow motion – despite the fact that the heads were the original ones.

Remote control didn't work: This machine's remote control system didn't work. The IR commands were being inhibited because the machine thought that the timer was on. Timer inhibit is introduced by the MA165 diode D7554 on the timer PCB. A check showed that the diode was leaky.

Panasonic NV830

No colour: A dealer asked me to sort this one out. For some reason there was no colour after replacement of the 12 V regulator Q1102. Why I do it this way I shall never know, but after proving that there was no fault in the colour circuits I discovered that the 12 V rail was low. Now this isn't the rail provided by Q1102. This was at TP1002, which recorded about 10.5 V. The base of Q1003 was at 4.5 V instead of 5.6 V while its emitter was held at a regulated 5 V, so it was off. This transistor's collector is connected to 45 V via 6.8 kΩ and 4.7 kΩ resistors, so there was no reason why the collector voltage read 11 V unless Q1003 was leaky, which it wasn't . . . In fact Q6013 on the system control panel switches the supply on and off by earthing the collector of Q1003 via D1004, which was leaky.

Overloading on E-E: It was not the modulator this time; the u.h.f. a.g.c. was ineffective. Replacing the BN5115B i.f. chip cured the problem.

No colour: The problem with this machine was no colour. We found that the switched 12 V line read 10 V. The switching transistor Q6013 in the syscon was leaky.

Panasonic NV8600

Pinch roller solenoid wouldn't pull in: These oldies were built well! Some look set to go clunking and twanging forever. One we had in for repair wouldn't play or record because the pinch roller solenoid wouldn't pull in. The solenoid would hold in when operated by hand and we found that the pull-in transistor Q622 was open-circuit. A BD139 transistor turned out to be a successful replacement, but we also checked the damping diode (D624) as a precaution.

Playback gives blank white raster: There was normal sound in playback. But there was no E-E picture and we couldn't get the test signal. I suspected the r.f. converter but as we didn't have one to hand I decided to start fault-finding. The first step was to check the supply to the r.f. converter. It was correct at 9 V. Next the converter's video input was checked. The manual says there should be a 1 V p-p signal here but there was a 10 V d.c. reading. This 10 V was traced to the buffer transistor Q313 which was short-circuit all ways. A replacement put matters right.

Sometimes wouldn't complete the threading process: Also, on occasions, the functions couldn't be selected as the keys were stiff. We'd have wasted a lot of time if we hadn't noticed the changing intensity of the light from the cassette lamp. The cause of the problem was that the cassette lamp lead-outs were intermittently shorting in the holder. Straightening the lead-outs provided a complete cure.

Panasonic NV870 (D1 deck)

No tuning memory: As a secondary fault the display went dim when the machine was switched on. Our decision to trace the cause of the memory tuning fault first turned out to be a good move. We found that the $-28\,V$ supply to the tuning memory chip was missing due to R7550 ($27\,\Omega$) being open-circuit. Replacing this cured both faults as the $-28\,V$ line also supplies the display driver buffer.

Occasional flicker from display: This was the only sign of life. There was less than $1\,V$ on the Reg $5\,V$ rail due to a $10\,\Omega$ short to chassis. Several plugs, sockets and links later I reached the operation display board and found that the earthed leg of C6512, which decouples the supply to IC6503, was pushed against the positive leg of the Reg $5\,V$ decoupler C6502, behind the digitron. With the leads apart life was restored to the machine. But no channels could be tuned or memorized for several minutes, after which this fault would suddenly clear. The AN5033 tuning chip was temperature sensitive.

Would unlace in 'review' mode: When the fault occurred the play symbol would appear in the display. Then, after realizing its mistake, the machine would lace up again, usually successfully but sometimes with tape chewing. Replacing the mode switch cured the fault.

Hi-fi sound lost intermittently in playback: The longer the machine was in use the worse the symptom became, suggesting that there was a thermal fault. Use of a hairdryer on the electronics failed to instigate the fault, however. Following a hunch, I decided to check on the mechanics and found that minute changes in the back tension coincided with the loss of sound. Back tension is applied via the tension arm I unit (VXL1157) which engages, on the underside of the mechanism, with the main cam gear (VDG0200). Close examination of this gear revealed that the Moriton grease used here for lubrication had become lumpy with age. This would occasionally prevent the back-tension arm engaging fully. The 'thermal fault' I had suspected must have been caused by the grease's characteristics altering with the temperature of the machine. Cleaning the cam and applying fresh grease cured the problem.

Panasonic NVD80

Audio record level VU meters didn't agree: With a mono audio source the left level was lower. We adjusted this as laid down in the manual, with VR4004, to obtain correct balance, but then found that the left meter displayed too high a level at the lower end and too low a level at the upper end. Correct conditions were obtained when the LED level meter unit (VEK3183) was replaced.

Dead machine: When the mains supply was connected the switch-mode power supply would give a quick whistle then die. There were no apparent shorts or overloads on the secondary side of the PSU. So attention turned to looking for the most singed and discoloured capacitor on the primary side. C1045 (47 μF, 35 V) won hands down. A replacement got the show back on the road.

Poor playback picture and noisy hi-fi sound: This machine came in for service with various complaints, the main ones being of a poor playback picture and noisy hi-fi sound. A new upper drum put that right. We then noticed that with SP recordings there were noise bars across the picture in the cue and review modes instead of noise-free horizontal 'cuts' – the LP results looked fine. The cause was discovered after some – searching around the signal paths in the video preamplifier area, there was a break in the print on the underside of the main PCB between pin 8 of connector BP3001 and pin 17 of the servo pack connector. This is a feed from the preamplifier called 'enve. select'. During normal playback there should be a voltage high here, becoming a square wave in SP cue/review. The square wave is used by the servo PCB to produce an output called 'h.amp switch' at pin 16. It returns to the video preamplifier PCB.

Mechanism problem: This machine would lose control over its mechanism, lapse into a sulk and power down. Moments later the fault would clear and everything operated normally again. The clue with this machine and indeed with most G mechanisms is that you should get a nice, satisfying 'clack' when a key is pressed as the mechanism solenoid engages and the capstan motor moves the mechanism to the selected mode. With this machine the solenoid was intermittently sticking. The system control then became confused and powered down. A new solenoid, part number VXA3735, cured the fault.

Intermittent E-E problem: The machine would sometimes display a ragged dark raster, usually after changing channels a few times. To obtain a clear channel you would then have to unplug the machine and start up again. Fortunately the fault was easy to cure. The outputs from the superannuated power supply module were all too high because C1012 and C1039 (both 47 μF, 16 V) on the primary side were low in value. Replacements cured the fault.

Panasonic NVF55

Monochrome recordings, unstable picture: On test, however, all that this modern Nicam machine would do was to display E9 in the self-diagnosis list. According to the service manual this means no serial data. The MN67431VREH systems and servo flat-pack chip IC6001 was totally inactive. A replacement got the machine working and we were then able to attend to the original complaint. This was dealt with by switching off the NTSC 4.43 switch, which is meant for copying NTSC tapes only.

No output from power supply module: Disconnecting the module from the machine then plugging it in again showed that the various supplies were at about twice the correct voltage. Checks around IC1103, which provides a stable reference for the power supply, showed that the earth pin was at about 19 V. The print had gone open-circuit around the nearest earthing connection to the screening can.

Search tuning fault: This machine would search but wouldn't lock on to stations. Checks on the sync low, a.f.c. defeat and a.f.c. feeds showed that there was nothing amiss to and from the demodulator pack, so out came this plug-in pack, revealing a surface-mounted diode (D6701, type MA151WK) with one end missing. A replacement cured the problem.

Blank, grey raster in E-E mode: In the E-E mode this machine produced a blank, grey raster with no VU meter display on the display tube. The M66006FP chip IC1701 enables the VU display and switches the AV1 and AV2 circuits: fitting a replacement cured the fault.

No normal sound in E-E mode: There was no normal sound or Nicam sound in the E-E mode. After making a few voltage checks we found that the audio defeat line to the Nicam pack was permanently high. The audio defeat line is produced by the M66006FP chip IC7001. A replacement removed the silence.

Mechanism operated very fast: Checks showed that there were no FG signals from the capstan motor to the systems control chip IC6001. The cause of the trouble was soon traced to L2001, which was open-circuit. It filters the 12 V supply to the capstan stator.

Coloured speckles on picture in the E-E mode only: The cause was dry-joints in the tuner unit.

Dead, following a power cut: There was h.t. across C1103 in the primary side of the chopper power supply, but there were no outputs on the secondary side. The cause was traced to R1103, which was open-circuit. It provides IC1101 with a start-up feed. Replace R1103 and R1133 (both 220 kΩ).

Panasonic NVF55B

Misloading of tape cassette: I've had a few of these machines throwing the tape out three times in four attempts. Check the tension of the cassette gripping flanges: slight readjustment is all that's needed.

Dead machine: We found that IC1102, part number. S13120C, was short-circuit internally.

Intermittent loading: When a tape was inserted, the capstan motor would judder around very slowly. If you were lucky, the tape might load. We checked the voltages around the capstan motor drive chip. They all appeared to be correct. So we tried a new stator, which made no difference. Time for some drastic action! We earthed pin 16 of the chip. This time the motor rotated but fast. Pin 16 is controlled by the syscon/servo chip IC6001. Replacing this cured the fault.

Panasonic NVF65

Would go dead during play: This machine failed when it was being installed. It would go completely dead, with no display etc. When it was powered in the workshop it ran all right for many hours before it failed. Failure eventually occurred during play: the tape remained laced up and it was just as if the mains plug had been pulled out. If the mains supply was disconnected then quickly reconnected a short buzzing noise was heard from the power supply. If the mains supply was disconnected for a couple of minutes before reconnection, however, the power supply would start up and the machine would work without any problems for perhaps an hour or two before it stopped again. Because of the disorderly shutdown a power supply problem seemed likely. A careful examination of the PCB revealed no suspect joints or breaks, but there was an interesting pointer. If the mains supply was connected to the power unit with no connection made to the rest of the machine the power supply wouldn't run: it just buzzed for a couple of seconds (unlike the G21 etc. which will work in this condition). As a check we fitted the power supply from another new machine and connected it up. The machine faulted again after several hours, so the fault wasn't in the power supply itself. We then found that the fault could be brought on by flexing the main PCB. Careful pressure in different parts revealed a sensitive point, down the left edge near the mechanism. As the fault occurred when the board was pressed down a break on the underside print was suspected. Ohmmeter checks on the print then revealed the cause of the trouble: a break in the print that connects R6036 to the base of the motor regulator transistor Q6004 near the front of the machine. To put matters right we connected a link across the faulty section of print. Presumably the loss of loading on the one power supply output caused the complete shutdown.

Capstan motor full speed in all modes – load, playback and search: Supply choke L2002 on the main PCB was open-circuit. It supplies 5 V to the capstan motor's FG stage. Because there were no FG pulses the servo ran the motor at full speed.

Wouldn't stop when search tuning: But if you tried tuning in the opposite direction the machine would usually (but not always) lock on a station. So checks were made for sync low and a.f.c. defeat switching at the pins of the demodulator pack. Normally when tuning the a.f.c. defeat voltage changes from 4.5 V to 0 V. In this case it remained at 2.5 V all the time. The MN12C261D front panel memory chip IC7502 is directly responsible for this and proved to be the culprit, a replacement curing the fault.

Sound would intermittently go off: This would happen in the E-E mode, and also with the customer's own recordings. We've had a few faults on the TV demodulator pack in this range of VCRs, so our suspicions were immediately directed to this area. Fault-finding on the upright panels, especially with an intermittent fault, can be nigh on impossible. Our suspicions were soon proved to be correct: we found another NVF65 and swapped the TV demodulator packs over, the fault transferring with the pack. No obvious cracks or dry-joints could be seen on the defective pack, but when the copper side was attacked with freezer we found that we could instigate the fault. The audio defeat transistor Q713 was being turned on in the fault condition, connecting the normal audio output line to chassis. We traced the reason for this to IC7651: voltage checks here showed that the 12 V supply to pin 3 disappeared. This was due to a faulty surface-mounted device, in fact a link. It's not shown on the circuit diagram and was going open-circuit intermittently. Once a proper wire link had been fitted in its place the machine worked normally.

Channel would jump: This machine worked perfectly when first switched on. But after an hour or so the channel jumped to the next one. As time went by the jumps got faster and faster. Each time it happened the machine beeped, so I thought I'd better cure it quickly before it drives me mad. Fortunately a squirt of freezer on the MN187125VFM tuning chip IC1 soon proved that it was the culprit. After fitting a replacement the machine never beeped on its own accord again!

Dead, squeal from power supply: When the main PCB was disconnected the power supply worked perfectly. So it was time to start disconnecting the loads to find out which one was imposing the excessive load. The 45 V line turned out to be the culprit. Now there's a rather unusual shunt stabilizer across this rail, based around transistor Q6021 which was being turned on hard. A fault in its base drive circuit was therefore suspected. But checks here proved fruitless. The cause of the problem was the fuse on the main PCB. It was dry-jointed. Resoldering it cured the problem.

Panasonic NVF65B

No signals: On the bench we found that although a test signal could be obtained and the record/playback/search functions operated correctly it was impossible to tune in any local channels. There was a lively E-E raster and as we had already confirmed that there was an r.f. output from the combined r.f. amplifier–tuner module attention was turned to the tuner voltages. The 12 V regulated supply was present at pin 2 (BM) and the tuning voltage at pin 7 (BT) cycled nicely through its range in the preset/tuning mode. But no pictures appeared on the screen. What was missing was the u.h.f. enable at pin 8 (BU). This should be at 11.5 V, the supply coming from pin 12 of the AN5043 chip IC7652. When pin 12 of this chip was isolated and 12 V from a separate source was fed to pin 8 of the tuner signals could be tuned in although the machine wouldn't lock on to them when set to scan through the band. A replacement AN5043 was obtained and fitted but made not one jot of difference. Feeling somewhat miffed, we delved further into the circuit and eventually followed the Band-U feed to pin 10 of IC7652 back via the audio board to the timer panel, where IC7502 (MN12C261D) was found buried under the display. Amongst its other functions this chip passes the v.h.f./u.h.f. switching from the timer/control micro-computer chip IC7501 to the TV demodulator. Although the voltage at pin 5 (U out) was correct at 2.1 V, as there is nothing else between IC7501 and IC7502 we decided to obtain a replacement MN12C261D. Fitting this restored correct tuning. It's a pity that the manual for this machine is rather vague regarding the functions of some of the control chips.

Very noisy rewind: We found that the supply and take-up spindles were as dry as a bone. Lubrication silenced the noise.

Low E-E output: Don't assume that the cause is the r.f. converter. It's more likely to be the M51292FP switching chip IC1300. Replacement should cure the problem.

Panasonic NVF70

Completely dead, mains fuse OK: This machine was completely dead. The mains fuse was intact so the fault was in the switch-mode power supply, where the primary side wasn't starting up. C1109 (1 μF, 400 V) should provide a pulse to the chopper transistor within IC1101 (STRD6008X) to get things going but there was no pulse at pin 2. The IC was found to have an internal short-circuit between pin 2 and pin 4 (ground), a replacement restoring normal operation. Note that there's an error on the circuit diagram, where the base and emitter of the chopper transistor are shown shorted together.

Power supply problem: This machine was virtually dead, the characteristic squeal at switch-on being rather muted. We found that rectifier D1111 in the power supply was short-circuit. This surprised us as it's a very large device that looks as if it's capable of carrying several amps.

Mechanics degrade with age: As these machines age, particularly the mechanics (G mechanism with review motor), various squeaks and rattles can arise. This is particularly the case with models, such as this one, that have jog shuttle dials. A delightful series of shrieks, whines, rattles and buzzes can be demonstrated. Replacement of the tension unit (part number VXA3516) and the soft brake (VXL1873) usually restores relatively quiet operation.

Dead or reluctance to start: The likelihood is that the kick-start capacitor C1109 in the power supply has dried up. It gets very warm in its little box. Replace C1114 while you are at it. To prevent premature failure of the replacements make sure that you use 105°C rated components in both positions.

No Nicam stereo: Checks at IC7901 (TA8662N) in the Nicam section showed that there was a digital audio output at pin 29 but no clock waveform at pin 27. X7902 wasn't oscillating and as pin 24 of IC7901 read 10 Ω to chassis I decided to replace it. This restored the Nicam sound.

Panasonic NVF75

E-E picture rolled and pulled: Video playback was fine. A scope check on the video output from the demodulator pack produced a nice waveform with plenty of sync pulse depth. This went into the luminance processing chip IC301 where it retained its purity. But the output from the M51292FP switching chip IC3901 on the back panel was a sorry picture indeed. Replacing this chip cured the problem.

Function display faulty: This all-singing and dancing machine worked perfectly unless you took notice of the function display – the usual display was pause, with no counter display in any mode. Our first checks were on the serial data and clock lines between the syscon chip IC6001 and the timer and display chip IC7501. As data signals go, they appeared to be all right. To eliminate the front panel timer and display circuits the front panel PCB from a nearby NVF70 was borrowed. This showed the same errors, so back to the syscon circuitry. Comparative checks with the NVF70 showed that identical data left the syscon chip. Only an inverter circuit centred on transistor QR6017 was left to check. It was working as well as I could tell but the culprit turned out to be C6011 at the base of QR6017. It's a tiny surface-mounted type capacitor and was open-circuit, thus apparently corrupting the data signals to IC7501.

Unreliable operation: This machine would power-down frequently and ignore all operational requests. When the machine could be induced to show some life the multi-function display produced all kinds of random indications. After much hair tearing I eventually discovered that the 1000 pF capacitor C6011 at the base of the serial data inverter transistor QR6017 was open-circuit. Replacing this surface-mounted capacitor cured all the problems.

Super-still/super-fine slow problem: This machine worked perfectly except for one small detail. While editing tapes, which the machine's owner was inclined to do, there was some flickering blue disturbance in the background with super-still or super-fine slow. Fortunately the cause of the problem was not as obscure as it first seemed. Replacing C1022 (47 µF) in the power supply cleared the fault.

Panasonic NVFS1

Tuning drift: The tuning on all channels would drift after several hours of use. When we'd removed the complete main circuit board assembly to make measurements around the tuner's BT supply circuit we were faced with what appeared to be a different fault: there was no or very weak vision in the E-E mode. A check on the buffer transistor Q703 at the output of the TV demodulator CBA showed that it was reverse biased – its emitter voltage was slightly higher than that at its base. As the collector resistor of this emitter-follower is only 180 Ω you'd expect the value of the emitter load resistor to be fairly low. In fact the reading was about 200 kΩ. No emitter resistor is shown on this page of the manual (3–94/5/6) so we had to trace the route the 'video out' takes. From the TV demodulator block CBA it passes through the sub-Nicam decoder section (and the 12 Ω resistor R792) then through the TV demodulator interface CBA to the channel select area, from there to the luminance/chrominance area and on to the input select section. It's here that the two series 75 Ω emitter resistors R3906 and R3909 are mounted. They go to a connection marked 'GND (TUNER, VIDEO)' which is pin 2 of the pack connection to the main circuit board. In fact this ground is actually a black lead which isn't connected to the main CBA ground but ends in an eyelet. This should be connected to a chassis point on the TV demodulator interface CBA but is disconnected when you remove the screw that attaches it to ground as this also holds the CBA. With the eyelet soldered to the board we were back with the original fault. When the tuning drifted, there was a slight change in the tuning voltage BT. This is produced by the channel select pack from the digital DAC signal that comes from the tuning/timer chip IC7501. We noticed at once that the BT voltage at the tuner was about 2 V lower than that at pin 8 (BT output) of IC7551. The drop was across a series 10 kΩ resistor R7555, one end of which is connected to pin 8 of the chip and the other via a 1 k resistor to the output from the pack (connection 8). A 0.01 μF decoupling capacitor (C7555) is connected from the output end of R7555 to chassis. It was leaky – around 120 kΩ – and it was this that caused the drifting.

Loss of audio in E-E mode: In fact when a phono lead was inserted in one audio input socket we found that this input was permanently selected. We also found that the input select switch (S video in/tuner/line) did nothing – the tuner's picture stayed there. The reason for this was soon discovered. The input and simulcast switches on the front panel provide highs and lows to switch between modes. We found that the high levels, which are derived from the 12 V line, read only 2.4 V. The cause of the problem was in the power supply can: the unswitched regulator transistor Q1004 (2SD638) was open-circuit base-to-emitter and the associated zener diode D1012 connected between its base and chassis was open-circuit.

Intermittent colour with machine's recordings: We found the cause to be dry-joints around the luminance/chroma pack, at pin 34 in particular.

Intermittent operation: Sure enough we found that it failed to function mechanically when cold. Unfortunately any attempt to take the top off the machine cured the fault! We had it on soak test for many days before discovering that the multi-voltage regulator in the power supply was the cause of the trouble. There were several dry-joints at its pins.

Would not accept a tape: When a cassette was inserted it would be accepted, taken into the depths of the machine then quickly spat out again. We checked the mechanism and mode switches for correct alignment, but everything was OK. Attention was next turned to the systems microcontroller chip, were we found that only one of the mode-switch tristate position signals changed state with the mechanism. The other one remained at 5 V all the time. R1501 (47 kΩ), which is across the mode switch, was open-circuit.

Panasonic NVFS100

Playback picture dark and blurred: In both the S-VHS and standard VHS modes the playback picture was just a dark, blurred mess. After much chasing of signals around this complex machine attention was focused on the subluminance pack, in particular the 1 H delay system. Checks showed that the video output waveform at pin 1 of IC3504 was badly distorted. The culprit was eventually found to be C3510, which decouples pin 5 of this chip. It was open-circuit, a new 3.3 μF, 25 V capacitor curing the problem.

Mechanism problems: There were problems with this S-VHS machine's mechanism, but nothing that you could really put your finger on. The machine would play a tape all right, but when going from play to rewind or fast forward there would sometimes be problems: after briefly spooling backwards and forwards the machine would lapse into standby. Our first move, more in hope than the expectation that this would work, was to replace the mode switch. We then carried out a full check on the mechanical alignment. All to no avail. The G mechanism in this machine is operated by the capstan motor, through a gear train which is switched in and out by a solenoid. Kick and hold circuits control the solenoid. What was happening was that the kick circuit was operating weakly, sometimes not at all, because C6017 had gone low in value. As a result the pulse to the kick circuit was of low amplitude.

Snowy E-E and r.f.–r.f. outputs, no reel counter operation: Voltage checks at plug P1101 in the power supply showed that the non-switched 12 V output at pin 1 was missing. Q1102 was found to be open-circuit base-to-emitter.

Faint diagonal patterning in playback: C3311 in the HQ pack is the usual suspect when this symptom is present. On this occasion it was OK. Deep in the bowels of the machine you will find the 1 H CCD delay line pack on the subluminance and chrominance board. Several small capacitors here can die: C3501, C3506 and C3516 often fail. This time C3509 (3.3 µF) had expired. After replacing it we had a very good, clean picture.

Problem with recording: Although the recordings were good, the real-time counter didn't work and, when it appeared, the 'write' signal was frozen on the display. The real-time counter worked in the playback mode, but the index function didn't. After some checking around in the servo circuitry I found a section of open-circuit print between D2306 and pin 12 of IC2201. The cause of the print problem was corrosion because C2311 had leaked on to the adjacent copper track.

Panasonic NVFS90

Poor picture quality: An abundance of faint, flickering lines were present in the background of the playback picture. The condition was even worse in the LP mode. I decided to check on the HQ pack. As I removed the screening can X3301, the crystal associated with the 1 H delay CCD chip IC3302, came out of the board. Resoldering it back into the board restored the normally excellent playback picture.

Refusal to playback S-VHS tapes: This all-singing, all-dancing editing machine would refuse to play back S-VHS recordings after about half an hour. Checks in the S signal channel brought me to IC303 (part number VEFH05BT) which proved to be heat sensitive. A replacement restored the excellent picture.

Pulling, colour phase shift and horizontal tearing at top: As the owner copied tapes from his camcorder he was concerned about the overall picture quality. There was pulling and colour phase shift at the top of his copies, and horizontal tearing. In this situation it's highly probable that there has been head and tape path wear – the VCR had been in use for a couple of years. So the owner agreed to replacement of the video heads, the tension band and the pinch roller. The resultant tape path improvement made the horizontal tearing much worse, however. We discovered that the vertical sync pulses had noise at the bottom: so too, on closer inspection, did the horizontal sync pulses. This noise turned out to be some type of carrier signal. It took a couple of hours one evening to trace the cause of the fault to a CCD delay line circuit on the subluminance board – in fact it took some time to find the board! The culprit was C1 (1 μF) which couples the video signal into the CCD delay line chip. Once this capacitor had been replaced the picture was greatly improved.

Intermittent horizontal patterning: The complaint was of intermittent fine horizontal patterning. This type of fault is often associated with a CCD delay line. Sure enough we found that C3311, which decouples the oscillator circuit associated with the 1 H delay line in the HQ pack, was the culprit. A new 10 μF, 16 V capacitor cured the fault.

Various complaints: There were various complaints with this machine: picture problems, no VHS or S-VHS playback, and the owner mentioned fine lines across the picture when it was there. Lack of playback was soon narrowed down to the subluminance pack where we found that C3501 (1 μF, 50 V) which couples the input to the CCD delay line chip IC3504 was open-circuit. We then found that the restored picture was smeary, with the fine lines complained about in evidence. These faults were cured by replacing C3509 and C3516 (both 3.3 μF, 16 V) which decouple pins 6 and 15 of the CCD delay line chip. A check on the S-VHS playback produced a badly distorted, mushy picture. When IC303 was treated to a quick squirt of freezer the picture cleared up, proving that a replacement (VEFH05B) was required. Finally, while the machine was on soak test another fault developed: fine lines across the picture slowly became more distinct. The cause of this fault was traced to yet another electrolytic capacitor, this time C3311 (10 μF, 16 V) which decouples the output buffer for the 1 H delay line in the HQ pack. The machine was then pronounced fit to resume its duties.

No S-VHS playback or record: The standard VHS was fine. When I traced the luminance path through I found that the signal entered the hybrid chip IC303 (VEFH05BT) at pin 2 but nothing came out at pins 7 and 8. The cause of the trouble appeared to be that one of the aluminium can, surface-mount electrolytics in the module had leaked. A replacement i.c. cured the problem.

Dark smearing: The complaint was dark smearing to the right of any black image during playback. The main cause of picture distortion in these S-VHS machines is the 1 H delay CCD pack on the subluminance and chrominance board. Scope checks in this area led me to C3506, in the 9 V supply to IC3504. It was open-circuit. A new 10 μF capacitor restored an excellent picture.

Bad dropouts: When any tape was played back there were excessive flashing black and white lines. We eventually traced the cause of the fault to C3311 (10 µF) in the HQ pack – it was open-circuit. After fitting a replacement the picture was clear.

No playback/weak record luminance: I've had several of these VCRs with varying degrees of no playback luminance and weak record luminance. The cause of the problem is signal loss within the thick-film hybrid chip IC303, which incorporates surface-mounted can electrolytics. They leak – as they do in Sanyo camcorders.

Poor playback picture in both the S and VHS modes: Its recordings were fine when checked with another machine. Many readers will be aware of the luminance problems that arise when the hybrid chip IC301 fails, but visual inspection showed that there were no signs of leakage from the aluminium can surface-mounted electrolytics in the module. The luminance varied with temperature, from being virtually non-existent through no field sync to being almost OK but with tearing on highlights. When I traced through the playback path everything was fine up to the sub-YC pack. Everything was OK within it up to the luminance feed to the 1 H delay pack, at pin 12 of P3504. But there was complete loss of the sync pulses at the output. The incoming luminance signal passed through FL3504 within the 1 H delay pack, but was being lost at the 1 µF, 50 V electrolytic C3501.

Panasonic NVG10

Channel display comes on when the VTR switch is operated: But if there's no on-LED light, play or E-E, check the voltages around IC1001. If 13.5 V is present at pin 6 there's a good chance that this i.c. is faulty.

White lines/bars on play and record: There were white flashing lines and white bars that affected both playback and record. It seemed likely that the fault was in the luminance/chrominance section and as a start we replaced the hybrid luminance processing chip IC301. Fortunately this provided a speedy and complete cure.

Adding a scart connector: I found the use of switched BNC and phono sockets for the external video and audio inputs to my VCR, a Panasonic NV-G10, annoying. In the J30 series of machines the problem was corrected by including a scart socket. In devising a modification (see Figure 8) it was necessary to find a solution to two problems: how to switch between the internal and external signals and how to drive the switching. To keep things simple and cheap, a relay was used for signal switching. This keeps the circuit modifications to a minimum, as the relay contacts replace the switching contacts in the sockets. The neatest way that I could think of to drive the relay was to use the tuning voltage. This enables the external inputs to be switched in and out even when in the timer mode, as selection is just like that of a u.h.f. channel. The very simple drive circuit consists of a comparator, using half an AN6914/μPC358 operational amplifier chip. A 0.6 V reference voltage is applied to the non-inverting input while the tuning voltage is connected via a 270 kΩ resistor to the inverting input. Thus when the tuning voltage is below 0.6 V the chip's output (pin 1) is high: conversely the output is low when the tuning voltage is above 0.6 V. This circuit works well, being able to distinguish between an external input and channel 22. The tuning voltage feed couldn't be taken straight from the tuner's BT terminal as the 10 kΩ series resistor present resulted in slight tuning drift. It was taken from a point further back in the circuit, therefore. The components are mounted on a small piece of strip-board. Most of them came from the PCBs in a scrapped NV333. The relay is the capstan motor direction switching one while the chip is half of IC6006. The transistor was found in the reel motor drive area. Diodes and resistors, also short lengths of screened lead to connect the small piece of strip-board to the main PCB, were also taken from the panels. I insulated the relay board and fitted it below the main PCB, with the connecting leads taken up through the gap around the right-hand edge of the main board for connection to the top.

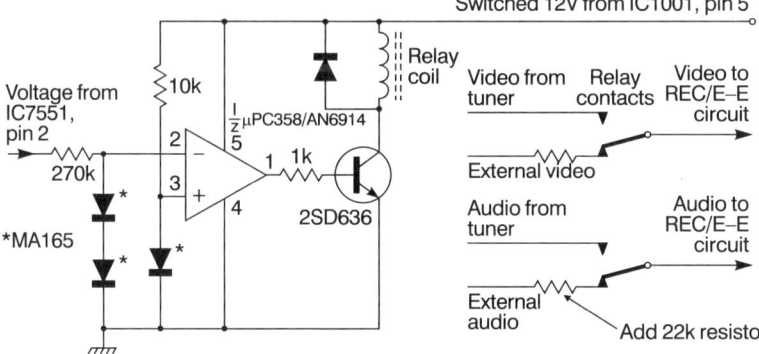

Figure 8 *AV switching circuit for the Panasonic NV-G10*

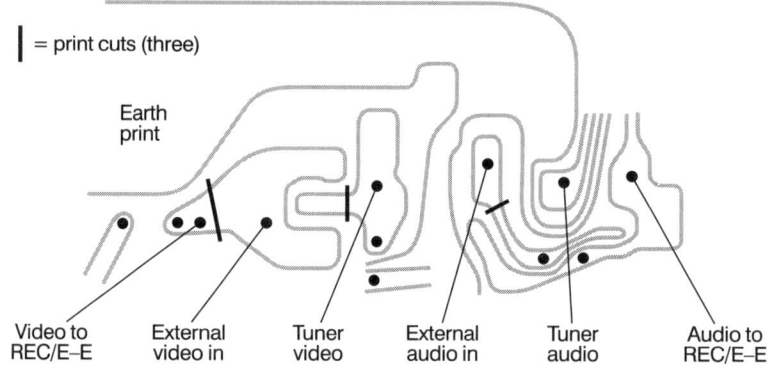

| = print cuts (three)

Earth
print

Video to   External   Tuner   External   Tuner   Audio to
REC/E–E   video in   video   audio in   audio   REC/E–E

Figure 9 *Connections to the main PCB, at the right rear*

No playback colour: We traced the chroma signal as far as C8002 (0.01 μF) which couples the signal to pin 31 (playback chroma input) of the luminance/chrominance pack. There was a signal at one end of C8002 but not at the other. A new capacitor restored the colour.

Intermittent mechanical fit: The mechanism would shuffle around, the machine then powering down. Selecting rewind or fast forward would sometimes produce a quick shuffle. Then the command would be ignored. A machine of this age is usually in dire need of maintenance kit VUD4103KIT, but the part needed to cure the symptoms just mentioned, the mode switch, is not included. Its part number is VSS0135.

Panasonic NVG12

Tuning problems: Tuning problems due to IC7551, D7555 and Q7551/2 going short-circuit have been a problem with these machines. Some guidance has now come from Panasonic. It seems that all machines with the serial number prefix E7 can be affected. Resistor legs around the i.c. were bent over during assembly, before soldering, and they can intermittently touch lands on the PCB. The cure is to straighten suspect legs. Some machines have already been done by Panasonic themselves.

Drum motor didn't move: When play was selected the drum motor didn't move so the machine cut out. This was quickly traced to an open-circuit fusible resistor, R2012 (6.8 Ω, 0.5 W), which is connected from the unregulated 14 V supply to the common connection of the three-phase DD drum motor. The resistance to chassis on the motor side was only 7.5 Ω, this turned out to be the resistance of one phase of the motor. A check at connection P201 revealed where the short was – on the winding connected to pin 1. When the head amplifier was removed to check the leads the reason for this was found. One of the ten leads was trapped between the right-hand head amplifier support bracket and the main chassis. It must have been like this since the machine was made some 18 months ago.

Very slight knocking noise: It was slight too, but once you'd picked it up it was annoying. The cause was a dent in the tyre of the reel clutch (VXP0599). Changing this is an involved job.

Some tapes' tracking lines visible: The picture was OK with some tapes but on others there were tracking lines. No amount of tracking control adjustment would cure this. We found that playback of the machine's own recordings was perfect, and that these were OK with other machines. Someone had a go and tried almost everything – head cleaning, drum assembly, guides etc. After much resetting and realignment the fault was still present. Fortunately another one of these machines then turned up. It had a broken display although it worked well. We decided to sort this out by swapping over the timer panels. The first machine then worked perfectly with all tapes. So it was the timer panel! In fact the (–) tracking control was permanently shorted and as a result the tracking was set at its most negative end.

No sound in any mode including E-E: A sound signal was going from the TV demodulator PCB to the audio pack, where all the audio goes. It wasn't being switched out of the BA7752LS chip IC410, however. The supplies and the d.c. switching conditions were all correct so it seemed that the i.c. was faulty. A replacement restored the sound.

No record: This machine would enter the record mode and proceed as if everything was OK, but when you rewound the tape and replayed it the previous recording was still present. Although the machine entered the record mode it didn't produce the delayed record 12 V supply. There was no 12 V at the emitter of Q6005 as Q6006 was switched off because its base voltage was high. Q6006's base voltage should be pulled down by the record prevent switch line from the microcomputer chip IC6001. The voltage was correct at IC6001 – the cause of the fault being a break in the print between the cathode of D6005 and R6041 as the print winds its way through the pins of the audio pack.

Various symptoms: These were as follows: there was no rewind, forward wind or play; the capstan motor was shunting; and the drum motor ran at full speed. A quick check showed that the 4.43 MHz clock signal from the video pack was missing. This signal is produced by IC8001, which on inspection was cracked. A replacement restored normal operation. We find that this fault is now quite a common one with several Panasonic models, in particular the NV830/NV870 series.

Loading arms out, mode switch fault: When this machine was brought in the loading arms were out. Then it unloaded, leaving the supply reel turning slowly – all this with no cassette inserted. Replacing the deck mode-control switch restored normal operation.

Would sometimes unlace immediately: This machine would sometimes unlace immediately because the drum failed to rotate. The reason for this was a dry-joint at the motor drive plug/socket.

Panasonic NVG21

Cassette wouldn't eject: In fact the machine was very sluggish in lacing the tape. If the capstan motor was given a helping hand everything worked. A new capstan motor put matters right.

Knocking noise: This fault could theoretically occur with any machine that uses the G mechanism. There was a knocking noise in play, record, fast forward, rewind and search. The cause was the pulley at the bottom of the capstan motor rotor/flywheel – it drives the loading and reel mechanism via the timing belt and was running eccentrically. There's considerable tension in this belt and the noise didn't occur with the motor running off load.

White line at the bottom of its own recordings and would intermittently roll and mute the sound: The audio/control head was at the wrong height, the control head section erasing part of the video track.

Creasing tapes: This machine came in with a stock fault – creasing tapes and indeed the pinch roller was of the unmodified type and faulty. Having replaced this, and its securing cap, I switched on to test and noticed that the capstan shunted then, when a cassette was inserted, the capstan rotated in the wrong direction while its clutch squealed very loudly. Switching off and on again cleared the fault, but after a few minutes it was back again. Previous experience with these machines suggested that a new mode switch would put matters right. One was soon fitted and after this we had no further trouble.

Squealing noise from machine: After removing the top cover we noticed that the capstan rotor was vibrating. A check at the torque control input pin of the BA6430S motor driver chip IC2001 revealed some bursts of 1.8 V spikes that sat on a d.c. level of 1.5 V. The timing of these bursts seemed to coincide with changes in pitch in the noise coming from the motor. After checking with another machine, however, we found that these bursts are quite normal. We also discovered that the noise would fade away when the machine had been on for a few moments. A quick spray of freezer on the motor driver chip brought the fault back again, proving that the i.c. itself was the cause of the problem.

No display, no LED indications: This machine showed very little sign of life. In fact all that worked was the r.f. amplifier. The cause was quickly traced to the power supply – there was no 6 V and hence no switched 5 V output from the STK5338 multi-voltage regulator chip IC1001. It was the second machine we'd had in a single week with the same fault.

Dead, multi-regulator failure: This machine appeared to be completely dead. There were no mechanical operations, no displays and no noises except for the characteristic power supply start-up squeal. When the PSU can was removed and disconnected all the supplies were found to be present and correct. The regulated 5 V and 6 V rails were extremely low, however, when the power supply was connected to the rest of the circuitry. There was no excessive loading on either rail: the STK5338 multi-regulator chip was faulty.

Poor sound on playback: The playback sound gave a fair impression of a washing machine and a tumble drier working in unison, drowning the recorded sound with buzzing and spurious oscillation. Voltage checks around IC4001 in the sound section produced no clues except that the noises were reduced when the meter's probe touched the input pin. So attention was turned to the input circuitry, where R4021 (47 Ω), a surface-mounted resistor connected to a filter network, was found to be open-circuit.

Machine hummed and buzzed alarmingly: This machine buzzed and hummed alarmingly when it played back a good tape. The dealer concerned had replaced the audio/control head and the audio chip (IC4001). Nothing was revealed by carrying out careful voltage checks, but the audio input to IC4001 seemed to be abnormally sensitive. There's a fairly severe input filter between the audio head and IC4001. Resistor R4021, which is a 47 Ω surface-mounted device, is part of this filter and was open-circuit. An ordinary 0.25 W, 47 Ω resistor fitted neatly on the board and cured the trouble.

Machine dead: We quickly found that a 5 V supply was missing. We removed the power supply can and then, with some difficulty, took off its covers. After this it was a simple matter to discover that IC1001 (STK5338) was faulty. Why do manufacturers fit the wire-ended/push-in type connectors when a plug/socket would surely be a more practical solution? We've had many of this type of connector produce intermittent results in various machines.

No audio playback: We found that the audio mute line was high, the cause of this being absence of the CTL pulses. We checked back to the servo pack where C240 in the 'AGC CTL' circuit was defective.

Lift jammed half way, machine dead: There was no clock display or power-on LED indication. A check in the power supply showed that the 1 Ω safety resistor R15 was open-circuit. This was the cause of the dead machine, and I assume that the jammed lift was responsible for its failure. Once the lift had been extracted I found that various plastic tabs had broken off, so a new side plate (part number VXA2677) was obtained from SEME. If you have to fit one of these I recommend that you read Nick Beer's classic article (*Television*, May 1991) on servicing this model: without his advice I'm sure that I would still be trying to align the gears.

Intermittent tape take-up or the tape being caught in the carriage on eject: This can be caused by a faulty play arm or a stiff mounting post. The remedy is to replace the play arm unit and lightly oil the play arms' mounting post.

Tuning drift: We found that a tape could be played but when stop was selected the machine would shut down, tripping. On examining the power supply we found that C12 and C39, both 4.7 μF, 16 V, were faulty. After replacing them the machine remained dead! Other capacitors in the power supply were then found to be either open-circuit or of incorrect value – C18, C22 and C25, all 4.7 μF, 50 V and C2, 1000 μF, 6.3 V. (power supply type VEK3254–2).

No playback colour and white lines: This machine had come into the workshop on several occasions, each time with a report of a different, niggly fault, none of which was ever really cured. Very rarely could we get the machine to show one of these faults. This time the complaint was of no playback colour and lines on the screen. We confirmed that there was no playback colour and found that the drum speed was varying slightly, as a result of which the picture shifted a little on the screen. The cause was easily found with the aid of a hairdryer and freezer: C23 (1000 μF, 10 V) in the power supply was leaky. Replacing this cured the latest problem and all the other ones that plagued the customer seem to have gone away.

Machine dead: This was because C1018 (47 μF, 50 V) which provides the supply for IC1001 was open-circuit. It's best to replace C1022 and C1025 as well. C1022 can be responsible for patterning on the screen.

Patterning on screen: As mentioned above, C1022 can be responsible for patterning on the screen.

Attractive swirling pattern on playback picture: This machine's playback picture had an attractive swirling pattern superimposed on it. The cause, not unusually for a machine that's as long in the tooth as this one, was a capacitor in the power supply. C1022 (47 μF, 50 V) was the culprit. For good measure C1118 and C1023 were also replaced.

No functions and no display, just a ticking from the PSU: Replacing the STR1006 chip in the power supply restored normal operation.

Advice on replacing post P5: When replacing post P5, make sure that you use the correct part for the machine on which you are working. After fitting a new post we found that it caught on the securing screws for the capstan motor. This resulted in a tape loop on eject. The new post looked identical to the old one but was about 1 mm thicker. It would seem that different ones are used in the deck.

Various mechanical problems: This machine originally came in for a service. In addition to replacing all the gears, the pinch roller and the mode switch we had to replace the side plate and connection gear. Two months later the machine came back, the complaint being that when it was switched on the carriage moved forwards then ejected, repeating this until the machine went back to standby. You often get this fault when there's a worn carriage mode switch or a bent lever. Not this time, however. The cause of the trouble was that all the joints of the new carriage's connector were dry-jointed. Re-soldering put matters right.

Badly distorted sound and diagonal lines across the picture: Anyone optimistic enough to try to play a tape would see that the drum and capstan speeds were incorrect. In a machine of this age the initial checks should be carried out in the switch-mode power supply. We found that C1019 (1000 μF, 25 V) had died of old age.

Dead, no functions: When I checked the power supply outputs I found that the regulated 5 V and 6 V supplies were far too low. Replacing the STR5338 regulator chip restored normal operation.

No display: This machine worked all right for a 10-year-old VCR, but there was no illumination from the multi-function display on the timer board. The cause of most faults with these machines can usually be traced to world-weary capacitors in the power supply, but in this case the chopper transformer T1001 was defective; windings S7 and S8, which should supply the filament voltage, didn't.

Unreliable tape unlacing action: The cure was to replace the play arm unit, part number VXL1490.

Intermittent loss of action: The fault with this machine was very intermittent and it took us a long time to track down the culprit. When the fault was present there was complete loss of action: even the clock display went out, leaving only the green cassette-in indicator (near the standby switch) alight to indicate that power was reaching the machine. The culprit turned out to be the STK5338 regulator chip IC1001 in the power supply can.

No on/off LED indication: A few of these machines have come in recently because of no on/off LED indication, although the switching works, indicated by the appearance of the counter display and the beep. The other problem has been no deck functions. The cause of these symptoms is that the 12 V output from pin 6 of the STK5338 regulator chip IC1001 has fallen to about 7 V. Replacing the chip puts matters right.

Dead, except for cassette-in LED on: One of these machines came in dead save for the fact that the cassette-in LED was on. Checks in the power supply showed that the unregulated 45 V line was low at around 25 V. The 47 μF, 50 V reservoir capacitor C1018 was open-circuit. As many of you will know, C1023 (1000 μF, 10 V) in the power supply commonly fails, causing various servo and chroma faults. These are sufficiently severe to lead to a service call, but if the capacitor is left to deteriorate the display and other features will be lost and regained rhythmically.

Intermittent loss of playback colour: Check C1023 (1000 μF, 10 V) in the power supply. It may be open-circuit or just dried up.

Panasonic NVG25

Various faults: Having cleaned the heads, a field engineer brought this one in for the following faults: the noise bars on cue and review were wider than normal (they are usually almost invisible) at 1125th speed every other frame was obliterated by noise, odd frames were noisy with the 1/5th and 1/10th speeds; and the still frame was very poor. On examination we found that the heads were faulty. When I used the special Panasonic tool – wonderful machine – to replace the heads I noticed that all the relevant adjustments (tracking fix, head frequency response etc.) had been set at one end of their tracks. Presumably the original head had been a borderline one.

No video in E-E or playback: The card just said 'faulty' . . . The fault showed up when the machine was unboxed. Before removing any covers I connected a scope to the video output socket and tuned the machine in, using the sound output from the monitor. Video was thus present up to this point so the trouble had to be in or around the r.f. converter. In this machine the video is fed from the converter via a buffer on the input selector then back to the converter. We found that the video went in at pin 12 but nothing came out at pin 14. The 5 V supply was present at pin 11. Once this was established it was not difficult to trace the trouble to Q3501 (2SC2206 but a 2SD636 will do) which was open-circuit all ways round and D3502 (MA27W) which was open-circuit. To gain access to this panel it must be unsoldered from the mother board.

Poor picture, intermittent low sound: We've recently had two of these machines with the same problems. Both were about a year old, the complaint being of a poor picture and intermittently low sound. When the video heads in these machines start to wear out you get a slight increase in overall picture noise and bands of noise in only the top third of the picture in the review mode, particularly near the beginning of a tape. The only cure as far as I know is to replace the heads. I did find that by adjusting the inlet guide by approximately a quarter of a turn the noise disappeared with a Panasonic tape, but the noise returned when a different type of tape was used. If more adjustment was tried the noise could still be cleared, but then the guide adjustment was wrong for normal playback. The low sound problem was due to audio head wear, so again replacement was necessary. The audio/control head assembly is now supplied as part number VBR0125 instead of the original VBR0116 – perhaps the new type will last longer.

Clock/counter display faded out: After a recent repair the customer complained that the clock/counter display faded out. Our field engineer (we call him Enid Blyton because of the fairy tales he tells us about the faults he sees) said he'd seen the fault. After many hours on soak test the display went dim, the final digit of the counter being even dimmer than the rest. The heater voltage was correct but the grid drive was incorrect. The drive comes from the MN15283VJU timer chip IC7501. Replacing this restored correct operation.

Problems with LP recording: There was a poor picture, poor sound and the speed varied. We found that the results in the SP mode were not very different and that the tape was riding high past the audio/control head stack then wrinkling between the capstan and the pinch roller. The problem was caused by a faulty pinch roller – the one that lowers itself into position. A new one and a clean up restored good results in both the LP and the SP modes.

Odd display: The display worked correctly but several of the unused portions were partially illuminated. We found that the voltage drop across the digitron's heater was correct but the actual feed voltage was high. The cause was traced to the supply's stabilizing zener diode D7502 (7.5 V) which was virtually short-circuit.

Intermittent loss of output: Premature failure of the heads is common enough with these machines and I didn't doubt for a moment that this was the cause of intermittent loss of the output from one of the heads on record or playback. When the drum was removed, however, the true cause became apparent, broken leads on the connection pins to the rotary transformer. We've had this on more than one occasion. Another of these machines would accept a cassette, thread it to the half-laced position as normally but would then unload and eject the cassette. The green cassette-in light never came on. The cause of the trouble was a faulty cassette insert switch – it's a slide switch on the side of the carriage.

Patterning on E-E, sluggish drum: The symptoms were alarming. There was patterning on EE, sluggish operation of the drum motor and, when the machine did play back, no colour. The dealer who brought it in said that the luminance/chrominance pack, the head amplifier module and the head drum motor had been changed.

After that his technician left, suffering from nervous exhaustion ...
Fortunately the problem was not as bad as it seemed. C1023
(1000 μF), the 14 V supply reservoir capacitor in the chopper circuit,
was faulty.

Playback picture noisy: It was also worse with its own recordings. On
test we found that the noise on the screen had a definite pattern and
wasn't random. As a start we checked the earth continuity between
the video head preamplifier unit and the main PCB and other
possible earth paths. All measured OK but the strange thing was that
the noise was made worse when an extra lead was connected from
the head preamplifier to the main PCB. A scope was next used to
check the noise on the various supply lines – the main connector
from the switch-mode power supply seemed a convenient place to do
this. Apart from the –19 V supply at pin 10 of P1001 the supplies had
very little noise. At pin 10, however, there were noise spikes of
around 1 V amplitude. When the power supply can was removed the
cause of the trouble was obvious: smoothing capacitor C1022 had
split its top and was open-circuit. A new capacitor greatly reduced
the spikes and cured the problem.

Intermittent shutdown: We found that capacitors C18–21 were all low
in value. Replacement cured all the faults.

Head clogged: No other problems were evident with this machine. A
few days later, however, the customer said that it had gone dead.
When it was back on the bench I found that the mains fuse had
blown. A closer look at the power supply revealed the cause. The
STR11006 chip had overheated so badly that its plastic clamp had
melted, allowing it to fall free of the metalwork and thus run without
a heatsink. D02 and Q11 were short-circuit, also D05 on the little
subpanel. R1009 on the subpanel was dry-jointed. I used a BYD33J in
position D02, a BC640 in position Q11 and a 1N4148 in position
D05. A new STR11006 and a 1.6 A fuse completed the repair.

Head drum motor won't start up: It twitches instead. The cause of
the trouble may well lie in the power supply, where the electrolytic
capacitors C18/22/23 are suspect. If replacing them doesn't cure the
fault, try replacing all nine miniature electrolytic capacitors on the
drum drive CBA under the drum motor – ensure that the types used
in positions C1, C2 and C3 are non-polarized. This note probably
applies to many models that use the Panasonic G deck.

Panasonic NVG40

Grainy r.f. loop-through: We've had several of these machines with faulty video heads when new, but the complaint with this one was very grainy r.f. loop-through. A check on the unswitched 12 V supply to the r.f. amplifier showed that this was low at about 5.2 V. Further checks indicated that the rail was not being loaded excessively so attention was directed to the power supply, which is usually very reliable. Regulator transistor Q1004 (2SD1330) was soon found to be open-circuit.

Dead, tape still laced up inside: The mains fuse had blown (blackened) and the bridge rectifier (D1004) was short-circuit. When this was replaced the fuse held but the power supply wouldn't start. The pulse was being killed by a primary short. Not, as had been the case on a previous occasion, a short in the auto voltage selection i.c. but one in the spike protection diode D1014 (VSD0002).

Digital bar scanner problem: A digital bar scanner is used for programming the timer. An LCD screen gives confirmation of the information read. In this case, when the information was read in the usual order the display would go dim and many of the previously lit segments would go out as soon as the date was reached. If programming was done out of order, with the date selected last, everything was OK. The cause of the fault was soon traced to two pins on the LCD panel being bridged with solder.

After 5 seconds the VCR switched off: When we saw the report on the job card our immediate thought was of a dodgy solenoid. In fact this wasn't the case. When the mechanism solenoid sticks the machine can't move the mechanism into the correct mode. It thus switches back to standby. The problem was that when the tape was moving in play, fast forward, rewind etc. it would run for only about 7 seconds before the machine went into the stop mode. This sounded like a reel rotation problem, so a check was made at pin 27 of the system control chip IC6001 (t.reel input). When play was selected, this input switched correctly between 0 V and 5 V a few times before sticking at the high level, even though the take-up reel was still moving. There's an operational-amplifier IC6002 between the system control chip and the take-up reel phototransistor. One of its inputs (pin 3) is held at 2.5 V by a potential divider while the other input (pin 2) is connected to the sensor. This latter input varies between about 1 V to 3 V to give a change of voltage at output pin 1. When play was selected we found that the input switched from high to low but the output remained high, proving that the operational-amplifier was faulty.

Fast flutter on playback sound: We were sure that this would be a mechanical problem, due to the capstan shaft or bearing. It turned out to be an electrical problem with the motor stator unit, however, part number VEK2944. Note that the capstan motor is not supplied complete, you have to order the various component parts to change the whole motor.

Refusal to keep a tape in: This machine uses the later version of the G mechanism, which is much more reliable. It would refuse to keep a tape in, however. The machine would load the tape then begin to lace it but the mechanism didn't click and engage half way through to allow lacing to be completed. An additional point is that the fault was intermittent. Experience of the earlier version suggested that the relay was probably faulty. Sure enough a replacement restored normal operation. The replacements now supplied are like those used in the subsequent L model number machines.

Refused to return tape: This machine would accept a tape and half lace it correctly, but when asked to give the tape back it refused. We found that when the machine began to unlace and the mechanism reached the point where it has to click the solenoid to move to the next stage, i.e. half way between half lace and the entry to the cassette, the mechanism stopped as the solenoid didn't move. Despite the fact that the solenoid worked perfectly on the outward excursion it was the cause of the fault.

No colour: The playback picture was clean and noise free but without colour. A recording played back on another machine proved that the fault was present in the playback mode only. The playback r.f. chroma is applied to the YC panel at pin 32. It was present here, although indistinct and of low amplitude. More alarmingly it sat on a d.c. potential of 3.32 V. If the feed from the head amplifier was disconnected the d.c. potential disappeared. We found that it was present at pin 13 of the chroma hybrid chip IC801 (part number VEFH04A). The same potential was present at pin 11 of the i.c. on the hybrid module, these two points being coupled by a capacitor which was apparently short-circuit. A new hybrid chip restored the playback colour.

'Noise like a machine gun': Sure enough the deck solenoid was firing continually. Quick checks in the power supply showed that the Reg. 12 V, Reg. 6 V, Reg. 5 V and power-off lines were all pulsing. I was about to look at the system control chip when I noticed from the i.c. block diagram that the Reg. 6 V supply isn't controlled by the power-off line. A new STK5340 chip was required.

Machine didn't play: Fast forward and rewind were OK. When we checked the operation of the mechanism we found that the P5 unit arm didn't come across fully in the stop mode. This was because the 'pull-out sector gear' was broken. We replaced the gear and also the mode switch as it's the most likely thing to have caused the gear to break.

Machine damages bottom of tapes: This machine made a real mess of tapes by damaging their bottom edge. The cause of the problem was a faulty pinch roller. It was quite an expensive repair as the pinch roller can't be detached from its drive assembly – the whole thing has to be purchased as a unit.

Playback and E-E pictures intermittent: But the owner said that the machine worked fine on its side! Tapping anywhere on the top main board affected the fault, so I scoped the video signal at input pin 3 of the luma/chroma subpanel. It was constant here, but at output pin 1 it fluctuated as the panel was flexed. When the subpanel was removed I saw that there were cracks around pins 1 and 2. Resoldering them provided a more permanent remedy than gravity!

Sporadic loss of sound on the machine's own recordings: The recordings themselves were OK, as a check with another machine proved, and playback of pre-recorded tapes was all right. The answer to the problem lay in realigning the audio/control head assembly in order to provide the microcontroller chip with better pulses. It was muting the sound of pin 18 of IC4001.

Intermittent loss of sound and counter in playback mode: This machine had been in several times but the fault wouldn't put in an appearance in the workshop. As the picture apparently remained perfectly OK, loss of control pulses, at least to the servo, was not the cause. This time, however, the fault was present, and the customer had been perfectly correct about the symptoms. There was loss of control pulses at the microcontroller chip IC6001 – in fact there was no activity at the relevant pin. The pulses come from the servo section on the sub main PCB via connection 11, where the pulses were present. The soldering on the wire hoop, so often dry, was fine. From here the pulses pass, via both sides of the PCB, to the base of transistor Q2003. We found that there was no output at the collector of Q2003, although it was not open-circuit. In addition the d.c. conditions around Q2003 and the following transistor Q2004 were correct. Careful checks showed that the pulses at the base of Q2003 were of about 35 per cent lower amplitude than those at pin 11 of the sub main PCB. This disparity was detected across the 10 μF, 16 V coupling capacitor C2022 which turned out to be low in value.

No signals: As there was not always 12 V supply this machine produced no signals, not even via the r.f. loop-through. Q04 in the power supply was open-circuit.

Tape counter stopped in record: Playback was OK but when the machine was asked to record the tape counter stopped after about 7 seconds and neither sound nor the control pulses were recorded. IC2101 was faulty.

No capstan phase lock in playback: Record was OK. Checks around IC2101 on the servo subpack produced the following results: the capstan speed duty cycle at pin 16 was correct at 50:50, although the d.c. voltage was high at 5 V instead of 2.5 V; the capstan phase duty cycle at pin 17 was wrong at 1:99, with the d.c. voltage at 0 V instead of 2.5 V; the d.c. voltage at pin 2 (tracking MMV) was low at 0.3 V. The latter seemed odd, as this is a simple d.c. control voltage obtained from the 5 V supply via the tracking control. Herein lay the simple answer – there was no 5 V at the top end of the tracking control as the grey single wire to the control sub-PCB from the timer PCB (pin 4 of P7503) had broken off. It had never been secured by tape or glue.

Panasonic NVG45

Drum speed variations: This machine was faulty when taken from the box. The playback picture would come and go, due to drum speed variations that could also be heard. When we turned the machine on its side to remove the bottom cover the fault cleared. While looking for a loose plug/socket connection we removed the drum cylinder and found a crack almost half way across the double-sided stator panel, between the socket and where the panel enters the drum unit.

No play or record: The reason for this was that the drum didn't start up. It just kicked backwards and forwards. Experience has shown that with the type of motor used here the cause of this condition is usually something to do with the rotor position sensing, since this controls the current that's switched to the three pairs of drive coils. A single Hall-effect device is used with these drums. The connections are to pins 2 and 6 of plug/socket P2001. With this particular unit we got a reading of 2.7 kΩ between these pins. When we checked a G40 drum the reading was 400 Ω. Unfortunately the Hall device is not available as a separate item. You have to order a complete motor assembly, which costs around £60. We had a scrapped rental NV430 in the workshop, however – it had been written off due to liquid spillage. It's drum motor has two Hall-effect devices, one of which was removed. Before removing the cup-shaped rotor of the G45's motor we marked on it the positions of the centres of the two securing screws in the two adjusting slots. The Hall-effect device was then removed and the replacement from the NV430 was carefully fitted in exactly the same place. The rotor was then refitted, with the two securing screws just tight enough to allow adjustment of the rotor's phase, and the drum was secured back in the machine. A standard tape was played and the head-switching point was adjusted visually before tightening the rotor screws fully and carrying out fine electronic adjustment of the head-switching point.

Whine from machine at switch-on and it wouldn't play back correctly: The reason for these symptoms was that the drum motor ran flat out. When tests were made we found that the voltage at the torque control input pin of the drum driver chip IC2901 (pin 7) was incorrect. In the stop mode (machine on) the voltage at this pin should be 3.8 V. It was low at 1.4 V. With the pin disconnected from the PCB the motor didn't start and the voltage rose to around 3.6 V, indicating that there was some loading on the line. It was due to a leaky capacitor, C2117 (0.47 µF electrolytic), which filters the speed and phase control voltage from the operational-amplifier IC2103.

Sometimes eject tape during rewind: The customer's complaint was that this machine would sometimes eject a tape during rewind. Our field engineer had tried the solenoid, resoldered the servo pack connections and noted on the job card that he'd not seen the fault. When we tried the machine it worked OK but then failed to half lace when a new cassette was inserted. The solenoid wasn't being energized, the most likely reason for this being a faulty mode switch. This would also account for the original complaint. Fitting a new one cured both problems.

Erratic capstan: So it proved to be on test, speeding up and slowing down at random. On this machine (G deck) the capstan also drives the mechanism operation, so you have to be careful with capstan speed faults because you can damage the mechanism. The fault would show up when any point in the capstan circuit was touched, but it seemed logical to start by checking the capstan FG waveforms. Sure enough we found that the FG waveforms fluctuated up and down wildly at the capstan FG amplifier IC2104. The key seemed to lie with R2184 at the input to the FG amplifier, as there was no waveform amplitude variation at one end of this component and a very erratic variation at the other end. We removed this surface-mounted resistor and tacked a conventional $1\,k\Omega$ resistor in its place. The capstan circuit then operated perfectly in all modes.

No video head drum servo lock in the record mode: A tape provided by the customer showed that there was no video head drum servo lock in the record mode. The symptom shows as the head crossover point wandering up and down the screen. Additional symptoms were poor capstan servo lock with sound pitch variation. Neither the dealer nor ourselves were able to confirm these symptoms. Fortunately Panasonic were able to come up with a circuit upgrade to deal with the problem. A $150\,k\Omega$ resistor and a $15\,nF$ capacitor are added in parallel across R338 on the Yc panel – this appears to be part of the sync separator circuit. Closer inspection of the results produced by the owner's tape showed that servo lock was lost when a rapid screen change from very dark to light occurred. This probably made the video signal ride up and down on the d.c. level, affecting the sync slicing.

Severe drum twitching: The sync separator modification detailed in the previous fault report can be improved by adding an extra $47\,k\Omega$ resistor across the components that Panasonic supply. I'd found this to be necessary because users complained about severe drum twitch and HSP running through the picture when recording from early Bush/Alba satellite TV receivers (the ones that used to drift off tune). I had heard that the video output from these units was suspect and that there was a modification for them. But customers who rented G40s and G45s from us preferred us to take action.

Capstan would run fast: This machine had an intermittent fault. The capstan would start to run fast in all modes, including lacing and unlacing. When the fault was present in the play mode the machine was toggling between SP and LP. We suspected a fault in the capstan FG pulse feedback circuit and sure enough found that the FG signal was going missing at pin 3 of IC2104 on the sub main servo part of the PCB. The only components in the feed here are R2184 and C2185. Scope checks at the resistor showed that the signal was present at one side but not the other. It's a surface-mounted component, with a value of $1\,k\Omega$. We cured the fault by fitting a standard eighth watt resistor.

Wouldn't play back own recordings: Pre-recorded tapes were played back correctly but the machine wouldn't play back its own recordings properly. A check with another machine proved that the cause of the fault was in the record circuitry – the fault was also intermittent. The record track wasn't being recorded as pin 1 of IC2102 in the servo section was dry-jointed.

Failure to accept tapes: This is often caused by a faulty timing belt. In one case recently, however, the flywheel was at fault: the collar that provides the belt drive had become detached. Check the flywheel by replacement.

Machine would accept a cassette half way then eject it: This was because the cassette flap opener had become dislodged from the carriage assembly, it didn't lift the flap as the cassette was being lowered. Simply clicking the opener back into place put matters right, and numerous test runs proved that all was well. When the machine was returned to the customer's home the teenage son decided to load a tape. He did this by placing the tape in the slot then, sitting on the floor with his back against a settee, pushing the cassette home with his foot – which he also used to operate the function buttons. I pointed out to his father that the guarantee terms were subject to 'normal' use, but this was greeted with a grunt. Back in the workshop the copy invoice was retrieved and a note was made about this in case of a guarantee claim in the future.

Panasonic NVG7

Would accept cassette then immediately eject it: Unless the end sensors are covered, immediate ejection of the carriage is normal with a Panasonic front loader when it's fooled into loading the carriage without a cassette in place. With this information in mind we suspected perhaps a leaky end sensor, and as another of these machines was available we swapped over the loading carriages. The fault persisted, however. We next compared the voltages at pins 20 and 21 of the system control chip with those in the working machine – these are the end sensor inputs. The readings in the faulty machine were 5.5 V against 3.2 V in the other machine. The supply voltages for the end sensor phototransistors, which are connected to pins 20 and 21, are obtained from a potential divider network (R6009/10/11/12) across the 6 V rail. A check revealed that there was no voltage drop across R6009 (1 kΩ) at the top of the network and an ohms check showed that there was a dead short across it. The resistor itself was blameless, as the short persisted when one leg was unsoldered. On closer inspection we noticed that R6009's other leg was very close to an adjacent print track. In fact the leg was touching the track, as a result of which R6009 was shorting itself out. When the resistor's leg was prised up the short disappeared and the machine accepted cassettes readily.

Dead apart from clock: This machine was dead apart from the clock display. The digital counter display appeared when the on/off button was pressed, but there was no channel number, no on-LED display and no mechanical functions worked. A faulty STK5331 main regulator chip (IC1001) was responsible.

Mode switch problem: The mode switches used in various Panasonic decks can be troublesome. If you replace one in the D1 deck (NVG7, NV-G10 etc.) don't use the VSS0110 type, which you may have in stock for earlier models. It looks similar and fits perfectly, but electrically it's quite different, giving rise to some peculiar deck behaviour. The correct part number is VSS0135.

Vertical rolling in playback: This would appear when a known good tape or a self-recorded one was played back. In addition the tape would be crinkled along its bottom edge. The pinch roller was glossy and tape ridges were evident. A new roller cured both problems.

Tuning storage problem: Stations could be tuned in and stored, but on channel change they disappeared. We suspected the MN1220 memory chip, but checks took us to the $-30\,V$ supply which was rather high at $-57\,V$! Q1101 and D11 in the power supply were found to be short-circuit. Despite this high voltage the memory chip was perfectly OK.

Capstan motor at wrong speed: The cause was faulty CTL pulses. C248 $(1\,\mu F)$ had failed.

Panasonic NVHD100

Recordings spoilt by twitching: It was as though there was a 'glitch' on the drum drive. The cause of the fault turned out to be the impedance roller next to the full-erase head. A replacement, part number VXP1402, restored flicker-free operation.

Occasionally slows down and stops: This would happen usually after many hours of operation. When the fault eventually put in an appearance we were able to condemn the XRA6439P capstan drive chip IC2505.

No microphone sound on audio dub: The customer had succeeded in blowing up the mic sound amplifier chip IC7701 by feedback from his TV set. Replacing IC7701 put matters right.

Intermittent shutdown: Any attempt to get the tape out would be frustrated until the machine had been left disconnected from the mains supply for a few hours. After many hours had been spent head scratching we eventually found that the mechanism loading motor, part number VEM0427, was the culprit. A replacement restored the machine's good nature.

'Bad picture': Sure enough the playback picture had very bad dropouts, with lots of black flashes and glitches. Clearly the dropout compensation circuit wasn't working. Checks in the video processing circuitry showed that there was no 5 V feed at pin 44 of IC301 – this pin supplies the dropout compensation part of the chip. The cause was coil L304, which was open-circuit. A much cleaner playback picture was produced when a replacement coil had been fitted.

Very intermittent play fault: The owner of this machine had included a recording to illustrate the fault: about once every half hour there was an interruption – a missing word, a flick of the picture or a jump in the scene. We came to the conclusion that the fault was probably caused by the capstan motor stopping briefly or slowing down every so often. Inspection of the XRA6439P capstan drive chip showed that its heatsink was very loose. For good measure we replaced the chip and refitted the heatsink securely. So far there have been no further interruptions.

Panasonic NVHD90

No E-E picture: Checks showed that the voltage on the 12.3 V line was low at about 5 V. As disconnecting the power supply made no difference to the voltage the cause of the fault was clearly in the power supply itself. C1130 (1000 pF) was eventually found to be leaky.

Glitch in reception: A transient flash muted the sound momentarily and made the stereo indicator flicker. Since this looked bit like tuner flashing we tried fitting a replacement. The glitch returned almost immediately. After much scoping and measuring we traced the cause of the fault to C1130, a leaky 1000 pF ceramic capacitor in the power supply. A replacement restored the tranquillty of the 12 V supply.

Only records two-thirds of screen: This machine recorded the head-switching point two-thirds of the way up the screen. Suspicion naturally fell on the integrity of the field sync signal that's used to lock the phase of the head drum in the record mode. It was completely missing, the cause being a defective video processor chip (IC301, part number VEFH29H).

Intermittent failure to load: This machine would sometimes accept a tape, try to load it, fail, then eject it. My first suspect was the mode switch, but it proved to be innocent. Eventually arm P5 was found to be the culprit. Because it was slightly distorted, it would sometimes jam as it attempted to load. A replacement (part number VXL2306) provided a reliable cure.

Panasonic NVJ30

After 3 hours' running the machine would cut out: It would remain powered, whatever mode it was in. Restarting it would provide a few more seconds of action before it once more decided to have a rest. We suspected loss of reel tacho pulses from the ON2170 opto-interrupter IC1501 beneath the take-up reel. The pulses were in fact there but were of only 250 mV peak-to-peak amplitude – a rather inefficient use of the 5 V supply! The supply was found to be low at only 2.2 V, however. It's developed from a 12 V feed by a regulator in the syscon section on the main PCB and goes to many areas. When the feed to the opto-interrupter was disconnected the voltage rose to 5 V. A new ON2170 restored normal action.

Intermittent capstan servo problems: This machine suffered from very intermittent capstan servo problems. After about 3 hours of operation the sound would start to wow and tracking bars would flash and flicker across the picture. Unfortunately the symptoms were erratic and couldn't be relied upon, so fault-finding of any sort was fruitless. Convinced that the cause of the fault was in the main systems and servo circuit we changed many i.c.s and capacitors in this area, but the fault continued to recur. After the machine had spent many weeks on the soak test bench the cause proved to be C1122 (330 μF, 10 V) in the power supply module. It was apparently going open-circuit when hot.

Machine would not tune in E-E mode: Because of an instruction from the timer microcontroller chip IC7501, which handles tuning, the machine was locked in the Band I mode. The instruction comes via

IC7502 (MN12C261D5) which was faulty – it provides parallel switching lines from the serial data fed to it.

Tape modes normal except for play: When playback was selected the machine laced up but the capstan, and thus the picture, ran too fast – at roughly the same speed as fast search, the monitor's screen remaining in the E-E mode. Then after about 5 seconds the machine entered the stop mode and shut down in the half-laced position. To cut a very long story short, the culprit turned out to be the mode switch. For those wishing to make a note of this the mode-switch voltages, measured for convenience at pins 73 and 74 of the microcontroller chip IC2001, should be as follows:

Pin 73: stop/half load position 5 V; rewind/fast forward 5 V; play 0 V.
Pin 74: stop/half load position 2 V; rewind/fast forward 2 V; play 5 V.

In the fault condition pin 73 remained at 5 V.

No E-E playback or test signals: After carrying out waveform and voltage checks we came to the conclusion that the r.f. converter module was faulty. A replacement restored the signals.

Dead power supply: The usual culprit is C1109, but not this time. Checks showed that the power supply worked initially but swiftly cut out. We eventually found that D1110 on the secondary side of the power supply was leaky. A new MA185 diode was all that was required to restore full operation.

Display problems: Although the machine worked perfectly, it was difficult to operate because the channel information and cassette functions were not shown on the front display panel. As the deck functions were not affected, it seemed logical to assume that the system control chip IC2001 and the timer and front display control chip IC7501 were both OK. So a thorough check was carried out on the serial data connections between these chips. This revealed that R6044 had risen in value from 220 Ω to over 900 Ω. The problem was cured by fitting the correct value resistor in this position.

Top half of E-E picture distorted when switched on from cold: The distortion would gradually decrease until, a few minutes later, the picture was fine. As with all such ephemeral faults, it took many days of soak bench testing to find the culprit, which turned out to be C768. This 10 μF capacitor decouples the 12 V supply on the demodulator board.

Dead, but would respond to heat: This machine was dead but was restored to life when its power supply was heated with a hairdryer. When it cooled down it returned to the dead condition. There was obviously a faulty reservoir/smoothing capacitor somewhere. It turned out to be C9 (1 μF, 400 V) which, when checked, produced a reading of about 0.03 μF. Where would we be without tins of freezer and a hairdryer?!

Dark/pulling E-E and playback picture: Check for dry-joints at the r.f. converter.

Dead remote control/scanner: As the combined remote control/bar scanner handset (part number VEQ1107) for these machines is an expensive item, repair is usually economic. This one seemed to be dead, although the chips were being supplied. There was no clock signal at pins 15 and 16 of the microcontroller chip IC1, however. Crystal X1 (3.52 MHz) was faulty.

Mechanical problems: I bought this ex-rental machine untested from a wholesaler. It came with a bag of bits that contained the pinch roller, a new unwrapped tuner/booster and the demodulator PCB. The tuner was fitted to the PCB, which was then fitted to the main board. On power-up, however, the machine was dead. A new 1 μF, 400 V capacitor in the power supply cured that. When a tape was inserted the solenoid clicked but the capstan wouldn't rotate. The cause was excessive clearance between the capstan's rotor and stator. After setting this up I had a working machine.

Panasonic NVJ35

Switch-mode power supply problem: This machine was dead and checks in the switch-mode power supply showed that D1113 was short-circuit. It's the 20 V protection zener diode connected across the 14 V line. So it seemed that the 14 V line was going way above its normal level, well above 20 V in fact. We suspected a fault on the primary side of the circuit and started to check the components here. This paid off: when C1114 was checked it was found to have fallen in value from 47 μF to 0.5 μF. It's a decoupling capacitor that's connected between the chopper transformer and the regulation control pin of the chopper chip. A new 47 μF, 16 V capacitor enabled the VCR to breathe again.

Maximum record time just over an hour: The customer card said 1 hour 20 minutes but when I tried it the machine just stopped after an hour and 12 minutes. It then resumed and recorded for about half an hour before again stopping. The further along the tape it got, the shorter its record period became. A scope check showed that the reel pulses that reached the syscon chip were at about 3.5 V. In my experience they are usually more than 4 V peak-to-peak. When I checked the opto chip from which these pulses are derived I found that there was a fair covering of fluff on the reflective surface of the reel drive gear. This has alternating black and mirrored portions that generate the reel pulses optically. Needless to say cleaning the surfaces cured the problem.

Record problem: Playback was perfect, but when record was selected the machine would run for a few seconds then return to the stop mode. L4002, the choke in the l.t. feed to the audio bias oscillator, had gone open-circuit because a solder blob inside the oscillator transformer T001 had provided a short to chassis. When L4002 had been replaced and the solder blob had been removed the record function worked normally. A microcontroller chip pin monitors the bias oscillator: if no oscillation is detected the deck is returned to the stop mode.

Machine refused to accept tapes: This machine refused to accept tapes, throwing them out immediately. We found that the cassette housing was misaligned with the main deck. Realigning the housing rack gear with the main deck metal drive gear provided only a temporary cure – the machine relapsed into its previous state with a clattering of slipping gears. Inspection of the right-side section of the cassette housing showed that there were two broken plastic retaining lugs. The result was too much play on the rack gears. A new right-side assembly (part number VXA3153) cured the problem.

Jammed mechanism (G deck): It's important to check the rack assembly on the right-hand cassette housing side. With the arm in the down (horizontal) position, the arrow on the nylon gear should line up with the one on the rack. If it's out by just one tooth you can get nasty crunching noises when ejecting the cassette because the switch on the side piece is in the wrong position and the capstan motor isn't switched off in time. As a result it tries to force the housing beyond its stop, crunching the gears. This occurs with any machine that uses the G deck. With this particular machine the rack was two teeth out. This is the reason why a complete rebuild was required. The right-hand side piece is also prone to damage: it's available as a complete assembly.

Died after 2 seconds: The deck made its usual whirry noises when the machine was plugged in but after 2 seconds it was dead, with no display or deck functions. Checks in the power supply department showed that all the output voltages were present and that during the brief period after connection to the mains the switched supply lines were also correct. So why no display? Over to the display section to check around IC7501. Its 5 V supply was correct and the resets were OK, but there was no waveform at the osc. 1 pins 2 and 3: one pin was at 5 V, the other at zero. Replacing crystal X7501 restored normal operation.

Nasty capstan fault: This machine had a nasty capstan fault, with bad wow on sound and tracking bars that jerked down the screen spasmodically. The capstan drive chip and stator are prime suspects when you get symptoms like these. On this occasion they were both innocent, however. We next changed the servo and system control chip, as the capstan drive seemed to be abnormally high, but again the verdict was not guilty. As things were now looking desperate the oscilloscope was wheeled into action. We were surprised to find that a check on the capstan error voltage produced a very corrupted digital waveform, even when the machine was in the stop mode. In this machine the capstan FG signals are amplified before being fed to the servo and systems chip IC2001. We found that there was a large spiked waveform sitting on the input at pin 15, which supplies the digital speed control circuit. Two operational-amplifiers feed this pin, from the capstan FG2 buffer amplifier: this is where the additional waveform was being added. High-frequency noise was being picked up and amplified. Where was it being generated? The power supply of course. C1118 had gone low in value, leaving a small high-frequency ripple on the 5 V line. A new 330 μF, 10 V capacitor put matters right. C1122 (100 μF, 50 V) in the 45 V supply had also fallen in value, and this is the more usual cause of the problem. While we were in there we also replaced C1109 (1 μF, 400 V) and C1114 (47 μF, 16 V) on the primary side of the supply as they can also give trouble.

Picture wobbled, wow on sound: The capstan jerked round, the picture wobbled and the sound had lots of wow on it. So we checked at pin 13 of the capstan drive chip and found that there was 2 V of ripple on the 5 V supply. There was ripple on the other supplies as well. Replacing most of the capacitors in the power supply restored normal operation.

Erratic mechanical operation: The mechanism solenoid on this machine would sometimes chatter and often disengage before an operation was complete. Our first step was to check the supply to the solenoid for ripple – we suspected a faulty decoupling capacitor. As the supply was OK we replaced the solenoid, but the fault was still present. We subsequently discovered that R6022 in the solenoid drive circuit was dry-jointed. A bit of resoldering was all that was required.

Wouldn't eject a cassette: On test it wouldn't power up. C9 in the power supply was open-circuit. After replacing this I found that the capstan speed was erratic and there were tracking problems on the picture. C18 and C22 in the power supply were low in value.

Counter had gone 'beserk': On inspection I found that in the search mode the tape moved up the head. The cause was post assembly P5 which was bent. Replacement restored normal operation.

No results: This is a very common fault when the mains input has for one reason or another been removed from this model. The auto-selector or power regulator chip can be responsible. Alternatively C1109 (1 µF) can, as here, fall in value or go open-circuit.

Clock display problem: Several segments were permanently lit up. Our experience has been that this fault is usually caused by the display itself or one of the diodes connected to it. Not on this occasion, however: the microcontroller chip IC7501 was faulty.

Noisy (lumpy) capstan motor operation when it's running slow: This occurs during play/record and loading. The cause is loss of capacitance in C1122 (330 µF). As a result the supply is affected by noise.

E-E picture blank, erase flickered: There are days when it doesn't pay to get out of bed. This machine was the cause of such a day. The E-E picture was blank, and the erase symbol flickered on and off intermittently in the fluorescent display. In addition the mechanism (type G) had shed a few gear teeth and refused to accept a tape. While puzzling over the former symptoms I realigned the mechanism, replacing the usual gears (VDG0343, VDG0346 and VDG0448). Once this had been done the machine accepted and loaded a tape – but wouldn't stop loading. I hastily unplugged the machine and replaced the mode switch (VSS0175A). This failed to cure the fault as the print to one of its pins was open-circuit. A good, clear playback picture appeared when this had been repaired. I started to check around the fluorescent display and timer circuitry. IC7501 was replaced and a new display was fitted, but the erase symbol still flickered on and off erratically. As all the diodes, resistors and capacitors on the timer board checked out OK, I decided that I should be looking for the cause of the other fault. When I checked through the E-E picture circuitry I found that the EE video stopped at the input/output AV switching chip IC3901. Checks in this area failed to reveal any obvious problems, but one of the switching lines was permanently high. As the signal on this line comes from the MN15522VMS subsystems control chip IC6801, I started to carry out checks here. It seemed odd that nearly every pin of this chip was at about 5 V. When I removed the supply and the reset to this chip, the E-E picture came up. Inexplicably, the display problem had also been cured. Replacing IC6801 completed the marathon job.

Wouldn't stop in tuning search mode: This machine would go into the tuning search mode but wouldn't stop when a channel was reached. The cause of the fault was cracked print leading to pin 2 of the tuner/demodulator pack. A link across the crack solved the problem.

Dead machine: The dead machine symptom with any model that uses this power supply can be caused by C1119 (680 μF) on the secondary side of the supply being short-circuit. This is unusual, however: the usual cause is C1109 (1 μF) on the primary side of the supply. The difference is that with C1119 short-circuit you get a slight chirp at switch-on.

Panasonic NVJ35B

No record chroma: Scope checks showed that everything was in order up to the 1.6 MHz LPF section of the multi-filter package FL801 (ELB4W002). It was open-circuit.

Dead machine: That there was no operation on the primary side of the power supply was indicated by the absence of squeal at switch-on. We found that diode D1104 (type AP01C) in the snubber network was short-circuit.

Capstan motor failure?: This machine gave every indication of capstan motor failure. There was very noisy operation and sluggishness whilst loading, and if loading was completed there was severe warble in play. Operation even sounded 'metallic', as if bearing failure had occurred. A colleague had fitted a new motor, but to his dismay this had no effect on the symptoms. Now electrolytics are often a problem with Panasonic equipment, so we carried out some checks in the power supply and discovered that C22 (330 μF, 10 V) was very low in value. Several other capacitors were checked as well and found to be wanting.

Panasonic NVJ40

Tuning problems: Sometimes no stations could be tuned in, but more often the mid-band stations were crammed at one end. Attention was focused on the plug-in demodulator pack and IC7652 (AN5043). This provides the tuner unit with the relevant tuning voltages etc. The voltage at pin 3 of IC7652 was low and varying, so I replaced C7666 (0.01 μF) which decouples this point.

Machine drifted off tune on any channel above 59: The cause was a faulty tuner.

At switch-on the cassette carriage tried to load although there was no tape: We've had this fault twice recently. The machine soon powered down as the system control detected that there was a problem. Checks showed that the capstan drive chip was operating at full tilt all the time, irrespective of the control signal at pin 16. The cure is to replace the BA6435 capstan drive chip.

No E-E sound or no tuning lock: This is a tale of two videos, one with no E-E sound, the other with no search tuning lock. No apparent connection but read on. We tackled the one with no sound first and soon found that no sound left the demodulator pack because Q713 (audio defeat) had a high at its base. There are two feeds to Q713, a mute from the syscon department and an interstation mute that's generated by IC7651. During search tuning IC7651 receives a video feed: it generates a sync-low signal to stop search tuning when a station is found. With a station already tuned in there should be nothing at IC7651's sync-low pin 9, but there was 5 V here. A new AN5421 cured the fault. The problem was the same with the other machine but as there were no previously tuned-in stations there was no search lock. A new AN5421 restored full tuning capability.

Blue and red patterns with new recordings: No full tape width erase was the trouble with this machine. The symptom occurred because the chroma from the previous recording wasn't being erased. The cause of the fault was simple. The full width erase head plugs in. During manufacture one pin had bent over when the plug was inserted and thus failed to make contact. With these later type G decks the erase voltage is fed via a ribbon cable across the top of the cassette housing, along with the end sensor supply.

No playback for first half hour, then bad patterning: The job card said 'no playback for the first half hour, then bad patterning'. It turned out to be an accurate description. Checks showed that from cold transistor Q3204, which provides the 'except record 5 V' supply to the head amplifier playback circuits, wasn't fully conductive. The supply would gradually increase from about 2 V (no picture) to 3 V (poor picture with lines across) then 4 V (reasonable picture with patterning). After much investigation in the switching and biasing circuits, all to no avail, I finally found that C1127 (330 µF) in the power supply was the cause of the trouble. It decouples the 5 V feed to the system circuit.

Nothing worked and the display showed E9: When the machine was plugged in the clock flashed 0–00 for 2 seconds then the display produced E9. We checked for oscillation at pin 82 of the surface-mounted chip IC6001. There should have been 10 V peak-to-peak but this was missing. The cause of the fault was Q6101 on the fold-down PCB.

Intermittent power-down when a tape was inserted: Preliminary checks showed that the head drum didn't start as the tape was loaded – in this machine the tape is fully loaded round the head in the stop mode. Inserting the tape a couple of times would get the machine to work. As the condition and alignment of the mode switches seemed to be OK suspicion fell on the systems and servo control, in particular the microcontroller chip IC6001, but again the verdict was not guilty. I still felt that there had to be a problem with the mode switches. A look at the carriage mode switch, which I'd already replaced, showed that it seemed to be less securely mounted than usual. Holding it securely with my finger as I loaded a tape proved the point: the machine worked faultlessly. A replacement right side cassette housing, part number VXA4468, cured the problem. The cause of the trouble was the excessive play in the mode switch mounting.

Tape stuck inside: The cause of the trouble turned out to be the capstan motor. Tape loading was OK, driven by the capstan motor, but once loading was complete and play commenced the motor wouldn't run at the slower speed.

Wouldn't eject tapes: This machine wouldn't eject tapes because the release spring had parted company with the release lever. Refitting the spring and retiming the mechanism cleared the fault. We fitted a new mode switch for good measure.

No recorded chroma: Playback was OK. After carrying out scope checks around the VEFH14D hybrid chroma chip we came to the conclusion that it was faulty. But in view of its price, about £42 plus VAT, we decided to give Panasonic Technical a call before ordering a replacement. A nice man agreed with our diagnosis, and a new chip cured the fault. Phew!

Intermittent stopping: The machines in this range and subsequent ones incorporate an internal fault diagnosis system. Pressing eject, fast forward and rewind puts the machine in the fault diagnosis mode (with shuttle search machines press eject and hold shuttle forward). The VCR will then recall from its memory the last registered fault unless it has been unplugged from the mains supply since the last fault occurred. The fault with this machine was intermittent stopping. Unfortunately it occurred once every two weeks. But the machine said E2, which according to the manual means reel stop problems. So we turfed out the two ON2170 reel sensors and cleaned the reflector surfaces. Our sincere thanks to the diagnostic system – the machine hasn't darkened our doorway since!

Interference, intermittent operation: The customer's original com-plaint was about interference. She also complained that occasionally, during playback, the machine would stop and come back on again. As it looked as if the cause of trouble was ripple from the power supply, I took the machine back to the workshop – where it performed faultlessly. When it was returned to the customer both faults showed up almost immediately. The switching on and off was a power supply fault: IC1102 was dry-jointed. The interference seemed to be caused by some form of interaction between the TV set and the VCR. Disconnecting the aerial from the VCR failed to cure it. Adding an attenuator made it worse. Slight adjustment of the r.f. converter frequency plus retuning the TV set eliminated the problem.

Blank screen in visual search mode: The customer complained that when he used this machine in the visual search mode all he got was a blank screen. Various checks were made, which brought us to the lower drum. This seemed to be worn. If the tape tension in the search mode was increased the picture reappeared. As the drum unit is so expensive we tried to revive it by polishing with Brasso and Duraglit. Needless to say, the lower drum soon found its way to the bin. A new drum restored perfect operation.

Deck jammed, tape wrapped around the head: There was nothing unusual about this, so the upper and lower deck service kits were fitted and aligned. I find the G deck alignment instruction video that's available from Charles Hyde and Son very helpful: it takes you through the procedure step by step – you simply use playback pause while carrying out the task described, then continue. When the service kits had been fitted everything worked normally apart from the fact that there was no auto stop at the end of rewind, play and fast forward. The microcontroller chip was calculating the speeds of the spool carriers accurately, because the tape speed was being reduced as the tape neared its end. But when the tape ran out the mechanism struggled to keep it going for a few seconds then the VCR entered the stop mode. This would not do the mechanism any good, and was probably the cause of the misalignment in the first place. A scope check across the infrared 'lighthouse' transmitter produced a 2.5 V peak-to-peak pulse at 25 msec. When we fitted a replacement the reading was 0.8 V peak-to-peak and the auto stop worked correctly.

Panasonic NVJ42

Search tuning fault: There was no lock in either direction. Sync low and a.f.c. defeat were normal but because there was no a.f.c. feed from pin 6 of the plug-in demodulator pack the front panel microcontroller chip had no information to work on. Pin 6 of the demodulator pack proved to be the cause of the problem: it was open-circuit to the demodulator pack plug itself. Fitting a new socket cured the problem.

Difficult to get cassette ejected: Although this machine would accept a cassette it was difficult to get the cassette back and the mechanism spooled backwards and forwards a great deal, rarely performing any function correctly. Checks soon showed that the solenoid which engages the mechanism was operating erratically. Instead of a satisfyingly solid clunk when the operation buttons were pressed only an anaemic click was heard. The solenoid drive system has two parts, a kick and a hold circuit. D603 in the kick section was open-circuit, a replacement restoring normal operation.

No record or E-E video: The sound was OK (use the test signal switch to get sound with self-muting TV sets). This machine is very hard to work on, with limited access to the luminance/chroma PCB. Replacing IC302 cured the problem.

Intermittently shut down to stop: This was a true video nasty! The machine would run for days without misbehaving, then suddenly shut down to stop, giving us a diagnosis time of perhaps 2 seconds. But we solved it! The waveform produced by the supply spool rotation-detector optocoupler was of low amplitude, just borderline for tripping the operational amplifier switch that's connected between it and the microcontroller chip. A new optocoupler solved the problem.

Did not stop rewinding: This is one of those Panasonic models that can increase its wind/rewind speed as it spools through the tape, slowing down before the transparent leader that activates the end stop is reached. This machine did not slow down when it rewound a tape, although it did when it wound one. Thus a tape that was being rewound had to stop very suddenly – sometimes this would tear the leader off the tape ... The systems and servo microcontroller chip has to monitor the two reel sensors to determine the type of tape and thus calculate when it should speed up and when it should slow down. If the tape is not recognized, usually because it has a non-standard length and/or hub size, or cannot be spooled evenly, it will not be wound or rewound at top speed. Now back to the fault. The amplitude of the output from the take-up reel sensor was lower than that from the supply reel sensor: perhaps more importantly, there was also a glitch at the bottom of the waveform. A replacement take-up reel sensor, part number ON2170, cured the unfortunate tendency to separate leaders from tapes.

Panasonic NVJ45

Displays flashed rhythmically: When this machine was tried out all the displays lit up at once and flashed rhythmically. Checks showed that there were no abnormalities in the filament and dynamic drives from IC7501. But R7504 in the grid supplies had rather a large voltage drop across it for a $100\,\Omega$ resistor. When it was measured out of circuit the reading was about $1\,\mathrm{k}\Omega$. A replacement restored normal operation.

Would cut out in record: This machine would cut out after a few seconds in the record mode. A check on the main PCB showed that the delay record 12 V (D Rec. 12 V) supply was missing. The 2SB1321AR transistor Q6203 turned out to be faulty.

When record selected some TV channels had diagonal patterning: There was an unusual complaint with this machine. When the record mode was selected, some channels on the TV set were affected by heavy diagonal patterning. On a hunch we checked the luminance and chrominance pack carefully for ageing capacitors. Sure enough when

C831 (4.7 μF) had been turfed out and a replacement fitted the problem had gone.

No colour with own recordings: If the drum servo is hunting, check whether the 4.43 MHz NTSC switch on the front panel is switched to 'on'. It should be in the off position – the switch should be in the on position only in the dubbing mode.

No E-E tuning: Loop-through and playback were fine. The tuner used in these VCRs is known for problems, but not of this sort. Checks showed that its 12 V BU supply was missing. This comes from the adjacent band-switching chip, which is largely redundant in the UK version. Pin 12 of this chip produced a low reading of 0.2 V, although its supplies and controls were all fine. A resistance check here produced a reading of 1 kΩ to chassis. So I tried, to no avail, a replacement chip from a scrap unit. I then noticed that the tuner had been replaced, and that one lug of its case is soldered very close to the print to the BU pin. There are a couple of surface-mounted links on this print. One was touching the rather bulbous joint on the tuner's lug – or was it? The resistance check had produced a reading of 1 kΩ: it now read short-circuit! Tidying up the joint restored the 1 kΩ reading and the tuning. But I didn't trust the original chip.

Panasonic NVJ47

Erratic mechanism, no playback: The mechanism had a lot of movement but it would rarely reach the play position without sighing to a halt and shutting down. The cause of the problem seemed to be the capstan stator. It had very little torque and emitted strange whistling noises intermittently. When the capstan rotor was removed to gain access to the stator we found that the soldering to the stator coils could have been better. In fact resoldering the stator coils and the Hall i.c.s cured the trouble.

Whistling or buzzing when the machine is switched off: The cause of this is excessive mains voltage, the cure being to add a 0.01 μF, 50 V capacitor (part number ECUM1H103KBN) across pins 3 and 4 of Q1103 in the power supply. However, the symptoms can be fixed by a squirt of sealant on T1101 in the power supply.

Broken subloading arm: When a new one had been fitted the deck worked fine – until you rewound the tape back to the start, when the end sensor didn't operate (hence the broken subloading arm). The waveform across the tower LED was low: tracing back, I found that R6612 (22 Ω safety resistor, part number ERD2FCVG220) was open-circuit. Replacing this restored normal operation.

Colour phase fault with playback: It would intermittently become black and white. It looked like a delay line fault, but the cause was IC302.

Dark flashing bars on E-E picture that seemed to vary with the sound signal: After a quick check on the power supply lines to confirm that they were all up to specification we moved to the luminance/chrominance subpanel. The main item here is IC302, which processes the luminance and chroma playback, record and E-E signals – and is expensive! Fortunately C318, a small 22 µF capacitor connected to pin 9 of the chip, turned out to be the culprit. When this had been replaced we had clear E-E pictures.

Head drum very reluctant to rotate: The head drum was very reluctant to rotate. It would eventually do so on about the third attempt. The result, however, was a picture that readily smeared across the screen as the drum servo struggled to maintain lock. Checks around the drum drive chip showed that C206, which is connected to pin 12, was defective. A new 0.1 µF electrolytic capacitor cured the fault.

Intermittent switch-off: This machine kept on coming back to the workshop because it would intermittently switch itself on or off. The cause was simple: the VCR button on the front panel was gummed up. As a result it randomly switched on the power. A good clean-up cleared the fault.

High-pitched noise in standby: The problem occurs only where the mains voltage is very high – 250 V or above. Transformer T1101 then buzzes. During a chat with Panasonic I was told to add a 0.01 µF, 50 V capacitor across pins 3 and 4 of optocoupler Q1103 in the power supply. There's a part number for this capacitor: ECUM1M103KBN.

Panasonic NVL20

Back-tension arm stiff on its pivot: We've had a couple of these machines with the back-tension arm stiff on its pivot. The effects are failure to erase previous recordings, picture rolling, and virtually unwatchable own-recordings. The machine usually works fine when it's stood on its left-hand side! The cure is to remove the lever, check that the box section at its pivot end is square, and reassemble with a drop of lubricant.

On/off button didn't work: When the machine was connected to the mains supply the clock display flashed as usual. Key scanning is carried out by IC7501, which is also the timer/display driver chip. It appeared to be working all right since pressing the timer button produced a warning that the clock wasn't set and that there was no tape, i.e. there were beeps and the timer and tape symbols in the display flashed. On a couple of occasions we found that the machine seemed to be on when connected to the mains supply since a channel number appeared in the display it could be changed up and down with the front keys. A check was made on the serial clock and data lines between IC7501 and the system control chip IC2001 (pins 44 and 45). Both were high (5 V) with no pulses. Since IC2001 produces the serial clock signal its reset and clock inputs were checked. There was no clock input from the osc. pack CBA, a small PCB beneath the main panel. The single transistor on this PCB was at fault. It's Q6101, type 2SC2206.

Dead machine: After an initial burst at switch-on there was no output from the switch-mode power supply. Diode D1103 on the primary side of the chopper transformer was leaky.

Refused to load a tape: This G mechanism machine refused to load a tape because it had damaged gears. So out came the main cam gears (VDG0343 and VDG0346) which both had damaged teeth. The subloading cam gear (VDG0448) and arm (VXL1857) were also broken. Replacements were fitted, along with a new deck mode switch (VSS0175A), in case this was the source of the problem. In the test mode with the carriage out everything worked correctly. After refitting the carriage, however, I managed to jam the cassette across it. Instead of the machine stalling and ejecting the tape there was a raucous clattering of disintegrating nylon teeth, after which the machine sullenly powered down. Once more to the gear drawer. In these machines the mechanism is driven by the capstan motor via a simple gearing system that's engaged by a solenoid. To prevent the full capstan torque damaging the nylon gears there's a friction clutch on the bottom of the capstan rotor: it's designed to slip before the gears are damaged. In this case the clutch was the cause of the problem – it was too tight. A replacement rotor and another set of gears restored normal operation.

Very poor E-E pictures: The fault could be cleared by tapping the tuner and r.f. converter. As MCES were on holiday I removed the tuner and converter and had a look inside. I found a large number of dry-joints along the output pins in both units and very carefully resoldered them. When the units had been refitted the machine worked perfectly and didn't even need to be retuned.

Would power down intermittently: This fault had been very intermittent and didn't show up in the workshop until it was provoked. The complaint was that the machine would stop during playback or record then power off. We found that pins 14 and 15 of connector P2001 were dry-jointed. These are connections to the capstan motor. When we flexed the joints during playback the capstan motor started to make a knocking noise then stopped, after which the machine tried to unlace then powered off.

Poor review picture, damaged tapes: There was a very poor picture in the review mode while, more seriously, the machine would sometimes throw out tape into the mechanism, much to the detriment of the tape. A careful check on the tape's progress along its path failed to reveal any obstructions or resistance: all the brakes and soft brakes were working faultlessly. The culprit was eventually found to be the head drum entry guide, which had seized. As the guide didn't rotate, the tape's progress was somewhat unstable as it stuck and jumped randomly over the guide. A replacement guide, part number VXP0863, restored correct operation.

Intermittent flashes on E-E picture: As I was unable to tune in any of the local channels I first checked the tuner's BT supply, which was low. It eventually transpired that C706 (0.01 μF) was leaky, compressing the tuning band at one end and probably causing some instability from cold.

In forward search mode picture broke into lines: This machine played back and recorded all right but in the forward search mode the picture broke into lines. The cause was a worn lower drum assembly.

Machine completely dead: There was rectified mains voltage at the bridge rectifier's reservoir capacitor but no start-up ripple for the power supply because C1109 (1 μF, 400 V, 105°C) was open-circuit.

Unable to tune channels: This was because the VL and VU lines were wrong. These two signals are set by the front panel microcontroller chip IC7501, determining the tuning bands. A replacement M37422V4AF microcontroller chip produced the correct band.

Noisy rewind: This range of machines uses a very slightly improved and modified version of the G deck. On test it wasn't noticeably noisy but a slight knocking was just discernible. As it causes the problem in the earlier versions I first replaced the main pulley (VXP0917). This time, however, the cause was the intermediate gear (VDG0546) which transfers the drive from the centre pulley unit.

No vision: The fault tends to be exceptionally intermittent and affects only the r.f. output. This is the big clue, but it takes some time to establish this when the fault occurs for only 2–3 minutes a week! The exact symptoms are akin to a TV set being off tune – the output seems to move through the tuning point but won't lock. The raster is dark or affected by hum, with exaggerated chroma present and distorted sync. The cause of the trouble is the ENC17952 r.f. modulator.

Power supply wouldn't start: After a mains supply fluctuation caused by a storm the machine's power supply wouldn't start. C1109 (1 μF, 400 V), which provides a start-up pulse, had gone low in value. The 12 V supply was also missing because Q1102 was open-circuit.

Panasonic NVL25

'Write' and 'Release' displayed: This seemed to be a very strange fault at first. The customer said that the two words 'write' and 'release' would intermittently appear in the display and that when this happened nothing could be done with the machine. Unusually for an intermittent fault it showed up straight away for us, and we found that by lifting the front right corner of the machine slightly it cleared. As soon as the top cover screws had been removed, however, the fault wouldn't occur, no matter how the machine was flexed. With the top cover removed the cause of the fault was spotted straight away. A flat 14-way connecting cable goes from P7401 on the main PCB to P7501 on the display PCB. It hadn't been bent down far enough and the insulation had been cut through at connection five (serial data line) by the front of the metal top cover. This connection was thus earthed when the cover screws were fitted.

Intermittent operation: This is often a cue to resolder X6101, the main system control and servo crystal. It didn't help with the two machines on my bench, however. After a few hours the capstan motor would sometimes cog and judder to a halt. Any attempt at approaching these machines with test gear provided a complete cure until the following day. I decided to remove the main PCB and inspect it for any suspect wiring or soldering. As I came to withdraw the capstan motor plug P20011 I thought that it came out rather easily. When I turned to the other machine I found that, sure enough, the plug wasn't fully inserted. Refitting the plug provided a complete cure with both machines.

No 'power-on', timer flashed zeros: A curious set of symptoms greeted me with this machine: no 'power on' and the timer flashed zeros, but above these were the words 'write' and 'erase' – it was not something I'd ever seen before. The machine would accept a tape, but as the controls were inoperative it wouldn't return the tape. This suggested to me that the systems and servo chip IC2001 was probably all right. I'd also no good reason to suspect the timer chip IC7501. Voltage checks between these two chips showed that the serial data and serial clock lines were the source of the trouble. The serial data line was sitting at about 1 V while the serial clock line, at around 4 V, was closer to the normal condition for these lines. Disconnecting IC7501's serial data pin removed the 'write erase' from the display, but the line's voltage remained low. Disconnecting the other two chips connected to the serial data line, IC2001 and IC6801, also had no effect – the voltage remained low. The only other possibility was the 270 pF capacitor C6012, which is connected between the line and chassis. When I eventually found it (it's a small, surface-mounted capacitor tucked away at the edge of the main board), it turned out to be leaky. A normal-sized replacement restored correct operation.

Sullen refusal to power up: A sullen refusal to power up and the legend 'write erase' displayed above the flashing timer display were new symptoms to me. The machine would accept a cassette but wouldn't return it. In fact the only control that operated was the timer on/off button. This persuaded me that the systems and servo chip IC200I and the operation and timer chip IC7501 were working. Checks on the serial data line showed that the amplitude of the serial data was low at about 1 V. Even with all the serial data ports disconnected and only a pull-up resistor and a 270 pF decoupling capacitor left in circuit the data line sat at about 1.5 V. Sure enough the capacitor (C6012) which is a surface-mounted type was leaky, a replacement providing a complete cure.

Very intermittent servo lock: The machine would play all right for hours then, suddenly, the capstan motor would rapidly speed up, causing sound distortion. At the same time the drum speed would go way off lock. The result on the picture being like loss of line hold. The fault would last for about 10 seconds after which everything returned to normal as suddenly as the fault had appeared. After much head scratching and component changing we found that the cause of the fault was the STK5392 regulator chip in the power supply. When the fault condition was present the regulated 5 V rail rose to 6.2 V and became unsmoothed with h.f. pulses on it. Because of the very intermittent nature of the fault it took several days of testing and probing to find the cause.

No results: Because of its cause the fault had been present for some time, unnoticed. C1109 (1 μF) was open-circuit. As long as the machine remained plugged into the mains supply it was all right. When the mains supply was disconnected then reconnected the power supply wouldn't start up.

Power supply problems: D1113 on the secondary side of the supply was short-circuit. This device is present to protect the machine from excessive supply voltages. If the supply lines rise above a certain level it goes short-circuit and brings the excess-current trip into operation until the power supply is reset. Unfortunately when D1113 had been replaced there was still no power – until C1109 (1 μF, 400 V) in the start-up circuit was replaced. But there was still a problem: C1114 (47 μF, 63 V) was also open-circuit, and D1113 again blew. When D1113 and C1114 had been replaced the machine worked normally.

Intermittent operation: After a few hours' use the mechanism would just shuffle, cut out, then power down. A replacement deck mode switch, part number VSS0175A, cured the problem.

Counter stops in forward search: On inspection we found that the tape was riding up the audio/control head. As the pinch roller was worn we fitted a replacement, but this made little difference. Next comes post P5, which will produce the condition even if only very slightly bent. But it was OK. Attention was next turned to the AC head, which was worn. Replacing this and realigning the machine cured the fault. The upper drum was also worn. The customer accepted a further estimate, and once a new upper drum had been fitted the machine worked faultlessly.

Wow on sound and picture flicker: Our first steps were to replace the pinch roller and stator unit, but the fault persisted. It was cured by replacing the XRA6435S capstan drive chip IC2101.

Mechanical and tuner faults: First, this machine required a new carriage RHS, connection gear and retiming of the gears. Secondly, the booster circuit in the tuner block had low gain. We usually have to fit a new tuner, but these are expensive. On inspection, however, we found that there were several dry-joints in the booster section. Resoldering cured the fault.

Would not go into standby: When this machine was put into the standby mode the vision remained and the red LED stayed alight! The cause of these interesting symptoms was a splash of solder that shorted the cathode of D1002 to an adjacent wire link. As a result the switch line to the power supply didn't change state.

Panasonic NVL28

Intermittent operation: This one had three reported faults. There was intermittent operation of the functions via remote control, the VTR power button mounted on the fold-down control panel didn't always work, and lastly the most significant symptom – when the control panel was in the upright position the tracking shifted off to one end of its range. All these problems were caused by one fault. The control matrix PCB is connected to the machine via a flat, 15-way flexible lead that plugs into both the front PCB and the control PCB. The cause of the fault was a very small conductive strand, which bridged between connections 14 and 15 of the FPC at the control panel end. As a result tracking minus ('Set Down' in the manual) was selected momentarily, or all the time with the control panel closed up. If any of these keys is held down no other key will be detected, including the VTR power switch, and it will also lock out the remote control input.

No digital functions: If picture-in-picture was selected, followed by the swap function, in the playback mode the machine would toggle between tape playback and E-E but without the inset screen. We removed the digital box and its covers and made checks around the digital microcomputer IC9007 but couldn't find anything wrong. We then noticed a couple of VCO/phase comparator chips, IC9003 and IC9004 (type MN6790), which are controlled by the microcomputer chip. They both receive line-frequency inputs, which are used to lock an internal 21.4 MHz VCO. A divided-by-two output derived from this oscillator appears at pin 8. When a digital function is selected, pins 9 and 7 are taken low. This should switch on the oscillators. The output from IC9003 was fine, but only 5 V appeared at pin 8 of IC9004. Replacing this chip restored full operation.

Intermittent playback of a mixture of E-E and off-tape pictures: This was not something I've seen before, but this machine sure did it. Checks on the video and E-E switching showed that there were no problems here, and I couldn't see any other likely cause of the fault. But this machine has an extra digital pack for picture-in-picture effects. This was where the cause of the problem lay. Two wires in this pack were shorting intermittently. Trimming them provided a complete cure.

No colour with pre-recorded tape: This machine would record and then play back perfectly but was unable to produce colour with a pre-recorded tape. Further checks showed that 'super still' was also very poor with pre-recorded tapes. My first checks were around the video head and drum assembly to make sure that the head hadn't been fitted out-of-phase – this can cause no chroma playback and still frame problems. But there was nothing amiss here. A look at the servo and colour circuitry suggested that the MN6740VCJK systems and servo main processor chip IC2001 might be responsible for the trouble as it feeds rotary signals to the colour circuits and many other signals to the slow and still servo circuits. Fortunately a replacement cured the fault. I then had colour and perfect slow and still when playing back pre-recorded tapes but any tape with copy guard on it caused pulling at the top of the screen. As a cure for this Panasonic recommend fitting a 22 μF non-polarized or tantalum capacitor across C9568 in the digital pack.

Dead: There was no power supply operation. Checks on the primary side of the power supply revealed that C1109 (1 μF, 400 V) was open-circuit.

Panasonic NVSD200 (K Deck)

Shuts down while accepting or ejecting a tape: A check on the built-in error codes told us that the mechanism was jamming whilst loading or unloading. As the tapes used for testing were known good ones, we turned our attention to the loading motor drive and the gearing from the loading motor. Checks at the loading motor connection PCB, which is on the back of the loading motor, revealed cold soldering at all four connections. After resoldering these the machine worked faultlessly.

Played back tapes at cue forward speed: The cause of the trouble was the BA6871S capstan motor drive chip IC2501.

Very low E-E audio: I found that one end of C0729 in the VIF pack was dry-jointed – it's the detector audio output coupling capacitor. Interesting that the other end had clearly been soldered manually!

Intermittently shows F04, F05 or F06: Check for dry-joints on the loading motor. In this particular case, however, the cause of the problem was a deformed contact on the mode switch plug-in connector, which stands up off the main board.

Panasonic NVSD25

No E-E picture: Checks showed that the 12.3 V supply was low at about 7 V. When the power supply was disconnected from the main board the voltage returned to normal. We eventually found that the UN2211 transistor QR1001, which buffers the power on/off switching, was leaky. A replacement transistor cured the trouble.

Poor recorded sound and slow wind/rewind: We decided to concentrate on the sound fault initially. There was some bottom edge damage with the tapes supplied: they seemed to play all right in the workshop, but when the tape was cued forwards it would run down the guide next to the audio/control head. We first suspected that arm P5 was slightly bent, as can happen with the K mechanism. But a replacement made no difference. We then tried AC head tilt adjustment, again to no avail. The cause of the trouble was eventually traced to excessive back tension. We found that the supply spool brake was permanently engaged because the brake lever assembly (VXZ0313) had a broken lever. After replacing this the machine worked perfectly, even in the wind and rewind modes. The worrying thing about this fault was that the tape remains threaded around the drum in the wind and rewind modes. With the tape tension so dramatically increased, how much wear had been imposed on the drum while the fault was present?

Cue function failure: This machine played back all right but any attempt to cue forwards or backwards would result in loss of line lock as the machine tried to default to the NTSC mode. A number of defects, e.g. worn video heads, poor tape path alignment, incorrect back tension etc., can cause this problem. In this case the cause turned out to be the capstan motor's top bearing, which was almost seized because of a build-up of dirt. Once the capstan spindle and bearing had been cleaned and lubricated line lock was maintained in all modes.

Panasonic NVSD30

Would lose capstan motor control: This machine would lose control of its capstan motor, frequently playing back as though in the cue mode. Suspecting that the main systems and servo chip IC6001 was at fault, we fitted a replacement. This of course made no difference. The culprit turned out to be the XRA6439P capstan drive chip.

Would fail to go into fast rewind: The complaint with this machine was that when certain pre-recorded tapes were rewound it would fail to go into fast rewind. This is a quirk of the machine, recognized by the manufacturer. The cure is to change the microcontroller chip to type MN67434VRSH.

Would accept tape, then eject it: Loss of capstan drive is the usual cause of this situation, but not on this occasion – the capstan motor rotated as the tape was being ejected. The capstan stator (part number VEK4097) was eventually found to be the cause. Presumably the FGs or PGs were confusing the systems control chip.

Intermittent tape damage: There was also a clicking noise between modes. On close examination I found that the slide cam was damaged. As a precaution I replaced the mode switch as well.

Problem in cue mode: Any attempt to cue forward resulted in loss of line lock because the drum speed changed incorrectly. Checks in the system control circuit showed that IC6001 was changing to the NTSC default condition, despite this machine not being fully equipped for NTSC playback. The cause of the problem was poor head-to-tape contact. Further checks showed that the performance in the review mode was very poor, the top half of the picture being covered with noise. A badly worn and very shiny pressure roller confirmed the diagnosis. The drum was badly worn: because of the relative newness of the machine this was something we hadn't considered initially.

Tape chewing: With the machine in the playback mode we saw that the tape was being chewed at its bottom edge. The back tension was very high, while the supply reel was very stiff. A careful deck inspection showed that the end of the take-up brake arm unit was broken. A replacement restored normal operation.

Panasonic NVSD40

Would accept a tape then power down: This machine would accept a tape then keep it: lapsing into sullen silence, it would power down then after a few minutes the power supply would cut out. The machine was brought back to life when I connected a battery to the loading motor to extract the tape. Checks in the loading motor drive circuit showed that the BA6219B chip IC6501 was badly overheating – presumably it was making the power supply cut out. Fitting a replacement cured the fault.

Would not retain cassette: Everything seemed to be normal when the tape was being loaded, but after a pause it was ejected. My first checks were around the mode switch and the systems circuits, but nothing seemed to be amiss. After much hair tearing and grinding of teeth I discovered that the BA6439P capstan drive chip was faulty. Presumably the system control section checks for capstan operation before lacing up, to prevent tape damage.

Lines across screen in play mode: It looked as though the loading arms were misaligned but inspection in this area showed that there was a circlip stuck in the way of one loading arm. When it was removed the machine worked all right. It didn't take long to discover where the circlip had come from and fit a new one.

Horizontal swaying on playback: Voltages were measured, oscillograms were examined and hair was torn from heads! IC6001 was replaced, then IC2505. Capacitors were checked. Eventually we found that the cause of the trouble was the regulator transistor Q2505, which supplies 12 V to the capstan stator: it seemed to have some sort of internal leak. A new 2SD601 cured the fault.

Would eject tapes when inserted: On investigation I found that the capstan had seized: because 'thrust screw UMT' was loose, the capstan flywheel was rubbing against the motor PCB. The problem was cured by adjusting this screw (it's the large white plastic screw on top of the capstan spindle).

Wouldn't accept a cassette: Replacing the BA6219B loading motor drive chip seemed to cure the fault, but the machine failed again on the soak test bench. Eventually we found that the cause of the problem was a dry-joint at one of the loading motor pins: it sparked intermittently, blowing the drive chip.

Wouldn't record: Playback of pre-recorded tapes was fine, but playback of one of its own 'recordings' produced just snow. Modulated r.f. reached the head amplifier, so the problem had to be within the head amplifier's screening can. We found that several of IC501's pins were dry-jointed. Resoldering them provided a complete cure.

Intermittent failure to record: At very rare and erratic intervals this machine would fail to record the picture. The effect on playback was a screen full of snow, with the sound continuing normally. Surface-mounted transistor Q3007 turned out to be the culprit: it had an intermittently open-circuit base-emitter junction. Its job is to switch the operating voltage to the video record amplifier.

Wouldn't accept tapes: On checking it we were perhaps lucky to find the cause of the problem first time: there were dry-joints at plug/socket P1506. It's connected to Q1503 (2SB941) which is mounted on the diecast chassis.

Tuning drift on one or all channels: Alternatively there could be a flashing or flickering picture that varied when the tuner/i.f. panel was tapped. We eventually cured the fault by removing the 0.1 μF chip capacitor C706, cleaning the paint mark off the PCB and then refitting it.

Poor playback/visual search: When we checked the machine in the playback mode the performance was very poor. A further check showed that the upper and lower drum assemblies were both badly worn. There were just lines in the forward search mode, as if the scanning wasn't locked. The cause of all this was excessive back tension because the take-up brake arm unit (part number VXZ0313) was broken. To restore good playback we had to replace this item and the complete drum assembly.

Low gain tuner/i.f. pack: This VCR was brought in because the tuner/ booster pack produced a low-gain E-E output. This is not unusual. After fitting a replacement, however, the machine wouldn't tune in any stations. A check on the tuning voltage showed that the sweep was normal, so we assumed that the new tuner was faulty. It wasn't. This model has a band-switching circuit, which is controlled by IC6710 on the syscon board. It should supply a low output to the band-switching chip, but the output was high. Replacing IC6710 restored correct operation. A quick check for this is to short the relevant pin of IC6710 to chassis and see whether you can then tune in stations.

Rejected video cassettes: This machine rejected any video cassette it was offered. When it took the cassette in it would pause briefly then throw the cassette back out. This is a classic result of no capstan motor operation. But in this case the capstan motor was going too fast, as there was no 12 V supply to the capstan motor's stator. The 2SD601 voltage regulator transistor Q2505 was found to be open-circuit.

No E-E or playback sound from modulator: The sound was OK at the AV connector. We found that the 1 kΩ chip resistor R7005 was open-circuit. It couples the sound signal to pin 4 of the r.f. converter.

Loss of E-E and playback signals: This machine had been to another 'repairer'. When it came to us the fault was loss of the E-E and playback signals. The TV demodulator pack is connected to the main PCB by several plugs and sockets, PS701, PS702 etc. Their contacts can break internally if care is not taken when the panel is removed. Replacing PS701 and resoldering the other sockets where they are connected to the main PCB restored normal operation.

Machine wouldn't record: Although it entered the appropriate mode, the expected recording didn't take place. A previous recording would not be affected, which proved that the full erase circuitry wasn't being activated either. Some basic checks in the record-switching circuitry revealed that the 2SB710 surface-mounted transistor Q3007 was faulty, a replacement curing the trouble.

Panasonic NVV8000

Looked like dirty heads: However, no amount of cleaning would restore the picture. Because of the price of the heads for these machines I checked for life around the head amplifier module, where the 5 V supply was found to be missing because of a loose plug on the chroma/luminance board. Refitting P3001 restored a perfect picture.

Playback and E-E pictures distorted: This impressive looking machine produced a less than impressive picture. This occurred with both S-VHS and normal VHS operation. With this model and its lower specified relatives the NVFS100 and NVFS90 you tend to get capacitor trouble in the small pack that houses the CCD delay line. The CCD delay pack is on the YC separation board in the NVV8000. An excellent picture was obtained when C3506/7/8 had been replaced. They are all 3.3 μF capacitors rated at 16 V.

Mechanism fault: Unfortunately these machines have a dual VHS-C and standard VHS cassette loading mechanism, which means that the cassette carriage assembly is fairly complicated. It would jam – it seemed to stop between the two modes, after which the machine would power down. The main mechanism within the machine stayed in the VHS-C mode. The cassette carriage was removed after a long struggle. I then realigned the many gears and levers. After that the machine worked consistently. I was unable to find any damaged gears or bent levers that could have been the cause of the misalignment, and was just grateful that the machine now worked.

Philips

Philips DMP series decks

Servicing tips: When taking the top plate from the mechanism be careful as you remove the erase plug: if you don't release the clips first you can easily break off part of the erase head mounting. This happens to be where the 180° roller adjusting screw operates, the result being that the tape path goes way off adjustment. Usually the engineer glues the erase head back on but leaves the L-shaped lug off, thinking that it has no effect – it comes into use only when threaded up . . . To do the job properly a new scanner ring is required.

Changes during production: When ordering parts, look at the paper label on the inside left-hand side of the metal chassis – the label is easier to read with the tray in the lowered position. Note the type number (DMP 2–2) and the week number (WD . . .) and check in supplement 4822 726 14564 whether the part you want has been changed – some have been modified twice. The IDM series deck is similar to the DMP type in appearance but many of the lift and threading parts differ and are not interchangeable. There's a different manual for this deck.

Take-up spool would cease to rotate: On going from rewind search to playback the take-up spool would sometimes cease to rotate and the machine would then stop. When play is selected the brake magnet normally holds the brakes off, but sometimes the magnet let go and the fault would occur. The cause of the problem was a dry-joint on the brake electromagnet.

Damage to lift guides: A lift guide repair kit for the Charlie range of decks is now available from Konig (through Wiltsgrove), the part number being VID1534. Damage to the lift guides previously meant either a replacement chassis or a bodge with Araldite, but this kit enables a satisfactory and neat repair to be carried out in only a few minutes.

Cassette ejected whatever deck function was selected: We had two machines with this symptom. In both cases the capstan motor had seized due to a build-up of sticky 'gunge' on the capstan shaft in the upper bearing. Dismantling and cleaning provides a cure. Some older DMP/IDM series decks are developing cracks in the top rails of the plastic racks that guide the cassette lift on its way in and out of the machine. If they actually break you have to fit a very expensive half-chassis subassembly (but see the fault report above about the Konig repair kit). To guard against having to do this we run a layer of hot-met glue along the top surface of any cracked racks we find, forcing it tight into the angle between the rack moulding and the metal wall to which it is fixed.

Problems of pinch roller mechanism: A common failure with these decks is the rather unusual pinch roller assembly. The rubber roller seems to decompose rather alarmingly. Symptoms vary – poor or varying sound or tracking, for example, or tape damage.

Philips Turbodeck VCRs

Poor tracking or slow auto-tracking: Check that the control head is clean. These and other half-loading decks seem to be prone to dirt build-up at this point.

The new style head doesn't fit: There have been reports that in VCRs which use this deck the new style head doesn't fit the lower drum spindle. In severe cases, dealers have been replacing the lower drum as well. Here's a tip that may save some money. When the old head has been removed, try using some rubbing compound on the lower drum spindle. Duraglit or some other metal polish may even work. It's possible that the cause of the problem is slight corrosion because of migration from the old head. If you are careful, the new head should now slide on to the shaft easily, using the special tool.

Dealing with broken plastic lugs: Plastic lugs that break off have been a difficult problem to deal with – until now! The lugs that break hold down the record-protection switch spring, pivot the brake pulse lever, and pivot the brake pulse roller. You no longer have to practise your modelling skills with Araldite. Just order kit part number 4822 256 92316 and you get the parts to repair all three trouble spots.

Philips VR202

Chopper transistor short-circuit: This machine was dead, with the BUT11AF chopper transistor short-circuit and the $3.3\,\Omega$ surge limiter resistor open-circuit – the fuse was intact. Nothing unusual here? But the replacements again blew, which was unusual. I should have checked the other transistors, shouldn't I? One of the drivers, 7126, was short-circuit. In went a replacement, along with another BUT11AF etc., and the machine was then powered via the variac. It worked and produced the correct voltages, but the BUT11AF was getting very hot and obviously wasn't going to last long. A scope check showed that the drive waveform was low while the frequency was much too high considering that there was no load. Problems like this are usually cured by replacing all five transistors in the chopper circuit – but not this time! Eventually C2127 ($330\,nF$) was found to be open-circuit. Incidentally, don't worry if you cannot get the BUT11AF's drive waveform to look like the illustration in the manual: it's the correct size but is drawn upside down! Press the invert button if your scope has one.

Microcontroller problem: The deck carried out the command when play was selected but there was no picture or sound and no other deck commands would be accepted until the mains input was interrupted. It seemed that the microcontroller chip was crashing during its programme: anyway a new P8052AH JSTD1–1U solved the problem.

Two vertical bars superimposed on picture: This machine worked perfectly in the play mode but on E-to-E or record two vertical bars were superimposed on the picture. These bars were not present at the output from the i.f. strip but were present at the output from the 4053 video switching chip IC7550. When we removed IC7550 we found that the /PBV (playback switching) line was pulsing in the E-to-E and record modes. The SAD1009 chip was faulty.

Controls would lock up: When any key on the control panel was used, the machine would lock up totally and the command that was entered would be totally ignored. On investigation we found that the main data line became continuously busy. After trying the obvious chips we moved to the on-screen display IC. When this was replaced the machine was back in full working order.

Philips VR231

Monochrome picture with field jitter: When a known good tape was inserted and play was selected the display consisted of a monochrome picture with field jitter. If forward search was tried the fault cleared and the display remained OK when you went back to play. Scope checks around the LA7191 luminance/chroma chip IC7051 showed that the video signal was being corrupted by the CCD delay line chip IC7504. The video input at pin 6 was all right but the output at pin 4 was 'chopped up'. The CCD clocking signal at pin 7 was similarly chopped up. It comes from IC7051, where the VXO crystal 1601 wasn't producing a clean oscillation. A new crystal solved the problem.

Machine dead: If the power supply is dead and the start-up voltage for the control chip is low, check whether diode 6115 is leaky. The type fitted in this position depends on the model. If it's a UG06B, the part number is 4822 130 83307.

Poor picture quality: While the cause of this can be worn video heads, there are other possibilities. Here are a few things to check first: (1) check that the tape guides, all heads and the lower drum knife-edge are clean; (2) check that the back tension is correct; (3) check that nothing in the tape path causes tape creasing – check especially the fixed guide between the ACE head and the pinch roller; (4) check that the tape moves across the CTL head correctly and is not riding up or down. Test by selecting reverse search then play. If in doubt, scope the CTL pulses. If the power supply is the type that uses an SPH4690 i.c., add a 33 μH coil (part number. 4822 157 53006) in series with inductor 5204. This should reduce the dots in the dark areas of the picture, and help the auto-tracking system to find the off-tape f.m. among the power supply hash. Watch out for poor electrolytic capacitors as these machines age. Put the machine in the service mode (method varies with the make and model) and check the X position of the ACE head as laid down in the manual. If the drum has already been replaced, use the mylar shims supplied with the new head as a feeler gauge to check that the upper drum to lower drum gap is correct. When replacing the drum, beware if the old one is tight on the shaft. If the securing screw has been overtightened, the motor shaft can be scored with the result that the new head jams as it is fitted. Make sure that the shaft is clean and undamaged. If there is slight damage, careful application of fine abrasive paper may remove the scoring. Otherwise a new lower drum/ motor will be needed. Later video heads have a triangular circlip at the top and bottom of the drum instead of the single-screw fixing. A special tool (part number. 4822 395 90977) is required to release this type of drum. If the tracking is OK at the top and bottom of the picture but the FM dips at the centre of the screen, the lower drum is probably worn. As long as the machine is picking up the CTL pulses all right, the auto-tracking should search three times then go into play in the best position it has found. If the off-tape FM is low or the CTL pulses are intermittent, the tracking comes to the wrong conclusion. If it goes past the best position, pressing play when the best place is found during auto-tracking will result in play continuing at that tracking value. Remember that these VCRs use narrow video heads that are optimized for LP use, not SP.

Philips VR323 (Charlie Deck)

Mechanism was jammed: When the jam had been rectified we found that the machine still wouldn't work because IC40 (SAA1310) was short-circuit internally.

Scrambled text: The cause was the SAA5231 teletext processing chip.

Tape damage: Because the tape rides up or down the capstan damage is caused. This is a routine fault with the older Charlie range of VCRs.

The cause is often a defective pinch roller, but lately I've had several cases where this has not been the cause. The take-up torque has been excessive because the coupling (item 214 in the exploded view of the deck) is faulty. This item and the pinch roller are included in service kit 4822 310 31803. Take care with the back-tension band when fitting a service kit to a Charlie deck. Make sure that it is clipped into its holder. I've come across cases where this has not been done. The resultant low back tension produces spotty playback pictures with the tape running up or down the capstan.

Philips VR422

Tape could be loaded, nothing else: This was one of the few faults we've seen on the TurboDrive deck. A tape could be loaded and would go to the cassette-down position but nothing else would happen. What should happen is that the tape should go into the subload position, i.e. the tape is taken up to the audio/control head. Curiously, if the carnage was loaded without a tape the machine did go into the subload position when the cassette-down position was reached. After first, incorrectly, looking for a brake fault we soon spotted the defective part, There's a V and a notch in the outer cam track of the main cam wheel on the upper side of the deck. When the subloading movement arm reaches this point the subloading arms are taken up to their positions. What had happened was that the notch had sheared slightly. Thus when there was some tension on the arms the movement arm ran straight past the notch. A replacement cam wheel, which is supplied in a kit, cleared the problem.

Intermittent mistracking and generally very poor playback: As there were no obvious mechanical faults and the deck had been realigned a few times we suspected an electronic fault. A scope check on the drum PG/FG pulses at the head amplifier showed, when a comparison was made with the oscillogram, that there was no synchronizing pulse. The oscillogram shows this as a gap in the drum pulse waveform. After replacing the TDA5140 drum motor driver chip IC7301 and carrying out slight realignment of the deck we obtained an excellent picture.

No vision playback, sound OK: Checks with the scope showed that the video was being lost in IC7501 (LA7391A). The off-tape f.m. entered at pin 39, came out at pin 3, re-entered at pin 4 but then got lost internally. As voltage checks on the supplies and the record/playback switching didn't come up with anything we fitted a new LA7391A chip. This solved the problem.

No picture or sound: The mechanism was OK but there was no picture or sound (just a blank raster) in any mode. While tracing the video path back from the modulator, I soon discovered that the +12a supply was missing. Safety resistors R3151 and R3147 (both 6.8 Ω, part number 4822 050 26808) were open-circuit. No reason for their failure could be found, and the replacements didn't fail during a soak test.

Philips VR502

Intermittent remote control: Check for dry-joints on the infrared receiver chip IC121.

Loss of signal: After running for about 15 minutes this machine lost all TV programmes – it was as if the aerial lead had been pulled out. We found that the voltage at pin 11 of the tuner was being pulled down from within the tuner. A suitable voltage fed to pin 11 from a bench power supply showed that 22 mA was being sunk inside the tuner. A new tuner restored normal operation.

Intermittent capstan rotation: This machine had an intermittent fault: the capstan would occasionally start to rotate uncontrollably. If this happened when a tape was threaded, it would be moved although the drum didn't rotate and the rest of the circuitry would be shut down. If it happened without a cassette the lift mechanism would lock and the capstan clutch would slip. The fault condition would last for 5 minutes or so after which it would clear and the machine would be fine for days. Board swapping enabled me to prove that the cause of the fault was on the family board. With the machine in the service mode while the fault was present I found that the microcontroller chip read the deck state correctly but seemed unable to control the capstan motor. New tape servo and interfacing chips made no difference. Two signals, CAP and CREV, control the tape servo chip. Resistor/capacitor integrating networks are used to DA convert these signals. C2316, a 22 nF surface-mounted capacitor, was found to be intermittently leaky – it read 6 kΩ when faulty.

Philips VR522

Nasty noise from heads when a tape was loaded: I stopped the machine and inspected the heads. There was a very nasty groove in the upper drum. Wondering what could have caused such damage, I looked at the rest of the mechanism and noticed that the left guide arm was at a peculiar angle, pointing into the drum. It locked back into position with a push. I then saw that the impedance roller was loose. After tracing the holding spring and rebuilding the assembly the machine worked normally – except for the need for new heads as they were by now totally useless. I assume that the damage had been done by someone using extreme force to extract a jammed tape.

Faulty f.m. sound in E-E mode: Nicam sound was perfect but there was faulty 6 MHz f.m. sound in the E-E mode. White noise was all that could be heard when 6 MHz was selected. Demodulation of the 6 MHz sound is carried out by the TDA3867T chip fC7820 on the front-end PCB. A replacement chip put matters right.

Machine was jammed: A field engineer had removed the tape, only to find that the entry guide had parted company with the load arm. Refitting was all that was necessary – a quick and easy repair for a change.

Philips VR6180

Sound would cut out intermittently: This would happen after about half an hour's play. If you left it playing the machine would eventually stop and eject the tape. A dose of freezer and hairdryer heat isolated the fault to IC7140 (type MUP8051H1–2-D3).

Intermittently failed to accept tape: If the machine was put into standby before the cassette was tried the display would go bright, showing that the cassette-in switch was being sensed, but the tray wouldn't move. No supply voltage reached the control motor as there was a dry-joint on plug B2.

Colour crosstalk on playback: I hit a major snag when I looked at the manual – it had the 5 V signal panel and I didn't have the supplement for this. Anyway I decided to start by changing the delay line 5102, which turned out to be the cause of the trouble. Rock on!

Timer fault: It's rare to come across a VCR with a timer fault that's not due to pilot error. This model is an exception to the rule. If the customer complains of shutting off in the timer mode, the VCR going dead intermittently with the display going haywire, the cassette being spontaneously ejected etc. and the error memory is empty suspect a power supply fault. Before starting with the hairdryer and freezer, disconnect the output plugs P1 and P2. Connect a 47 Ω resistor across the 5 V output (pins 7–8 of socket P1) and monitor the 5 V line – the reading should be 5.2 V! – while heating and cooling the power supply. Likely causes of a varying output voltage are: the BZX79/B5V1 zener diode 6012, the TCDT1101 optocoupler 7103, and transistors 7001 (BC547B) or 7004 (BC548B).

No cassette lift operation: There was no cassette lift operation. When a cassette was inserted the system controller sensed the switch but the motor wasn't energized. The supply to the control motor was missing as R3002 and R3003 were open-circuit.

Cassette ejected at play: The cause of this was that the head drum didn't spin fast enough. When it was rotated by hand it seemed to be slightly stiff and a faint scraping noise could be heard. Someone had disturbed the drum and hadn't used the mylar spacers to set the gap between its upper and lower sections. Normal operation was restored when this was done. New Philips video heads come with a fitting kit that consists of the following items. (1) A plastic holder so that the head can be fitted without getting finger marks on it. This holder has a plastic rod that locates in the lower drum to align the head position. (2) A plastic spring-loaded rivet that fits through a hole to set the drum motor position with respect to the lower drum. (3) Two transparent mylar shims to set the gap between the upper and lower drums. The fitting kit can be removed when the head fixing screw is secure. Keep it in the workshop so that you can swap heads between machines if necessary.

Excessive wow and flutter: A word of warning with this one. The problem was excessive wow and flutter because of a worn capstan bearing, which is not uncommon. But it took us a long time to get the correct replacement because the mechanism is not the same as that referred to in the Willow Vale catalogue. I believe it's called the DMP4: the capstan assembly 262 is neither the 4822 520 10635 nor the 4822 520 10559 but the 4822 535 92909, which Willow Vale supply under code 164467CP although it's not listed in the catalogue. The problem is that this mechanism has the capstan FG head mounted on the exterior of the flywheel, so the magnetic area has to be on the outside. The other types supplied by Willow Vale have the magnetic area on the inner edge of the flywheel, the result being that there are no FG pulses when they are fitted in this machine. Thus the cassette is ejected whichever deck function is selected.

Changes to DMP2 and DMP3 decks: A number of changes were made to later versions of these decks, which have the capstan tacho head on the outside of the flywheel. The following parts were changed: top plate item 255, part number 4822 466 82467; scanner ring item 227, part number 4822 532 11776; erase head item 247, part number 4822 249 40252; capstan item 262, part number 4822 535 92909; tacho head and PCB, part number 4822 214 32587; threading motor item 252, part number 4822 361 21242; crank item 259, part number 4822 528 20593; control lever item 272, part number 4822 403 53744; in addition a washer and spring were added under item 256, washer part number 4822 532 11775, spring part number 4822 492 52095.

No playback f.m., just noise: The cause of this was the fact that the lower drum earthing bracket's securing screw was very loose. When this was tightened there was a nice clean picture but no colour. Recordings were fine when played back on another machine. I found that the f.m. from the head amplifier was about 50 per cent down and rather noisy: this was triggering the colour killer, though the luminance seemed to be perfectly all right. After much searching around I found the cause of the problem. Someone had soldered the leads from pins 1 and 2 of L6 to the print side of the PCB. Fitting them to the connector itself restored the colour. Perhaps a previous attempt to cure the loss of f.m. had gone awry?

Philips VR6185

E-E signal would fade leaving snow: After a few minutes' use the E-E signal would fade off, leaving just snow on the screen. A scope check on the frequency divider signal at pin 13 of the tuner showed that it gradually got smaller and smaller until it reached 0.2V, when the picture went off. The U744 tuner was faulty.

Would accept cassette, would change channel in E-E mode but no sound: With this one the keyboard and the syscon didn't talk to one another! The machine would take in a cassette and would change channel in the E-E mode, but there was no sound. When a deck function was tried the display showed that the command had been received but nothing happened. If the operation board was pressed in the right place the fault cleared. On inspecting the print we noticed that there was a crack by the side of the infrared receiver can.

Tape stuck in machine: Instead of calling me first, the customer had taken the top off and used a knife to get the cassette out. When I looked at the machine the lift was lying loose inside together with the remains of lever arm 238. According to the customer there had been intermittent picture rolling for some time, then the machine had continuously ejected tapes. The problem tape got stuck when he had physically blocked the cassette opening by holding the cassette down while the machine was trying to eject it. In spite of all this mistreatment, the rest of the mechanism seemed to be in reasonable order. The cause of the rolling picture was a disintegrated pinch roller. After splitting the deck and fitting a replacement and a new lever arm, then reassembling everything, I found that the deck seemed to work when run with a d.c. supply to the loading motor. But when a tape was tried it was immediately ejected. The service mode indicated that there were no capstan pulses. As the capstan was turning, I decided to check the sensor. Once again the deck was split, and when the sensor panel was removed one of the sensor's leads was seen to be adrift. On putting this right and reassembling the deck I found that the machine now worked mechanically. My problems were not over, however, as the customer had fiddled with the sync head. In fact he'd tightened the screws to the extent that the base of the head was bent. I had to fit a new head and set it up. So the original faults had been a worn pinch roller and a lead off the capstan sensor. What should have been a half-hour job had taken almost 3 hours.

Intermittent jamming: The symptom was very intermittent jamming when laced up in the play/record modes. After much investigation I found that a pivot lever (Philips ref. 260) hidden under the threading ring at the rear of the deck was worn/broken. Replacement with one from a scrap machine provided a cure.

Cassette stuck in machine: This machine came in with a cassette loaded but the tape not threaded up, the complaint being that the cassette couldn't be ejected and the machine would shut down. It seemed to initialize when reconnected to the mains supply, then the cassette ejected all right. I thought that the deck microcontroller had become corrupted, but the machine then again failed to eject. The problem was that the fault would occur only every so often, while in all other respects the machine worked perfectly. The service mode suggested that the threading mechanism might be too heavy. As I've had threading motor failure quite often with these machines I fitted a replacement. Unfortunately this made no difference. I eventually cured the fault by replacing the L293 loading motor driver chip IC7001, after discovering that pins 11 and 14 were sometimes both at 4.5 V in the eject mode (alternate pins should go low during threading or unthreading). Oddly, the cassette loading was never affected.

Philips VR6290 (Charlie Deck)

Would get stuck in pause: If you tried to go to 'stop' after this machine had been playing for a few hours it would get stuck in pause and couldn't be restarted, stopped or put into standby. Scope checks around the keyboard processor chip showed that the SDA2 and SCL2 signals changed from blocks of data to continuous signals. Replacing the keyboard module as an initial check made no difference. The SDA2 and SCL2 lines communicate with the tuner, so a new one was fitted. This cured the problem. It seemed an unlikely cause, but fitting the suspect tuner in a working machine took the fault with it.

Half accepted tape: This machine half accepted a tape then the carriage made a squealing noise and ejected it. Inspection of the carriage's operation with the top cover removed showed that the carriage attempted to go in all the way but the cassette flap didn't open and thus fouled on the left-hand guide roller. The cause of this was that the tape flap opener spring had become displaced. Simply relocating the spring into the opener cured the problem.

Cassette stuck inside: When I switched the machine on the tape wouldn't eject, and if any function was selected the clock display flashed on and off. I found that the power supply was pulsing because the CNX83A optocoupler IC7124 was faulty. A new CNX83A restored normal operation.

Philips VR6362

Broken cassette lift: This machine came in with a broken cassette lift and the complaint of no rewind or reverse picture search. In fact what happened was that when the VCR entered the rewind mode the tape tightened and was then unloaded. The drive coupling, item 214, was at fault.

Erase during playback: The symptom with this machine looked like faulty video heads – until I tried the tape in another machine and found that it had been erased! The +11.8d record supply was present all the time as transistor T7701 (BC328) was short-circuit.

Would not accept a cassette: This machine would switch on and off but wouldn't accept a cassette. When a 9 V battery was connected across the loading motor the mechanics would thread in and out correctly, thus exonerating the eject rack and differential gear assembly. The fault was being caused by the load motor driver chip which had a short-circuit output pin – the pin that drives the load motor in the threading in direction.

Playback drop-outs: It looked just like a worn head, but there was no improvement with a replacement deck. We found that the fault was on the luminance/chroma panel P306, where the dropout offset control (3304) was broken. This can happen if you forget to hinge up panel P306 when removing the front control panel, as the IR receiver can hit it . . .

CMOS RAM battery flat, no memory: One of the dealers we do work for regularly sends us repairs with no indication as to the fault. Luckily with this range of machines a look at the error memory usually gives a clue. Not this time, however – the CMOS RAM battery was flat! The machine played all right but when I tried to tune in a station the picture was overlaid with diagonal black-and-white bars (like a monochrome barber's pole). A panel swap proved that the fault was on the signals panel P306. We found that the play supply switching transistor T7304 was short-circuit.

Problems with the clock display: The clock display would either go off or only one digit would appear – it was an intermittent fault. When it occurred the machine didn't answer the keyboard. Crystal 1001 on the clock module was dry-jointed.

E-E picture was smeary but OK on playback: The E-E picture was smeary but was OK on playback. By substitution we narrowed the cause down to the P607 motherboard. Voltage checks around the 4053 video switching chip showed that the EXT switching line was at 2 V when it should have been at 0 V. The chip itself was leaky, a new one (IC7951) restoring normal service.

Tape would eject when play selected: A check on the error memory found that the capstan tacho signal was missing, although the capstan was seen to turn a few times when the cassette went in. The cause of the trouble was no supply to the tacho amplifier as C2206 on board P607 was short-circuit.

Would eject in fast forward/rewind: In rewind or fast forward this machine would stop and eject the tape after a few seconds. In the play mode it would lace up then unlace and eject the tape. All this was because the drum didn't rotate as the $3.3\,\Omega$ resistor R3142, which is in the 13 V feed to the drum drive circuit, had failed.

Would power down when tape inserted: The cause of this was the drum motor, which was very tight. No drum rotation switches the machine off of course.

Sound warble, and eject when any function was selected: The original complaint with this machine was of sound warble. When it arrived on my bench it would eject the tape whenever a deck function was selected. Now before it will allow any functions, the microcontroller chip requires sample capstan FG pulses. When a tape is inserted, the capstan does a shuffle for about half a second. In this machine the capstan turned for about 5 seconds. The capstan FG pulses are picked up by a tacho coil that's mounted close to the capstan flywheel. They are amplified by the tacho amplifier, which is mounted under the subplate. It receives an 11 V supply that's derived from the +13a rail via a 1 kΩ resistor with a 47 μF, 16 V decoupling capacitor, C2206. This last item was short-circuit. When a replacement had been fitted we had some FG pulses, but they were too low in amplitude for the micro to recognize them. After replacing the tacho amplifier PCB and the tacho head the machine worked normally – with no trace of any sound warble.

Philips VR6460

Would not play pre-recorded tapes: This machine nearly drove us round the bend. The customer's complaint was that it played back its own recordings quite well but with films from the hire shop there were wow on sound and noise bars on the picture. Our field engineer gave the machine a thorough clean-up, paying particular attention to the tape path and the control head, although the machine worked perfectly while he was there. A week or so later the inevitable repeat call came and the machine was brought back to the workshop, together with a film. Once it was opened up on the bench our first step was to try an MH2 test tape and see whether it would play this back correctly. It did. Several other tapes were tried and seemed to be OK, but the machine flatly refused to play the film supplied. The capstan servo seemed to lose lock, as if the control pulses were missing. As the tape was suspect it was tried on another machine, which played it back normally. So now we were thoroughly confused! After a lot of time was wasted scoping the control pulse outputs etc. and setting up the lateral position of the control head as per the manual, we discovered that if the back tension was reduced by resting a finger against the tension arm the fault almost cleared. A similar effect could be obtained by increasing the pinch roller pressure. The pinch roller and back-tension band were replaced and, you've guessed it, the fault was still present! We decided that the back tension was excessive. This was backed by our observation that the tape seemed to be very tight between the exit guide and the pinch roller. At this point the penny dropped. Could the lower drum be too shiny so that the tape was sticking to it? Changing the lower drum completely cleared the fault. The price of this unit was £220 from Philips, £150 from Panasonic. Fortunately the machine was insured!

Poor E-E picture: Playback was OK but there was a very poor picture in the E-E mode. There was no sync and the picture was pulling and shaking. C22 (1 µF) on the i.f. panel was open-circuit.

Dead: There was no clock, no deck activity, nothing. The AT supplies were present but there was no activity on the I²C bus data line. It was shorted to chassis (47 Ω), but the short cleared when the keyboard was unplugged. A new TMS3763ANL28 chip was required.

Dead: We found that the 10.2 V supply was missing. It comes from the L4811 regulator 7110. As this device had no 12 V input we moved back to the LM317 regulator 7105 which proved to be open-circuit. Do others find these Philips manuals almost impossible to follow or is it just me?!

Intermittent going into rewind: This machine would work normally for days or weeks at a time. Then it would 'hang up', going into permanent rewind no matter which button was pressed. Once the tape had been ejected it wouldn't accept another one. We were convinced that it was a mode switch fault and fitted a replacement. The machine worked for two weeks then the same thing happened again. Heating or freezing the servo board had no effect and another engineer had tried the microcontroller chip. We suspected plugs and sockets but couldn't fault them. By now the fault had once more cleared. When the fault next appeared I was ready to do battle! Armed with the service manual and a logic probe I set to work checking the input conditions at the servo chip IC7125 from the mode switch. The manual is very helpful, giving the logic conditions for all functions at pins 4, 5 and 6. All three inputs were high, which is incorrect. The reason for this was soon apparent as the earth connection to the mode switch also measured high! This connection goes to a plug and socket on a small PCB (P667) which is mounted on the front deck. The panel is earthed by a single screw and star washer that had worked loose. A screwdriver was all that was needed to provide a complete cure. I now check this on all VR6460s that come in.

Display lit but no other operation: If the machine was powered up it would immediately shut down. There was also no capstan motor shuffle when the mains voltage was applied. As a tape couldn't be inserted, I started by making some cold checks in the power supply. Basically the 12 V supply was missing, or rather it was being dragged down to approximately 1.2 V, because of a short on the audio board. The cause turned out to be C2024 (330 µF, 16 V) which was very leaky. Replacement of this item cured the power supply problem and restored normal operation.

Philips VR6462

Very low playback and E-E luminance: We checked the CVBS signal output from signals panel P302 and found that the luminance was missing. When we checked back to the TDA3740 chip IC7251 we found that there was no signal input. We moved back to the BC548 emitter-follower transistor T7301 and found that there was a signal at its base but not at its emitter. A check on the emitter voltage showed that it was high and unstable. Changing the transistor made no difference but when we checked the resistance from its emitter to chassis we found that the reading indicated an open-circuit instead of around 400 Ω (via L5201/2/3/4 and R3202/3). L5202 turned out to be open-circuit and when replaced we had normal luminance.

Jerky tape motion in reverse play and reverse picture search, with corresponding picture instability: This was traced to swivelling wheel item 264 which was slipping. The wheel tyre was found to have traces of dirt on it. Scraping carefully removed the deposits and eliminated the problem.

Rewind mode and cassette tray fault: This machine had two faults as follows: first, at power-up with no cassette the tray moved in and out more times than usually; secondly, the machine remained in the rewind mode and refused to do anything else. The cause of the first fault was that the threading motor counter signal was missing due to a faulty Hall-effect sensor on board P671. The second symptom was present because the head drum didn't rotate – IC7001 (L272) was open-circuit.

Test pattern with cassette in: If play is selected without a cassette inserted these machines usually provide the test pattern. This one produced the test pattern even with a cassette in! The test signal is enabled by the TPI signal on Bus C: it was high all the time because transistor 7508 was open-circuit.

Cassette lift failure: A common problem with this range of machines is failure of the cassette lift to operate – the cam turns but the lift doesn't move. This is often because the pin has dropped out of lever 242, which is hidden under the main cam wheel 247. Changing the lever isn't too difficult as long as you mark the relevant positions of the gears (in the stop mode) and don't disturb the other levers under the cam. When reassembling, make sure that the threading mechanism is fully unthreaded and that the back-tension lever is on the correct side of the left-hand slider. It's a good idea to remove the loading motor and worm before switching on so that the cam operation can be checked manually.

Wouldn't tune a signal: This machine played OK. We found that the tuning information pin 16 of the SAB3013 chip was at a higher voltage than it should have been as T7420 (BC547) was open-circuit base-to-emitter.

Intermittent loading problem: Intermittently one side of the cassette lift didn't go down as the cassette was taken in. Lever 209 had been bent when the idler had been changed at some time in the past. As a result it didn't locate in the tray very well. Something to bear in mind next time you change one.

No reel motor rotation in play: Fast forward and rewind were all right. Comparisons with another machine (as usual there were no voltages on this section of the circuit) revealed that the supply to pin 2 of the drive chip was missing. It should be 7.5 V in play and comes via the BAX18 diode D6147 which was open-circuit. In the fast-wind modes the supply comes via D6146.

'Goes berserk': The fault note wasn't far wrong! At switch-on the deck would initialize then the clock would go off, the deck would go into wind and wouldn't stop. All this was accompanied by clouds of smoke from the i.f. module. The smoke was coming from R3426 as C2422 was short-circuit. It's part of the tuning voltage generator. . .

Head rotated too fast: When play was selected the tape threaded but the head rotated much too fast. So the machine immediately unthreaded. This is usually due to an open-circuit head position sensing optocoupler, but not this time. The opto LED is in series with the tower LED (IR cassette bulb) and it was this that was open-circuit.

The clue is that the head optocoupler LED voltages are normally around 3 V but in this case were at 12 V.

Would play but not record: When record was selected you couldn't tune in a station. As another machine was handy we swapped the tuner/i.f. panel P104. The first machine then worked all right. Access to this panel in situ is very poor, but after replacing the machine's own panel we managed to check the voltages at the tuner. This showed that there was no voltage on the tuning line to pin 7. We traced the source back to IC7401 where there was no still output, but connecting the meter to pin 8 of this chip restored the tuning. This pin is the 30 V input from the regulator. We let the machine cool down then tried again. This time there was no voltage at pin 8 and on tracing back to the regulator we found that there was a 40 V input but nothing from T7601 (BC556A). Replacing this transistor restored normal operation. The second machine would jam a tape as it loaded. We found that the spring in the right-hand side of the cassette lift had come out. As a result the lift would jam half way down.

E-E signals, no modulator output: There were E-E signals but when play or the test pattern was selected there was no output from the modulator. The +12b supply was disappearing – check it at R3160. By disconnecting PCBs I was able to establish that the fault was on the P302 signals panel where C2329 (220 μF, 16 V) was short-circuit.

'No sound': The ticket said that the complaint was no sound – also that the machine had been to another repairer. Playback sound was OK but the E-to-E sound was weak with buzzing. A look at the sound subpanel showed that there had been a lot of soldering activity – also a new 5.5 MHz sound filter had been fitted! Fitting the correct 6 MHz type cut down the buzz while a tweak on coil S5 brought back the sound. Someone hadn't read the small print on the diagram '6 MHz for /05', i.e. for 6 MHz UK use.

Machine wouldn't accept cassette: If the cassette was placed in the slot, however, and a key such as play or stop was pressed, the tape went in and the machine worked normally. The cause of the problem was a shorted switch in the lift housing – as a result the machine thought that a cassette had already been inserted.

Lift problems: The tape wouldn't go in. We found that with a cassette inserted and then a key pressed it went down and everything worked correctly. The switch on the lift assembly has to be pressed and released before the lift will go down. If the cassette is put in and doesn't come out a little the lift will stand still. It all depends on the cassette being pushed out after being pushed in by the user. If you look at item 2421 on the lower side of the deck you'll see that there's an eccentric nut with a locking screw. Adjustment of this determines how far out the cassette is ejected after being inserted, and hence the action of switch 204 (lift switch tape in). The adjustment enables a small amount of mechanism wear to be taken up.

Dead, but remote control worked!: This machine appeared to be dead, with no clock display or deck functions. But the customer had noticed that it worked perfectly with remote control. A new TMS1934 clock display/function chip on the front panel put matters right.

Goes into standby: When reverse or play/search reverse was selected the supply reel would rotate slightly then the machine would go into the standby mode, with the tape still laced. If any key was pressed the machine would power up and unthread the tape. All the other functions worked perfectly. The cause of the fault was eventually traced to the control disc (large gear) on the underside of the deck. When we removed this we found that a small piece of plastic was stuck in the groove that operates the back-tension arm. The debris made the control disc jam before reaching its destination. The control system saw this as heavy running of the threading motor and thus shut down to standby.

No picture: Bench tests showed that the fault was no loop-through and no test signal. When checks were carried out in the power supply we found that there was a short-circuit across the 12 V line. The cause was traced to C2404 (330 μF, 16 V) on the audio panel.

No playback sound, E-E sound normal: I like to use a signal tracer. So I lifted out audio panel P502 to make checks. There was plenty of signal from the head, at the base of transistor 7010, but nothing at its collector. A few further quick checks showed that although there was 11 V at the top end of R3037 (3.3 kΩ) there were no voltages around transistors 7010 and 7009. The decoupling electrolytic C2027 (330 μF, 16 V) was dead short, a replacement restoring full sound.

No sound in any mode: The cause was traced to C2007 (330 μF, 16 V) on the audio panel. We've had a number of these blue Philips capacitors on the audio panel fail, causing a number of symptoms. They usually go dead short, so a faulty one is not too hard to find.

Capacitor failure: The 330 μF, 16 V capacitors used extensively in this and other Philips VCRs seem to fail quite often, usually going short-circuit. The complaint I had with one machine that came in recently was no sound. C2007 on the audio panel had gone short-circuit.

No signals, either E-E or playback: It wasn't possible to obtain a test signal as there seemed to be no output from the modulator. Mechanically the machine was OK. A substitute i.f. panel failed to restore the signals, so I checked the voltages at pins 4 and 6 of socket P5 on interface panel P005: these are the supplies to the modulator. The switched 12 V supply (12b) at pin 4 was missing. It comes from transistor 7002, which had correct voltages at its base and emitter but nothing at its collector. The 'on' line to IC7150 seemed to be working correctly. Component replacement on panel P005 can be carried out only after removing it. Remove the i.f. and chroma panels, then the three screws that secure the mains transformer. After unplugging the transformer, release the plastic clips that hold the panel, raising it gently as you do so. The panel can be worked on by resting it on its side, and you can plug the transformer back in. All deck functions will then remain operational. Once transistor 7002 had been replaced normal operation was restored. It's a BD678 Darlington-type transistor.

Insufficient record reel torque: The usual cause of insufficient reel torque in the record, playback and search modes in these machines is slippage (insufficient friction) in the 'flying-shuttle' reel idler. We've had one or two cases where the trouble has been due to too much friction, however. The idler is retained by an elliptical spring and a circular baseplate, upon which it revolves. Excessive friction can sometimes develop between the underside of the idler and the top of the plate – the friction can be sufficient to bring the reel motor to almost a standstill in any but the fast-forward and rewind modes. Once it has been confirmed that the reel brakes are not binding, i.e. the brake solenoid is not sticky, the point can be proved by pulling the idler clear of the motor drive boss, when the speed of the reel motor should zoom up for a second. Cleaning and polishing both face surfaces will cure the binding, but it's best to replace both the idler and the plate.

Insufficient reel torque: These machines use the early Philips VHS deck with conventional M loading of the tape. Many of them now suffer from insufficient reel torque in the fast transport modes – fast forward, rewind, cue and review. Typical symptoms are spillage with E180 or E240 tapes in the forward search mode and long rewind times. Some models benefit from the official modification – fit a 22 Ω resistor across

R3101/3103 in the reel motor drive circuit. Whether or not you do this, check for excessive friction between the reel idler wheel and the guide plate below it, and for gummy shafts on the spool turntables. To cure the latter problem, remove the turntables and thoroughly clean the holes in them, then clean and polish the shafts. Apply a tiny drop of light oil when reassembling.

No functions and no clock: This fault was cured by replacing the MAB8420 chip IC7091 on the back panel.

Philips VR6467

No playback video: There was a healthy f.m. signal output from the head but no video output from the Y/C processing board. The f.m. signal entered the TDA3730 chip IC7351 at pin 17 but didn't reappear at pin 16 as it should have done. This was not surprising as the supply to this chip was missing. It comes via a BC238 series regulator transistor T7304 which we discovered had gone short-circuit base-to-collector. This was not the end of the story since there was still no output when we'd replaced this transistor. It turned out that C2329 (330 μF, 16 V) was also dead short.

Periodic noise bar through picture: This occurred every 10 seconds or so. Scope checks showed that the amplitude of the control track pulse at pin 13 of 7551 was low at 2 V instead of 5 V. This chip (8051) was defective, with an internal pull-up open-circuit.

Sound faulty: Our field engineer brought this machine in with the report 'sound faulty, fit rack, suspect pinch roller'. Now if you don't know it this deck does tend to produce sound fading problems: the pinch roller hardens, pulling the tape down across the audio head with the result that the audio section is taken below the audio head. So, I thought, here goes, another rack assembly (although I find I'm not fitting as many now as in the past). But after fitting it the sound still warbled. The capstan motor and associated circuits were then checked and found to be OK. Back on the deck I released the tension on the arm bracket 268. This pushes the pinch roller into place and the warble stopped. So I replaced the bracket, again to no avail. Changing the capstan itself also failed to provide a cure, but slight pressure on the capstan holding assembly did the trick. So I took the white cap off the top, revealing the cause of the trouble: the top brass bearing had worn to an oval shape. A replacement cured the fault and occupied the rest of the morning's working time, although I'd learnt a lot more about these decks.

No sound and vision playback: This was just a blank screen and was due to absence of the 10 V supply at the head amplifier panel. Transistor 7607 (BC328) was open-circuit.

No playback or E-E video: After refurbishing the mechanism (rack slider kit etc., part number 4822 403 53377) the machine displayed only a test signal, i.e. there was no playback or E-E video. The 10 V supply was missing because the BC328–40 transistor 7607 was short-circuit base-to-emitter. A replacement restored the signals. Transistor 7304 on the chroma/video processor board causes a similar fault – it's also a BC328–40. Also check the electrolytic capacitor C2329. If transistor 7304 is faulty this capacitor will almost certainly be short-circuit.

Rolling pictures in all modes: For rolling pictures in all modes record, playback and E-E – check the 220 nF, 100 V capacitor that decouples pin 2 of the TDA3755 chip IC7451. You will probably find that it is dry-jointed at both ends.

Philips VR6468

No clock or controls: At switch-on the cassette carriage moved in and out as usual but the clock display was out and none of the keyboard controls worked. The +13a supply was missing at the keyboard panel as R3509 (15 Ω) on panel P607 was open-circuit.

Smeary, monochrome playback and E-E pictures: Panel swapping proved that the fault was on the on-screen display board. It was brought out for access by removing the PCB plug and soldering a ribbon cable between the plug and the PCB. Scope checks revealed that the signal was being dropped across R3122 as the d.c. voltage here was too high (8 V instead of 3.5 V). There was no base-emitter voltage drop at T7101 because its emitter was floating, the cause being a crack in the print by C2103.

No vision in E-E or play: The test pattern worked, however. The +11.9b supply was missing as C2329 on the signals board was short-circuit. The short had also damaged transistors 7607 and 7304 (both of type BC328).

Cassette would eject when play or rewind selected: This machine would accept a cassette normally, but if wind or play was selected the cassette would eject. A check revealed that the microcomputer chip thought the capstan wasn't turning even though it was! The tacho pulses were missing – a new P687 amplifier module put that right.

The same symptoms occurred on another occasion. When the service mode was selected the error was –2 (capstan not rotating). This time it wasn't loss of the tacho signal: the capstan motor had a dud spot.

Buzzing noise on sound, hot smell: The hot smell was coming from Tr7108, which is one of the drum motor drive transistors. A check on this transistor's drive waveform (HMC2) showed that there was an oscillation on it. Replacing Tr7108 and its driver transistor Tr7107 made no difference, in fact the oscillation appeared to be coming from the P8051–C21 D4 chip. A replacement cured the problem.

Philips VR6470

Wow on playback sound: This is often due to a faulty capstan motor or belt, but not this time. The speed servo loop includes the SDA1009 chip, the capstan tacho signal and the L293 drive chip's signals. The mark-space ratio of the drive signals varied, as you would expect, but they were otherwise OK. Phase control is via the SAB8051 chip, where there were too many control track pulses: instead of one every 40 msec, each correct pulse was followed by another one 20 msec later! The cause of the problem was that the control track pulse processor didn't ignore the negative control track pulses because C2326 (4.7 µF) was open-circuit.

No i.f. output from tuner: Checks showed that the tuner and SAB3036 CITAC chip supply voltages were OK but the tuning voltage remained at zero. 33 V was present at pin 9 of the CITAC chip but there was no output at pin 8. As the I^{2}C bus lines were OK we changed the chip. That did the trick, and the tuning points were still stored in the memory – all four channels were available straight away.

Chews tapes: When we tested the machine the tape played for a while then the take-up spool stopped. With the clear cassette in place we saw that the brake magnet released, the brakes then coming on. Transistor 7053 on the servo board operates the brake magnet. The transistor was OK, but the wire between it and the magnet was open-circuit: there was a dry-joint where the wire is connected to the magnet.

Philips VR6470 etc.

The machines covered by these fault reports include the Philips VR6470 and VR6670, the Tatung VRH8495TK and the Pye DV468, DV562 and DV761. They all use Philips mechanisms and electronics and there are probably other clones. We've serviced a hundred or so of them, and as a result have been able to note a few common faults. These are listed below.

13 V line missing: Replace transistor 7001 (BD436) and 1C7051 (LM393). Always replace them both before switching on.

Loud crackling noise while loading: This is usually caused by a broken tooth on the rack slider. Replacement is the only cure.

Noisy playback, rewind, fast-forward or no rewind or fast-forward picture search: This has always been found to be due to coupling 214. Repairing this will restore normal operation.

Tape jams in deck: Can be tape stop broken in cassette deck, spring 215 bent or cam 217, 206 broken.

Poor picture: Usually caused by poor video heads. The type of head used often seems to last for only about one year. On a few occasions the cause of the trouble has been the spring coming off back-tension lever 204.

Switches off when fully loaded: Check whether the capstan motor is trying to go in reverse. If so replace 1C7251 (L293B).

Tape loads then unloads: If this is not caused by a badly positioned cassette deck try replacing the tape-end sensors.

Tape loads only half way when play is pressed: The threading motor can have a tight spot.

Won't accept a tape: On one or two occasions, however, we've had a very weird problem. The complaint has been that at times the machine won't accept a tape. On each occasion when the machine has been put on the bench the fault has been present and no initial measures would cure it. The tape was then loaded half way using a battery connected to the threading motor. After connecting the mains the machine unloaded and after this the machine accepted a tape on all occasions. We could instigate the fault only by unloading the tape, switching off at the mains then reconnecting the machine. The cause of the problem was eventually traced to the fact that when IC7551 was being reset it held the threading motor on for 6 seconds longer than it should have done. Replacing IC7551 puts matters right. As a final point, when replacing the top plate make sure that the pressure roller assembly is located properly. Otherwise the teeth on the rack slider can be broken at the first attempt to load a tape.

Displays flash, no functions: We've had this fault on a couple of these machines. The symptoms are as follows. At switch-on from the mains supply the clock and counter displays flash at a very fast rate; there's no E-E display; and if any of the function keys are touched nothing happens. You then start to scratch your head and look at the circuit diagram. On looking up, the display seems to have cured itself and you now have E-E, but try a function and the display and E-E disappear, with no sign of anything working. Voltage checks will show that all four outputs from the power supply are low and fluctuating. Then the fault clears and the power supply is OK. The problem of course is whether something is pulling down the outputs from the power supply or the power supply itself is faulty. I cheated and fitted a new one – it's an all-enclosed unit. There were no shorts on any of the outputs so it was time to get out the scope and look for pulses. The only one that seemed to be missing was that at the collectors of T7001/2. When these transistors were checked out of circuit they seemed to be all right, but the fault was still there when they were replaced. Fitting new ones cleared the fault. A further check revealed that T7001 was actually the cause of the trouble, but it didn't measure faulty – and was the culprit in both machines.

Philips VR6490

E-E and record pictures smeary: However, playback of a pre-recorded tape was fine. Scope checks showed that the video output from the i.f. section was OK and remained normal right through the AV switching chip IC43. It was poor after R3A6 because Q3A3 was conducting when it shouldn't have done, placing C3B4 (0.01 μF) across the video signal – hence the smeary picture. There was 1 V at the collector of Q454 when there should have been 0 V. On test Q454 (2SA933) was found to be leaky.

Kept stopping in play: The reel rotation signal was intermittent, although the reel was turning all right. On investigation we found that the ribbon cable to deck plug P1504 wasn't clamped into the connector. A press on the locking bar was all that was required.

Dead: Check whether R102 (330 kΩ) in the power supply is open-circuit. It's a safety component, part number 4822 116 52272.

Sound problem: There was intermittent sound on the tape that was brought in with this machine. It came on and went off suddenly, as if the muting circuit was operating. When we tried the customer's tape in another machine there was a different symptom – the sound stayed on but there was a warble. I put the cassette back in the customer's machine and checked the control track pulses. They were weak and varied in amplitude. When the pulses were large the sound remained on – when they were small the sound was muted. All that was wrong was that the control track head was dirty. Cleaning its face and making a test recording proved that the problem had been cured.

Philips VR6542

Recordings had intermittent colour: It was to some extent signal dependent – a weaker signal was more prone to cause the symptom. Changing IC501 on the Y/C panel made no difference and we eventually found that the 627 kHz signal was off frequency by about 70 kHz. Resetting this produced reliable operation. We've had several of these machines that don't seem to like E240 cassettes – the tape commits suicide on the mechanism although there doesn't seem to be any mechanical fault. Strange.

Lost cam switch gear alignment: I recently had to sort out one of these VCRs for another local engineer who had tried to change the cam switch but had lost the gear alignment. The manual shows you how to align the gears using the triangular timing holes – the only snag is that the factory fitted gears don't always have them! Replacement gears do, thank goodness, so I cheated and fitted new ones. If you have to change the cam switch and the triangular holes aren't present, mark the positions of the main and brake cams by scratching marks on the metal with a jeweller's screwdriver, through the small holes. If one of these machines comes in with intermittent mechanical problems, to save time I change the loading belt, brake and master cams, and the cam switch. After doing this you should – as long as the reel idler is OK and the capstan motor hasn't got a dead spot – have covered the likely failure points. This has been my experience to date anyway.

Mode switch problems as Panasonic: This machine has a Panasonic deck and suffers from the same mode-switch problems as Panasonic models.

Machine was in permanent rewind: This Sharp-based machine was in permanent rewind. After checking the light sensors I removed the cam assembly to get at the mode switch and found that it had fallen apart. When a new mode switch had been fitted the machine would wind and rewind. But it wouldn't play as the capstan refused to turn in this mode. I made various checks and was beginning to suspect the system control chip IC801, although I've never known one of these to fail. There were some peculiar voltages around pins 25–28, they were varying slightly up and down. A look at the print side of the board showed that these pins are covered with a piece of sticky foam that's used to isolate a couple of capacitors from the PCB. I decided to remove the foam to check whether the chip's pins were dry-jointed. They weren't, but when I checked the voltages at pins 25–28 they were now correct. Not only that, but the machine now worked. I looked at the piece of foam and checked it with a meter. It had a resistance of a few kohms! There was another fault with the machine: the counter didn't work (although pulses were present) and it wouldn't change channels. The cause: you've guessed it! A similar piece of foam fitted to the back of the front control panel. Once this had been removed the machine worked perfectly.

Machine's recordings very poor: The pictures produced were dull and rolled. There was just as bad a picture in the E-E mode. A check showed that the waveform at the video output socket was badly cramped. The video input to the YC PCB was OK, but the output was cramped. C201 on the YC board was the cause of the trouble, a replacement putting matters right.

Philips VR6561

Tray problem: If it's powered up without a cassette this machine usually moves the tray in and then out again. With this one the tray moved in and after a few seconds the machine switched off. No, it wasn't a mechanical fault this time: the control motor drive chip IC7251 (L293b) was faulty.

Head did not rotate: This machine wouldn't play as the head didn't rotate. It would begin to turn but couldn't manage a full rotation. This pointed to one of the motor's coils not being driven. D.C. checks in the drive circuit revealed that transistor T7113 (BD135) was open-circuit. Another of these machines had the same fault but in this case it was intermittent. There was a broken wire in the loom from the PCB to the drum motor.

Idler wheel problem: If the idler wheel doesn't flip across to contact the reel discs, check that the block (item 257 4822 466 81643 that fits under the top plate) hasn't fallen off.

No sound and blank raster on playback: We found that the +10d (record) supply was present at all times as the RE2 line didn't go high in playback. It comes from the P608 board where T7701 (BC328) was open-circuit base-to-emitter.

Philips VR6585

After 10 seconds sound would go weak and distorted: The signal was OK at the P127 front-end module but faulty at the audio output sockets and of course the modulator. The cause turned out to be on the P524 f.m. audio board where decoupling capacitor C2211 was dry-jointed. For no sound at all check that the 80 mA Wickman fuse on this module hasn't blown.

Jammed mechanism: This machine uses the Panasonic G deck. I fitted a new gear set then tested the mechanism by turning the capstan motor by hand. As it went through the various motions without a hitch I powered the machine. It went into turbo drive, accompanied by some nasty crunching noises, then promptly seized solid again (I hadn't inserted a cassette). Clearly there was a power supply fault that had caused the original failure. I retimed the mechanism (fortunately no damage had been done) then borrowed the power supply from a known good machine. The result was perfect operation. I have to admit that I fitted a new power supply from stock. One day I may feel brave enough to repair the old one!

Mechanical misalignment: There is usually, but not always, a mechanical cause for mechanical misalignment of the Panasonic G deck. In this case the cause was electrical. After replacing the stripped gears and checking the operation of the mechanism by rotating the drive by hand I connected the power. The mechanism took the cassette in but didn't find the half-load position. After four or five attempts the cassette would be ejected. Sometimes the capstan would run at full speed when the power was connected: as the motor didn't stop at the fully threaded position you would have to repair the gears again if you weren't quick to disconnect the power! We fitted a new mode switch and a new tray position switch, but the fault was still present. Fortunately we had a new stock machine of the same type, so we were able to do some panel swapping to localize the cause of the fault. This proved that the cause of the trouble was in the power supply. The output voltages were correct, but they dropped when the deck was in operation. A new CNX83A optocoupler put matters right.

Blank screen in either E-E or playback: The cause was failure of the 315 mA wickman fuse T501. It's on the luminance/chroma panel, not in the power supply.

Sideways picture movement: This machine bounced back on me after I'd rebuilt the deck, which is the Panasonic G type. The customer complained of sideways picture movement. Small displacements were apparent, just about discernible on my 14 inch monitor. The effect was very much more noticeable on the customer's large-screen set. I checked the previous work, then investigated the drum circuit. Well, to be truthful, I started off here. This didn't get me anywhere and, having woken from my stupor, I then noticed that the fault was worse from cold. So I reached for the freezer. This led me to the non-polarized, 4.7 µF electrolytic capacitor C255, which turned out to be low in value. Two 10 µF electrolytics in parallel improved the situation but didn't cure the fault completely. As the lower drum was a little noisy I removed it and applied a drop of penetrating oil to the bottom bearing. This finally cured the problem.

Philips VR6660

Almost dead, drum rotating fast: This machine was almost dead – the only sign of life was the head drum which was rotating much too fast. All the supply lines were present and disconnecting various plugs and sockets as instructed in the I^2C bus fault-finding chart proved to be inconclusive. Whilst carrying out scope checks we discovered that the power-on reset signal was high all the time. Transistor 7151 was open-circuit.

No recording: This machine worked correctly in the playback and E-E modes. When record was selected, however, the monitor's screen went blank and nothing was recorded on the tape. When checks were made on the power supplies I found that the +12b line dropped in the record mode due to excessive current. Checks on the i.f. board led to the head amplifier can P400 where C2013 was found to be short-circuit. Along the way I had a wild goose chase around the P604 system control board which was fitted with a small subpanel. This isn't shown on the circuit diagram but is included in the manual for the VR6862.

Magnets of threading motor fall out: The result is that the deck doesn't initialize. If you can find the lost magnet, it does matter which way round it goes. You need to have the N pole pointing outside with one magnet and the S pole pointing outside with the other one.

Intermittently stopped playing: If the tape deck test was called up it said that one of the tacho signals was missing – although a visual check showed that the reels, the capstan and the head drum were all rotating when the fault occurred. Scope checks revealed that the wind tacho signal was going missing. When the reel was removed you could see that the reel magnet rubbed on the Hall sensor: as the magnet wasn't glued to the reel it would stop rotating while the reel kept going. The sensor wasn't sitting correctly in its slot as a wire had been trapped underneath it.

Couldn't remove test pattern: When the test pattern was selected it couldn't be removed except by interrupting the mains supply. When playback of a cassette was tried the tape would thread then immediately unthread. The head drum rotated even in the stop mode. I started with this last point as it seemed to be the easiest one to deal with. As the motor stop line (pin 25) didn't go low in the stop mode a new MAB8420C047 microcomputer chip was fitted. This cleared all the faults. Phew!

Flutter on sound: A panel swap proved that the cause was on the P604 servo board. In the past I've had the DAC cause this fault. As there was a 1 V peak-to-peak square wave at pin 13 of this chip I fitted a new one – to no avail. As I started to unsolder the servo microcomputer chip I noticed a solder blob that shorted pins 9 and 10 together. When the short was removed and the pins I'd unsoldered were resoldered the sound was fine. The soldering looked original, so why hadn't the customer noticed the fault before?

Philips VR6760

No E-E or playback picture: This was due to the absence of any 10 V supplies on the signal board (P306). Tracing back led us to an open-circuit transistor (7607) on the main board (P606). Removing the panel to replace this transistor is no easy task. It was even more frustrating when the replacement gave an impression of Vesuvius 10 seconds after switching on. Further checks revealed that there was a short to chassis on the signals panel. The +10c supply stabilizer transistor on this panel was found to be burnt up and short-circuit, but the short was still present after it was removed. It was found to be in C2329, and when this and the two transistors were replaced normal operation was restored.

Sound problems and distorted E-E picture: This machine came in with a long list of faults: no hi-fi sound on playback of any tape, no normal sound, and no E-E sound through the TV set with the picture distorted. The distortion took the form of a black-and-white bar that rolled up and across the screen, similar to sync bars breaking through from an adjacent channel. After chasing several red herrings we found that the culprit was transistor T607 on board P606. It was open-circuit, as a result of which the +11b supply was missing.

Wouldn't play old hi-fi recordings: This machine had come from another dealer. He'd cleaned the tape path because of intermittent hi-fi sound, but when he returned it the customer complained that it wouldn't play any of his previous recordings in hi-fi at all. New recordings were OK, as were pre-recorded tapes. We found that with the old recordings the hi-fi sound faded in and out although the picture remained perfect. On inspection we discovered a lot of tape oxide by the capstan, suggesting tape crinkling at the last guide, 256. Yes, it was another case of a faulty pinch roller.

Jammed rack: This machine had the usual jammed rack. So the deck was stripped down and the rack, pinch roller and coupling were replaced. After giving the mechanism a dummy run using a 9 V battery we tried the machine out for the first time for real. This showed that the power supply was dead, the cause being the BD436 transistor 7001 which was open-circuit. After replacing this we made another attempt. This time the deck initialized but a burning smell came from the 5 V section of the power supply. Someone had replaced the BYV10–20 with a 10 V zener diode. As there is normally a 14 V square wave across this diode it was working rather hard. Everything was OK when the correct diode had been fitted.

Refurbishment: While disposing a few wrecks at the local dump my eye was caught by a forlorn looking Philips VR6760 hi-fi stereo machine. I paid a small sum for it and when I got it back I found that someone had tried to fix the DMP deck and got themselves into an awful mess. When I tipped out all the bits it was virtually complete: a service kit then got the machine working nicely. Ensure that the record-tab sensor switch is fully down on the chassis before inserting a tape. It can collide with the bottom of the tape with the result that your carefully rebuilt deck jams. I also had a problem when I came to fit the cassette flap, which was missing. The one I ordered came with much larger square supports at the top than the round holes in the machine would accept. Willow Vale told me that the type with the small supports is no longer available and that I should open out the holes in the machine. Since there would have been nothing left of one of them I instead filed down the new and expensive flap I'd just bought. I was reluctant to do this, but it worked.

Philips VR6870

Buzzing noise from the power supply at plug-in: The clock-display segments lit at random and flashed on/off. Another dealer mentioned to us that he'd had a similar problem, which was caused by a faulty capacitor. We removed a small subpanel from the power supply can and used a digital capacitance meter to test the three electrolytic capacitors on it – C7, C11 and C27. All were very low in value. After fitting replacements the machine worked normally.

Slow drum: When play was selected the drum turned only very slowly. Then, after a few attempts, the cassette would eject. The drum drive transistors all tested OK and the loom to the motor was intact, so another motor was connected temporarily. The drum motor was faulty.

Machine dead, no displays and no deck functions: This is a nice stock fault. Go straight for C112 (3.3 µF, 25 V) on the stand-up subpanel in the power supply module. It goes open-circuit or changes value.

No functions, flashing in display: Putting the machine on test no operation was possible and there was random flickering within the display. A check showed that the power supply outputs were all slightly low. As we were carrying out checks we noticed that more display segments lit then, after about 10 minutes, the machine came into operation with all the supplies correct. When we applied freezer to the VA4006B power supply control chip the machine shut down, but a replacement produced the same results. More careful use of the freezer then showed that C2311 was very sensitive. When a replacement, 33 μF as fitted, was installed the machine would start up after about 90 seconds. Reference to the circuit diagram suggested that the correct value for this capacitor is 10 μF. Operation was normal when we fitted a 10 μF capacitor.

No E-E sound: This machine produced no E-E sound unless the audio select button was pressed. All became clear when the front was removed. The audio level sliders were not located on the controls themselves! Thus although the knobs were set at maximum the controls were at minimum. Fitting the front correctly and resetting the levels cured the trouble.

Machine dead: We soon found that the cause was the usual culprit – C2011 (33 μF) was leaky. After replacing this capacitor we left the machine on test for the rest of the day then returned it. Two days later it came back with a note to say 'no better, as before, worse'. It was in fact dead. Further checks revealed that two more electrolytics were leaky, C2006 and C2007. C2006 is 220 μF and is in the power supply on the main PCB. C2007 is 25 μF and is on the sub-PCB – the circuit diagram gives the value as 10 μF.

No displays, power supply 'noisy': Scope checks on its outputs showed that there was enough mush present for us to wonder about possible damage to the rest of the machine. The machine was therefore disconnected and cold checks were started. We soon found that D4, a 10 V zener diode, was leaky.

Distorted sound: When we tried it out we found that the sound was very distorted, it was rather like an output stage with no bias. There was perfect sound, however, when we checked at the scart socket. This simple test saved us a lot of time. Both linear and hi-fi audio are fed to pins 1 and 3 of the scart socket via a couple of 470 Ω resistors. As the sound was OK here everything up to this point, including the switching chip IC7061, could be ruled out. The sound feed to the modulator is via a couple of 100 kΩ resistors and a buffer stage with a single transistor, Tr7904. Checks showed that there was a clean signal at one side of the two resistors R3925/6 but a very distorted one at the base of

Tr7904. The transistor was OK but its 3.3 μF coupling capacitor C2917 had a 2 kΩ leak. A replacement cured the distortion.

Clock gained 1 hour per day: This machine had previously been in for replacement of the loading belt, which was worn and slipped. It was now back because the clock gained approximately 1 hour a day. We had to put up with the old story that 'it was all right before'. The cause of the trouble turned out to be a faulty 1.2 V NiCad battery. Strangely, the tuning data had been retained.

Philips VR727

Threading motor seized: This machine's threading motor had seized and its drive chip had burnt out. Unfortunately the chip is not listed separately in the manual: you have to replace the capstan drive board. Replacing this and the motor restored normal operation.

No playback or E-E sound: It' s not easy to trace the cause of this sort of thing where there is hi-fi, Nicam and linear sound. We eventually found that there was faulty switching within the TDA2518 chip IC7205.

Poor load or eject: It's worth checking the long pulley shaft that drives the main cam. The small end cog splits then slips when torque is applied. Replace it and check the other gears for damage: this should cure the problem. The pulley shaft doesn't seem to be up to the job.

Warning about deck spare parts: Take care when ordering deck parts for VCRs that use the Philips Turbodeck. Although they look similar, there have been changes in later versions – mainly in the braking system. Instead of using a trigger-operated brake, the main cam operates the brake directly. To do this, the design of the main cam has been changed (the cam has a sun or star shape moulded into its top surface), the worm shaft is simplified and the trigger components are omitted. Unknown to me, the machine that caused me problems had had the wrong cam fitted. The deck wouldn't initialize. When a cassette was inserted, the tape threaded up and unthreaded twice, then the machine powered down. A check on the error memory in the service mode showed that the microcontroller chip sensed a threading error, although visually the deck appeared to be operating correctly. The microcontroller chip senses the threading position by counting the pulses from the 'windmill' optosensor on the threading worm in relation to the moment when the INIT switch closes. Because the wrong cam had been fitted, the INIT switch closed later than it should have done. The microcontroller chip sensed this. A cam from the N kit instead of the A kit had been fitted. If you find that the marks on the main cam are different, or the threading worm is a different shape from the one originally fitted, check that you've got the correct part number.

Pioneer

Pioneer VR727

Switch-mode power supply fluttered: This range of VCRs uses a switch-mode power supply. The one in this machine fluttered audibly all the time the machine was on. We found that the outputs were low and fluttered in sympathy with the sound coming from the chopper transformer – there was a huge, triangular ripple voltage. If the load on the power supply was disconnected by withdrawing P023 the fluttering stopped and the output voltages settled down at the correct levels. Despite this the trouble was within the power supply, where C211 had gone low in value. It says $10\,\mu F$ on the circuit diagram but was actually $33\,\mu F$. Good old Philips!

Machine's power supply tripping: This machine's power supply was tripping. It's apparently a clone of the Philips VR6870. On investigation we found that all the power supply outputs were low but went back to normal when connector M8, the power supply output plug, was removed. Fitting another power supply from a working machine proved that the fault was in the power supply. The cause was eventually traced to C2011, which was open-circuit. It's on the vertically mounted switch-mode power supply control board, within the power supply can. The manual lists the value as $10\,\mu F$, but $33\,\mu F$ was fitted in the machine. We've had this fault on several machines now, so beware!

Power supply capacitor failure: Failure of C2011 in the chopper power supply is quite common, as highlighted in the previous fault report. Another, nastier failure in the same area occurs when the 6.8 V reservoir capacitors C2032 and C2033 dry up. They are both $680\,\mu F$, 16 V. The symptoms are several: the machine may take minutes or hours to come to life after being switched on: all segments in the fluorescent display may light up; or there may be intermittent or no sound in the E-E and playback modes.

Intermittent functions, going into rewind on its own: This machine had given trouble for some months. The first complaints were of intermittent functions and going into play and rewind by itself. On that occasion we were unable to find anything wrong, and after a long soak test returned the machine to its owner. It came back the following day with the complaint of no results at all. We switched it on and after half an hour it sprang to life and carried on working. So we replaced various components in the start-up circuit and returned it, once more with a nominal charge. It was recently back with the same complaint. This time there was a tape in the machine, in the stop position. So we plugged in and waited. As the machine sprang to life half an hour later it started to do some strange things. The clock display came on first, then the carriage tried to lift the tape about an inch or so then plonked it down again. It did this several times before going into rewind by itself. Then it was all right for the rest of the day. We switched it off for a few days and pretended that it wasn't there, like you do. Unfortunately it didn't go away. The next time I switched it on I monitored all the supply lines. Except for the 6 V and 5.1 V supplies they were all correct. The 6 V supply was low at 4.2 V while the 5.1 V supply was down at 4 V. A look at the circuit diagram showed that three 680 μF capacitors smooth the input to the BD434 transistor Q7008. They had dried up. Replacements rated at 25 V rather than 16 V were fitted. After this the machine worked perfectly.

Failure to load with mechanical noises: This should lead to a check on the pulley shaft that transfers the loading motor drive to the deck. The small plastic cog at the end splits and just spins on its shaft. To replace it you have to remove the capstan motor, the loading block, the deck and the cassette housing. The part number is 4822 528 81462 – it's item 47 in the exploded view. This is the only item that's available on its own. All other deck components are supplied in kits. This applies with all Turbo deck VCRs

Saisho

SAISHO VR1000
SAISHO VR1200
SAISHO VR1200HQ
SAISHO VR1600
SAISHO VR3400
SAISHO VR705
SAISHO VR805S

Saisho VR1000

No record, monochrome playback: The playback whites were clipped and the syncs crushed – this explained the absence of colour, as the sync separator couldn't cope with the distorted signal. The record signal disappeared half way through an i.c., and the white clip test point TP4001 had no signal on it. Also the PB 9 V line had just over half a volt on it in the record mode, suggesting a switching problem. Three i.c.s were ordered from Mastercare, an LA7031 f.m. demodulator, LA7034 f.m. record processor and an OEC2003 power rail switcher. Fitting the LA7031 restored normal playback levels and colour, but there was still no signal at TP4001 despite healthy signals at pins 19 and 20 of the LA7034. After much mucking about we discovered what had happened. The white clip level control VR4008 had been turned right down and the dark dip level control VR4006 had been turned right up – full white clip and full dark clip leaves you with a straight line! Some further time was then spent setting the record a.g.c., carrier level and deviation (using a method described in my book!). A lot more time was spent arguing with the customer that £91 for some 6 hours' work plus parts was justified, and that his complaint should be with the friend of a friend who had twiddled it in the first place . . .

Idler problem: Roughening the tyre with wet-and-dry has so far provided an effective cure as an alternative to a replacement idler.

Capstan permanently in operation: We had to obtain a manual from Mastercare – £25 for a poor quality photostat copy. It turned out that

C2039 (0.022 µF) was leaky. This took us a long time to find, not helped by the fact that the circuit is incorrect – C2039 goes from pin 3 of CD2003 to the base of Q2017, not its collector.

Intermittent failure to eject tape: This is a common fault, the result of carriage overshoot. Carry out the following modification to overcome this problem. Remove the blue lead from the cassette loading motor and replace it with a BY127 diode (cathode to the motor), with a 27 Ω, 0.25 W resistor in parallel with the diode.

No E-E video, but normal playback: C9, a 470 µF, 10 V electrolytic that couples the output from the i.f. section to the rest of the circuit, was leaky. This fault also occurs with the Amstrad VCR4600/4700 range.

Made intermittent squeaking noises in the play and record modes: Cleaning and lubricating the bottom flywheel bearing silenced the squeaks.

Saisho VR1200

Complete lack of sound: We traced the incoming sound signal from the tuner/i.f. section through to pin 8 of IC5001 (BA7751LS) with a scope. Thereafter there was nothing. As this chip does just about everything in the audio department it was replaced. Success! It's fiddly to get at, however. The control and bottom PCBs have to be hinged out as otherwise access to the top of the i.c. is limited by the plastic chassis/frame. In addition it's a 24-pin SIL with the pins quilled, so you need a fine-tipped iron. CPC's part number is HN107T67751L.

No functions: A faulty mode switch proved to be the cause of no functions. To replace the switch the carriage must first be removed. The switch can then be taken out by releasing the retaining screw and unsoldering the three leads that are attached to it. Reassembly is the reverse of this procedure. Take care to align the two slots on the switch.

Machine didn't always start: There would also be no clock display. Q2582 can cause the problem but this time the cause of the fault was dry-joints on the 6 V regulator. It's mounted on the chassis, to the rear of the head drum.

No play or record: On inspection we saw that the machine loaded and went into play but a limiter post prevented the pinch wheel from moving the tape. The limiter post lever assembly is made of plastic and the pin will drop out if it's broken.

Dead with no clock display: Sometimes the cause is a failed back-up battery – C821, 0.1 F. More often, however, Q02 has expired due to a heat stroke. It's type 2SD1207, but a BD131 runs cooler and is cheaper. It's wise to treat the printed circuitry in this area very gently – it looks as though it has been drawn by a gentle fairy with a mapping pen. Once it lifts reach for the aspirins!

'Recording problems': That is what the ticket said. 'Worn head', said the pictures on the monitor. So a replacement was ordered and fitted. This seemed to put matters right, and the machine was duly collected and paid for. Next day it was back again. 'It's just the same', protested the owner. 'It records for only five minutes, then nothing.' As I'd given the machine a full 3 hour record/playback test this seemed unlikely, but I checked it again. After 5 minutes the f.m. luminance record signal disappeared in a snowstorm. I found that the signal could be restored or killed by flexing the luminance subpanel on the YC board. As the print side of the panel is inaccessible when in situ I solder wicked all 29 pins, removed the subpanel then wired it temporarily to the print side of the YC board for detailed inspection. There are printed pads along the edges of several subassemblies on the luminance panel. These assemblies are inserted into the panel at right angles, then soldered. It's rather like a microscopic version of the old ITT CVC5 i.f. module. The assemblies suffer from the same problem too – hairline cracks across the thinly soldered junctions. When I'd resoldered actual and suspect cracks, using a fine tip, I found that the panel could be flexed without any faults arising. So it was refitted in the normal manner. This wire-looping technique may sound lengthy and laborious, but it's a useful aid to diagnosis with an inaccessible board. This particular job took less than an hour to complete.

Machine would accept cassette but no functions worked: When the bottom cover was removed we found that the main drive belt was broken. A replacement restored normal operation.

No functions except load: This machine had no eject, fast-forward or rewind operation, although it would load. We noticed that there was no capstan rotation, and a quick check showed that there was no 16 V feed from the power supply. This led us to the good old circuit protector ICP201, which was open-circuit. Replacing this and giving the machine a good clean-up completed the repair. I love the easy ones!

Very poor take-up: Rewind and fast forward were OK, and the clutch seemed to be working. When I removed the idler it was obviously too stiff. So I dismantled it and cleaned the spindle. On reassembly all was well.

Saisho VR1200HQ

Would accept a cassette: This was a NICAM (Nasty Intruder Caused Absolute Mayhem) job. The original complaint was that the machine wouldn't accept a cassette. Its owner had accepted the kind repair offer of a friend at the local Electricity Board. When she retrieved it some time later it was completely dead. A replacement 12V regulator transistor (Q2502) restored some life, but the machine still refused to accept cassettes. If a tape was wound in manually the start and fast-wind functions worked, but with no end-sensing operation. The left-hand PT361 sensor was open-circuit (the original fault?), but a replacement made no difference. The right-hand sensor is decoupled by a 10 nF capacitor (C1012), which had been carefully replaced by a 4.7 Ω resistor! The correct component restored loading and end-sensing, but with no playback picture. Meterman had removed the head amplifier module and refitted it with the PCB edge connector misaligned. After correcting this we had a working machine, the owner had a large bill, and I suspect that her SWEB friend was about to receive a shock.

'It won't stay on': When we find 'HQ' appended to this breed of VCRs we're inclined to take it as meaning highly questionable. 'It won't stay on', said the owner, the reason being that fuse F502 had blown. When our adapted Thorn 3500 cutout was connected temporarily the machine wouldn't accept a tape and the BA6239A loading motor chip quickly reached fried egg temperature. After using an external supply to check the loading motor we replaced the chip, fitted the correct fuse, and another Saisho limped off into the sunset.

Intermittent stop while in use: It didn't matter which particular function was asked of this machine, i.e. rewind, FF, play etc. Although we didn't really suspect the deck sensors they were replaced as strange things do happen. The supply lines were all checked, the mode switch was cleaned and aligned, and a thorough search for dry-joints was carried out. On test, however, the fault was still present. Eventually we found that the BU2716S (IC01) was the cause. Other Saisho models such as the VR1600 etc. could equally be affected by this problem as similar circuitry is used.

Failure of Q02 type 2SD1207 common: This also happens on Matsui VX820 and Hinari VXL35. We find that a TIP41C with a heatsink is a reliable replacement.

Would lace up and run for 3 seconds in the play and record modes: Since the machine would start to operate normally we suspected a false end-of-tape message from the left-hand end sensor. For once we were right first time!

Saisho VR1600

Slow drum rotation: The drum then coming to a stop. This is usually caused by IC01 (BU2716S) being faulty.

Plays, picture freezes, shuts down: These machines are now quite old, but still produce good results. The cause of the problem is usually the limiter post near the take-up reel. When you remove it you'll find that the pin is missing. A replacement will restore full working order.

No playback colour, E-E OK: Mindful of Panasonic power supply faults, I started off in this area. Bingo! C08 (100 μF) looked stressed. When it was replaced the playback picture had good colour.

Saisho VR3400

Dirty head symptom: This newer-generation machine came in with the symptoms of dirty heads. We cleaned the heads, the drum and the entire tape path. We then inserted a tape and pressed play. The tape laced up around the drum but the pinch roller failed to pull in on to the capstan. After a lot of messing about I found by accident that the machine has a quick-start function, similar to Akai machines. All that was necessary was to push the play button twice!

Take-up wheel wasn't being braked: The result was that when stop was pressed during rewind, tape would be spilled into the machine. We've had this fault often enough with the Sharp VC8300, but in this case we were unable to find the brake. On the underside of the deck the take-up reel is obscured by the master cam: it looked like a fair old job stripping the lot down. A check on the supply spool, however, suggested that the brake was on the inside of the spool. After removing the take-up and supply reel discs the cause of the problem was clear. A small spring was missing. We assumed that it had never been fitted. When we'd found and fitted a suitable spring the machine functioned normally.

Intermittent loss of luminance record signal: We eventually found that R52 on the YC board was going open-circuit intermittently.

Tape would lace up then eject: When a tape was inserted it would be laced up then ejected. The cause of the trouble was no capstan drive because the OEC9011 chip IC2001 had failed.

Saisho VR705

No picture and poor rewind: Many thanks for a simple job this time. Replacement heads and a new reel idler were all that were required.

No E-E or tape playback signal: The deck functions operated normally. The switched 9 V supply was missing. It's produced by Q507 (2SD1266) which had 20 V at its collector. Fitting a new transistor cured the problem.

Cyclical noise bars on display: The symptom could be mistaken for absence of the control track pulses but was caused by failure of the 2.2 µF, 50 V electrolytic capacitor C503 in the power supply. It smooths the regulated 12 V supply to the servo chip IC2007. A scope check showed that there was a 1 V square wave on this line. A new capacitor restored normal operation.

Saisho VR805S

Intermittently stopped during playback: The usual standard Sharp idler was replaced but the fault persisted. This time the cause turned out to be the cassette-in switch on the main deck. Cleaning it cured the problem.

Spurious white liny pattern on E-E video output: We tackled the fault by breaking into the E-E circuit at various points leading to the record/playback switching chip to establish where the interference was being generated. In fact we worked back until we were off the main board! The cause of the trouble lay in the u.h.f. module and was fully cured by those nice people at MCES.

Intermittent colour: This machine had been back to us several times with this same fault. Replacing crystals X3001 and X3002 and realigning the colour circuit usually provides a cure, but not this time. A new colour board, type PCB 301, put matters right.

Would accept a tape but no commands: A new cassette lamp put matters right.

Salora

```
SALORA SV6500
SALORA SV6600
SALORA SV8500
SALORA SV8600
```

Salora SV6500

Machine damaged tapes: The two tapes supplied by the customer had creased sections hanging out, both very near the beginning of the tape. When a workshop Panasonic tape was inserted, rewound and play selected everything appeared to be all right. We then tried the customer's Scotch tape. After a few seconds a loop of tape started to grow slowly between the pinch roller and the take-up side of the cassette. The tape was rewound to remove the loop, then taken out. We again tried with the Panasonic tape, in the playback mode. As the tape swung round the swing guide (the one between the pinch roller and the cassette) it could be seen to lift clear then tighten up again cyclically, but it didn't form a loop. The problem was that at the beginning of the tape the take-up drive wasn't going fast enough to keep the tape taught, although the drive could cope further through the tape. The cause was a defective reel drive motor. Interesting the different effects with the two types of tape.

Poor playback picture: After replacing the video heads the results were much better but the picture still didn't look as good as it should have done – in fact it was better with a pre-recorded tape than with one of its own recordings. When I tried to adjust the tracking control at the front of the machine I found that it didn't alter the tracking at all. The control consists of a $100 \text{k}\Omega$ slider with a centre notch for the normal setting. A resistance check revealed that in the centre position the potentiometer read around $1 \text{k}\Omega$ instead of $50 \text{k}\Omega$. The cause of the problem was revealed when the potentiometer was removed from the panel and taken apart. There are four connections to the potentiometer, one for each end of the track, one for the slider rail and one for the unused centre tap (see the Figure 10). Some material had

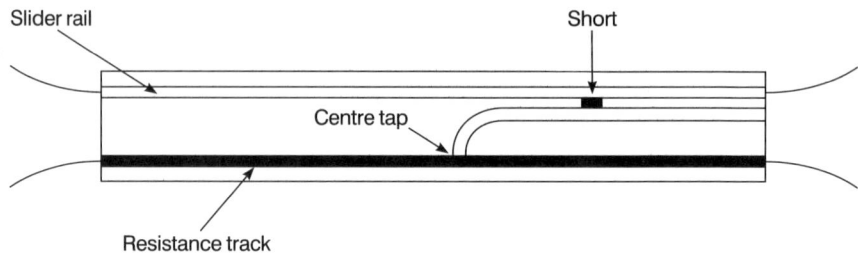

Figure 10 *Salora SV6500/Sanyo VHR1100 tracking control*

collected between the slider rail and the internal land, making a link to the centre tap. After removing this material with a cotton bud and refitting the control the tracking worked perfectly. The original heads were also found to be OK when they were refitted. The tracking control had been set at one end on playback, so that even with its own recordings the picture was poor – it just so happened that the pre-recorded tape was recorded with the correct tracking offset.

Capstan speed varying: The customer had previously taken this machine to a local cowboy who said that the capstan motor needed replacement and that this would cost over £100. On inspection we noticed that several chips had been changed. The capstan speed was certainly varying, but careful scope checks around the servo chip IC4002 showed that the tracking control was open circuit. The NCS (Genserve) part number for this $100\,k\Omega$ slider potentiometer is 6130010278.

Salora SV6600

Poor pictures, heads badly worn: When the heads were replaced things looked fine. While the machine continued to play the test tape I relocated the PCB that hinges over the mechanism: the drum and capstan servos then began to vary widely. By flexing the board both motors would stop and the machine would unlace. The fault could be provoked by prodding or poking in the servo area of the board to any degree. As the fault was so general I checked the servo d.c. supplies and found that the always 5 V and 15 V voltages disappeared between CN1003 and CN1004. The tracks that link these two connectors run right at the back of the panel, over the r.f. connection sockets, and were all broken in two places each. A point to note if you have a heavy-handed customer.

Intermittent tuning drift: The problem with this machine was intermittent tuning drift. As all channels appeared to be affected we checked the 33 V supply, which was slightly high. We also noticed that the panel (the timer/tuning panel) in the vicinity of the 33 V regulator IC6206 was brown and showed signs of overheating. The supply to this 33 V regulator chip is provided by a constant-current regulator arrangement on the power supply/system control panel. Note that there are two different circuit diagrams in the manual, this was the more complex one. The voltages around the regulator transistor Q5004 (2SA984) didn't agree with those in the manual, but the transistor, along with diode D5003 and zener diode D5004, were all OK. Resistor checks were then carried out. The emitter resistor R5010, which is used to sense the current, was found to be only 270 Ω instead of 560 Ω, the wrong value had been fitted. In addition R5015 was 560 Ω instead of 1.8 kΩ. Both resistors were original parts and had been in the machine for around three–four years. Maybe this was a one-off occurrence, or maybe more machines with these errors will start to show up soon.

Intermittent stopping in play or record: The machine was then difficult to get going again. We ran it in the workshop for four days and it never stopped once. When it was returned the fault immediately occurred – isn't that always the way? – so back it came. This time we were able to see the fault. When play was selected the machine would lace up and would then straight away unlace and stop. Take-up was present, the capstan turned, so did the drum. As the machine started to unlace just after the capstan and take-up started, leaving insufficient time for a cutout operation from either of these functions, I checked the head-switching waveform at pin 28 of IC4002 on the top PCB. It was absent. As it's fed to the system control chip to provide an indication of drum rotation (or lack of it) this was the reason for the machine stopping. The pulses from the PG coil enter IC4002 at pin 25. There were no pulses here either, because the pick-up coil was open-circuit – the slightest pressure would correct the problem. To restore correct operation we had to order and replace the complete stator assembly.

Intermittent stopping in playback, capstan servo warble and poor channel tuning: The tuning problem was drifting and hum in the E-E mode. We soon found that the voltage at the 33 V regulator transistor Q5004 was 47 V. Zener diode D5004, type GZA32, was zenering at 47 V! A replacement put everything to rights. The 33 V line is also used as a reference by the STK5482 regulator, hence the various fault symptoms.

Salora SV8500

Wouldn't record sound: This Mitsubishi clone would, however, erase and play sound. The trouble was traced to Q302 and Q304 which were both leaky. It's major job to dismantle this machine for service!

No chroma in playback mode: On removing the top cover I could see a lovely break in the print, bang in the middle of the main panel. Repairing this restored the colour. It's a very long stretch of print that eventually connects the anode of D2A1 to C6D1.

Intermittent sound in E-E and playback: This machine was being checked after coming back off rental. The customer had pointed out to the engineer who collected it that the sound was intermittent in E-E and playback. While thinking about getting to the heads to clean them I spotted the cause of the trouble – the r.f. modulator's audio pin had never been soldered!

Salora SV8600

Intermittent shutdown: The machine would also sometimes fail to power up from cold. After some searching we found that in the fault condition the switched 5 V supply was missing. We traced the cause to the regulator transistor Q9A2 which was dry-jointed on all three legs.

Stopped in the play mode: This machine allegedly stopped in the play mode. It ran faultlessly until I got in and provoked it. The pinch roller was distorted, apparently because of spillage that entered courtesy of the tape (the tape path was coated). A good clean and a new roller were all that was required.

No reel drive: brakes not released: There was no reel drive as the brakes weren't being released in the fast-forward and rewind modes – playback was OK. The cause of the trouble was that the brake release bar latching lever spring, beneath the mode motor assembly, was disconnected at one end.

Problems with the mechanism: This one came in from another dealer who was having difficulty with it – the cassette carriage was up, the guide poles were forward and the pinch roller wasn't engaged! I stripped down the mechanism, cleaned and regressed everything, put the lot back together again in the correct timing sequence and replaced the mode switch. The result: a working machine.

Samsung

SAMSUNG SI1240
SAMSUNG SI1260
SAMSUNG SI3240
SAMSUNG SI3240 AND SI3260
SAMSUNG SI3260
SAMSUNG SI7220
SAMSUNG SI7230
SAMSUNG VI375
SAMSUNG VI611
SAMSUNG VI616/626
SAMSUNG VI710
SAMSUNG VI730
SAMSUNG VI910
SAMSUNG VIK310
SAMSUNG VIK316
SAMSUNG VIK320
SAMSUNG VIK326

Samsung SI1240

Machine performs no functions and switches off after a few seconds:
The cassette-in symbol also shows on occasion even when the FL cradle
is empty. The cause of the trouble is failure of the KA8301 loading-
motor drive chip. An equivalent is the more common BA6209.

Failure of loading motor drive chip KA8301: The Samsung technical
information book, Volume 3, contains details of a modification to
overcome this.

In E-E mode picture was negative: In the play mode the video level was
fine. A check at test point TP3202 (video output) produced a reading
about twice the correct value (over 5 V) while the top of the waveform
was cut off. Adjustment of the E-E level control VR3205 made no
difference. This potentiometer sets the d.c. level at pin 19 of the
LA7323 video chip IC3201, but the voltage here didn't change when

VR3205 was adjusted. The cause of the trouble was R3218 (2.7 kΩ) which was open-circuit. When a replacement had been fitted and the E-E level had been set to 2 V peak-to-peak at TP3202, we had normal results in both the E-E and the playback modes.

Intermittent recording: Fortunately the owner supplied a sample tape, which showed an apparent loss of signal rather than recording. The suspect tuner was eventually persuaded to go intermittent in the E-E mode by being tapped. So we removed it and wired it back to the PCB temporarily for easier access and checking. The cause of the fault was then found to be a hairline crack around one of the SAWF's pins in the tuner/i.f. section.

Cutting out during rewind: On test we found that when a 3-hour cassette was rewound the tape would slow after the 1-hour point, the capstan would start to labour and the machine would then cut out. When the belt was removed from the capstan motor there was still plenty of torque. This was confirmed by the fact that fast forward was fine. As it's a full-lace machine I removed the tape from around the guides manually and let it run straight across the front of the cassette, to eliminate excessive friction in the tape path. This helped, but by no means cured the symptom. The cause of the trouble was excessive friction in the spools. Removing them then cleaning and lubricating the shafts cured it. All of the diodes were changed as they were the original ones. The FL motor chip was also replaced: this had been done previously, but the earth link modification in the power supply hadn't been carried out.

Random functions: It was as if the end sensors were faulty. With a cassette loaded we found that the voltage at one end sensor was 0.6 V and at the other 5 V.

Standby shortly after a function was selected: This machine would shut down in play, rewind, record etc., but the fault seemed to go away when pause was selected. This suggested that the cause of the trouble was to do with the reel pulse sensor or the reel turntable itself. Scope checks at the 2SC945 reel pulse amplifier transistor Q2601 showed that the pulses at its collector were only marginally greater in amplitude than those at its base. A new transistor increased the amplitude of the pulses significantly, but the fault persisted. The output from Q2601 is fed to the TC4021BP chip IC6201, which also had to be replaced. The two faulty components are mounted on a subpanel next to the loading motor.

Fast playback: Tape playback was at about the same speed as fast forward. I did not have a manual, but luckily spotted print corrosion at pins 5–11 of socket CN206. It's at the front right-hand side of the top main board, and can just be seen with the front fascia in position. A good scrape followed by resoldering restored normal operation.

Samsung SI1260

Machine records sound but not video: This machine recorded the sound but not the video – there was just snow. Playback of pre-recorded tapes was correct. Scope checks showed that there was low record f.m. in the vicinity of C3203, which when checked read 68 Ω. Strangely, its value is 68 pF. We subsequently had the same fault on two other machines.

Failure to record when warm: This is becoming a common fault. The first report you may get is of poor recording in the LP mode, progressing to no LP or SP recording. It looks as if no signal is being recorded, just noise – as though the machine is off tune. In fact the cause of the fault is lack of the luminance record signal: a scope check at TR3201 will show that it is almost non-existent. Replace C3203 (68 pF) which goes open-circuit.

Tape stuck inside: This machine came in with a fully laced tape inside. When it was powered it refused to unlace the tape. We eventually found that the cause of the problem was D212, which was open-circuit. It provides IC206 with a – 15 V supply.

Eject on power-up: As soon as the machine was powered the loading motor would run in the eject direction, stopping only when the syscon microcontroller chip decided that a fault condition was present and shut down the switched supplies. Attention was eventually turned to the mode switch, where the voltage at pin D never rose above 1.8 V regardless of the position of the mechanism. This pin is effectively the cassette-up switch – so the machine must have thought the cassette housing was down even though it wasn't. A new mode switch (part number 63579–101–026) provided a complete cure.

Returned to standby after a few seconds: This fault is beginning to show up regularly on these machines. There is no capstan motor rotation. The cause of the fault is always diodes D109, D110, D212 and D213. Always check this by measuring the voltage across the diode – with a good diode the cathode voltage will be 0.7 V less than the voltage at the anode. We have found that with some diodes there is as much as 4 V or more across the device, although the diode reads correctly when checked with a meter. Any diode from 1N4001 to 1N4007 will work all right in these positions. It's best to replace all four diodes to prevent comebacks.

No motor functions: This machine powered up and the clock and E-E system worked, but there were no motor functions at all. The always 15 V rail supplies the motor drive circuits via the 1N4001 diode D212 on the main PCB. It was open-circuit.

Blank raster and no playback: After removing the top cover we saw that there was neither drum nor capstan rotation: the blank raster gave the impression that the machine was stuck in the AV mode, although it wasn't. Voltage checks showed that the switched 5 V supply was missing at D109 (1N4001) which had gone open-circuit. A replacement restored normal operation.

Would not respond to key operations: This machine could be switched on and produced normal displays. It wouldn't respond to any key operation or accept a tape, however. IC206 has given trouble in these machines, so voltage checks were made here. A low supply voltage led us back to D212 (1N4001) which was open-circuit. Normal operation was restored after fitting a replacement.

Blank display: We connected a scope to the modulator/booster can's input, with the machine operating in the playback mode, and found that there was no signal. A check on the play mode supply line DLPB5V at pin 4 of CN3201, the 11-pin plug to the preamplifier module, produced a reading of only 3.6 V. We traced the source of this back to D110 (1N4001), which turned out to be faulty. It provides the PB5 supply from 'always 5' via switch transistor Q105.

Machine wouldn't load: So I got my meter and went to the loading drive i.c. to check the voltages, but it wasn't there! When the panel was lifted it was found at the bottom of the machine. Replacing it restored normal operation. Apparently a cassette had got jammed: it seems that the chip had so overheated while trying to load the cassette that the solder had melted. I fitted a new chip to be on the safe side.

Loud noise in rewind mode and would shut down in play: The cure is to remove the take-up spool and push the two halves together.

No supply to loading motor drive: There was no 15 V supply to the loading motor drive chip in this machine. After a few checks we found that D212 was open-circuit.

Tracking error caused by label: The playback picture produced by this VCR had a slight tracking error at the top, as if the back tension needed adjustment. When the top cover was removed and the tape path was examined we found that a sticky label, of the type used to index cassettes, was stuck to the entry guide. Removing it restored normal playback. I seem to recall having a similar fault with a Ferguson 3V55 some years ago. On that occasion, however, the label was stuck to the underside of the take-up spool. As a result the machine shut down because there were no reel pulses. In addition the label was not as obvious and not as easy to remove – the bottom of the machine and the reel sensor assembly had to be withdrawn in order to gain access to it.

No deck functions: The usual cause of this is loss of the power supply to the motor and servo sections because D112 (1N4001) has failed. Experience has shown that to ensure reliability it's wise to replace D108, D109, D110 and D123 as well. They are all type 1N4001 and are on the main, not the power supply, panel.

No drum and capstan rotation, tape chewing: You'll probably find that D109 on the main panel is open-circuit. This diode is in the 5 V supply.

No response when cassette loaded: A scope check on the logic condition at input pin 5 of the KA8301/BA6209 loading motor drive chip IC206 showed that it changed state. So we replaced the chip, fitting a BA6209 as it seems to be much more reliable. This brought the machine back to life. The BA6219 is also a suitable replacement – it's higher rated than the BA6209.

Samsung SI3240

No clock display: The machine worked correctly in every other respect. Checks showed that the 3.4 V a.c. supply across the end pins of the fluorescent display was missing. The cause of this was defective pads (10 and 11) at the mains transformer. Hardwiring these connections cured the fault.

Various intermittent problems: These problems were: no functions, tapes jamming, no or a flashing display and sticking in standby. We found that the always 5 V rail was intermittently low or missing. It took some time to discover that the print track below the metal plate which covers the mode switch had been sparking across to the plate (earth). Some PCB sealant was applied and the machine was then given a good test run. After this it was declared to be OK.

Tape stuck, quarter arm out of sync: Check the gears and arm for damage, replace any defective parts and align the mechanism. The most likely cause of the fault, however, is a broken brake trigger spring on the loading motor spindle. So check this as well!

Samsung SI3240 and SI3260

Would load, but unlaced when play was selected: The tape would be left hanging out of the cassette. Fast forward and rewind were OK, however. Very low capstan motor torque was the cause of the trouble. The torque control circuit consists of Q102, D108, D109 and D110. In the play mode power is fed to the motor via the three diodes. In the fast-forward and rewind modes the three diodes are switched out by Q102. The cause of the trouble was D109, which introduced a voltage drop of about 4 V although it tested OK on an ohmmeter. All three diodes are type 1N4001, and I suspect that they came from the same bad batch that affects Models SI1240/1260. Replace all three to avoid comebacks.

Cassette loading problem: This is quite common with these machines. There's a modified side plate for the carriage as well as a different connect gear (the front loading drive comes from the main mechanism loading motor). These parts are available from a number of sources – but beware they are sometimes up to 400 per cent more expensive than from Samsung, which charges just over a pound for that side plate!

Mechanical problems with the housing: Cassettes being jammed in the housing intermittently, poor eject, the housing going out of sync and other housing faults can be cured by replacing the whole right-hand side of the cassette housing with a new, modified version, part number 62203–0025–01. Don't mess about ordering new cogs for the old one: the whole side chassis costs just over a pound!

Poor E-E picture stability: For this fault, especially when the machine has just been plugged in from cold, replace C4112 (0.47 μF, 50 V),

C4110 (47 µF, 16 V) and C4120 (2.2 µF, 50 V), which are all in the i.f. can. The cause of the problem is that the i.f. chip runs very hot. Hence the three electrolytics dry out.

Samsung SI3260

Tuning had disappeared: When search tuning was tried it took a long time to search and when a signal was found it was very unstable with what looked like hum bars across the picture. In addition the tuning drifted. A check on the 30 V tuning supply produced a reading of only 16 V. The 30 V regulator is fed from a 40 V rail via R108 (1 kΩ) and was zenering at 16 V! A new 30 V regulator put matters right – we replaced R 108 as well as it had become discoloured.

Dead, but after repair it would shut down after a loading tape: This one made us long for the gift of hindsight. It was dead with no AL5 V supply at pin 5 of CN601. This is derived from the AL6 V (or AL5.8 V, depending on which page of the manual you look at) line via the 1N4148 diode D602 which was faulty. But the repair bounced. When the machine came back it shut down after loading a tape. There was also the strange symptom that operating the on/off button removed the channel number from the display but not E-E reception! Checks showed that power control pin 11 of the TC4094 expander chip IC606 was stuck at 1.8 V. A new TC4094 restored correct power supply switching, but the shutdown after loading still occurred. We eventually found that there was a hairline crack near a plastic locating peg hole on the top PCB.

No capstan drive: Rewind and fast forward were both OK, but in play and record there was no capstan drive. A check on the 12 V feed to the capstan motor in play, at pin 15 of CN201, produced a reading of only 2–3 V. Replacing L201 and D110 (1N4001) cured the fault.

E-E signal was blue mute raster: The machine refused to tune in any stations. A check at the collector of Q401, which supplies the tuner's VT (voltage tuning) pin, produced a reading of zero volts. There was no 33 V feed because D105 (1N4001) was open-circuit.

Customer had levered a tape out of this machine: Fortunately the carriage was intact, but its timing was wrong. When this had been corrected the machine wouldn't thread up. An external d.c. voltage fed to the loading motor with the carriage removed proved that the mechanism was partly jammed. The only solution was to retime the machine – the main cam was almost 180° out. Once the timing had been reset everything worked well.

Capstan didn't turn: This meant that the machine was damaging tapes. We found that D108 was going open-circuit under load. As a precaution we also replaced D109 and D110.

Samsung SI7220

Loss of E-E sound: The problem cleared when the bottom cover was removed, but could be made to come and go by slight pressure on the main panel. We soon found the cause. There was a 100 μF capacitor mounted on the underside of the panel, a production modification. Its negative lead should be connected to chassis but bridged from where it was soldered to an adjacent land. This land is the audio feed from the tuner/input select area via pin 3 of connector CN401, and was thus shorting out the signal.

Black and white playback with hum bar: This fault was a little disconcerting. When a good recording was played back the picture was in black and white with severe distortion in the form of horizontal pulling – like hum. The fault cleared when the YC panel was lifted to make measurements, and I soon found that messing about with connector CN3201 could provoke and clear the symptom. The cause of the trouble was a high-resistance connection to pin 5, via which the playback f.m. from the head amplifier passes to the YC processing section. A cure was achieved by soldering the lead to the board. This step was necessary because Samsung tell us that they don't stock as spares any leads/looms/connectors, and this one couldn't be repaired successfully. We've had similar problems with other Samsung models, but nothing that can be noted as a stock fault.

Vertical jitter on playback: The symptom suggested that the setting of the PG shifter VR201 was incorrect. When VR201 was adjusted the symptom varied but couldn't be cured completely. In fact VR201 couldn't be set correctly. This suggested that the drum motor was the cause of the problem. A replacement proved this to be the case.

Dead machine: Calls to a completely dead machine – no clock or anything are becoming common with this model. The cause is a locked-up microcomputer chip. Remove the mains supply for a few seconds then reconnect it and all will be well.

No r.f. output: The supply to the modulator was missing. This comes from the regulator transistor Q105 which was without its 15 V input. A choke and diode deliver this supply. The latter (D114), which is in the power supply section of the main PCB, was open-circuit. A new 1N4002 restored the signals.

Intermittent servo faults: The causes of intermittent servo faults can be frustrating and time consuming to trace. This machine was no exception. It came in with the complaint that the sound suffered from wow and flutter while tracking noise bars moved up the picture. After resoldering numerous suspect joints on the motor control/servo subpanel to no effect we decided to check back to the audio/control head. The screen on the connector plug from the head to the main PCB proved to be loose.

Samsung SI7230

Would lace then unlace in play: When play was selected the drum rotated and the tape laced up. It then unlaced and the machine shut down. There was no rewind and no fast forward operation. The basic cause of all this was no capstan movement. R244 (3.3 Ω) in the 15 V feed to the BA6209 capstan drive chip was open-circuit. A replacement immediately burnt up because the chip itself was faulty. A new chip and feed resistor restored normal operation.

Unreliable eject: After fitting a new loading belt and release belt I received a recall. The final solution was a new motor, CPC code number SS64769–052–140.

Always ejecting: Although there wasn't a tape in this machine it continuously tried to eject the carriage, which would have spent the rest of its life going in and out if it hadn't been for the carriage lock hook. This kept the carriage in the eject position. All deck functions worked normally when a tape was inserted and loaded by hand – until eject was selected. We then had a repeat performance. The cause of the trouble was the tape-in leaf switch that's mounted on top of the carriage. There was 5 V at both connections irrespective of the position of the contacts. Closer inspection revealed that one of the contacts was slightly twisted and was thus permanently closed. We were able to untwist the contact using small, long-nose pliers. This restored correct switch and machine operation.

Samsung VI375

Completely dead with a tape loaded in the mechanism: A check in the power supply showed that the 2.7 Ω surge limiter resistor R901 was open-circuit. The usual cause of this is failure of the STR11006 power regulator chip. When this chip and the 2.7 Ω resistor had been replaced the power supply squeaked loudly. Cold checks revealed that the 22 V zener diode ZD101 was short-circuit. A replacement immediately failed. When we removed the power supply from the rest of the machine and lifted ZD101 from the board we found that the 16 V supply had risen to 29 V. There was obviously another fault. The culprit turned out to be the 100 μF, 25 V, 110° electrolytic capacitor C110, which was leaky. If you have difficulty locating it, you'll find that it has a small rubber cap glued to the top. When this item had been replaced the 16 V line returned to its normal voltage. After replacing the zener diode and refitting the power supply the machine worked perfectly.

Dead with no clock or functions: Checks in the power supply showed that IC101 (STR11006) was short-circuit, R101 open-circuit and ZD101 short-circuit. As ZD101 provides over voltage protection, it seemed that the power supply outputs had gone high before these various items failed. The culprit was C110 (100 μF) which had fallen in value to around 10 μF.

No results: The 2.7 Ω surge limiter resistor R101 was open-circuit and the STR11006 regulator chip IC101 short-circuit. In addition the 22 V zener diode ZD101 on the secondary side of the chopper transformer was short-circuit. When these items were replaced and the machine was powered up they all failed again. Further investigation led us to C110 (100 μF) which was open-circuit.

Dead machine: There's a repair kit for the power supply, called the 6 WINNER1 kit – it's available from Samsung or CPC. When we fitted one the machine worked but the brightness of the clock display varied. In fact it sometimes almost went out. Replacing D109 and C118 in the power supply cured this problem.

Samsung VI611

Dead slow front loading: The fault with this machine was dead slow front loading and slow lacing followed by immediate unlacing. No loading motor voltages could be found in the manual, but comparison checks with another machine exonerated the electronics. Although the motor felt OK in operation fitting a replacement restored full speed loading.

No rewind or fast forward: The fault with this machine was due to a worn-out idler unit. Since the machine was a new one the fault could become quite common.

Would not change channel: This machine would not change channel and we discovered that there was no 12 V supply from the power unit. Transistor Q2 and the i.c. stabilizer were both overheating – Q2 feeds the chip, which had no output. Replacing these items restored normal operation.

Tape left out when the cassette is ejected: This looks at first sight like a case of poor rewind torque. Usually, however, it's that biggest blight of our lives, the mechanism state-switch. You'll find it hidden between the subdeck and case moulding webs.

Wouldn't record off air: I came across this machine in a friend's shop. It played all right but wouldn't record off air. The E-E display consisted of a blank raster with a murky bar near the top and the tuning was not precise. I suggested scoping the video detector's output but the scope was broken. So I injected a 39.5 MHz signal from a signal generator into the i.f. amplifier. This showed that the i.f. section was all right. The next step was to decouple the various feeds to the tuner. First the a.g.c., then the a.f.c. and finally, using a 22 µF capacitor, the tuning voltage at pin 2. This last action restored the picture and sound. When the voltage was traced back to source I found that C4 (47 µF, 100 V) in the 52 V part of the power supply was open-circuit.

No signals: Here's a quickie on this range of models. For no signals check R6 (3.9 kΩ) on the power supply PCB – it feeds the 33 V regulator. We've had this resistor go open-circuit on a number of occasions.

Loading fault: The loading motor tried to push the cassette tray out when it was already out with no cassette in it. After a few seconds the motor stopped and the power LED blinked. The cause of all this was switch 'a' on the lift assembly. It's the one beneath the stopper cam that prevents a second tape being inserted when one is already in. It also acts when the tray reaches its extreme unloaded position. As it was open it didn't tell the microcontroller chip to stop.

Hum bar in E-E mode only: We found that C4 (47 μF, 100 V) was open-circuit. As a result the voltage at pin 18 'PRST VTG' was low. It should be 33 V (this voltage is not shown in the service manual).

Tape would only run for a second in any mode: The machine would then go into the 'emergency state', with the standby LED flashing. The cause of the fault was in the rotation sensor section. As cleaning made no difference, we checked at pin 3 of the sensor – you are able to see it when the bottom cover is removed. A square wave was present when the disc was rotated, but its amplitude was only 2 V. A new sensor (part number 62309110243) produced pulses of greater than 4 V amplitude, the signal now reaching pin 55 of the syscon chip. This restored normal operation.

Machine would stop intermittently: On investigation we found that the take-up reel pulses were weak or absent. A new reel sensor cleared the fault.

Clock display intermittent: There should be –24 V at pin 17 of the power supply's output socket: when the display was out there was only –1 V here. The –24 V supply is derived from the mains transformer by rectifier D1 (1N4002) and its reservoir capacitor C3 (47 μF), with stabilization by Q1 and ZD2. As you often find in Samsung machines, some of the components are glued to the PCB. The trouble is that the glue becomes conductive. This was the cause of the problem – when the fault was present Q1 was bottomed. Removing the glue cured the fault.

Samsung VI616/626

Plays back with noisy chroma, and would record only in monochrome: This was a brand-new stock machine. I started out by assuming that one fault was the cause of both problems – not so! A scope check showed that chroma was entering IC0301 in the record mode, but it was not coming out at pin 1. Everything around the chip seemed to be OK crystals running, etc. – so the chip was replaced. This gave us colour recording, after slight adjustment of the record chroma level. The other fault was cured by replacing IC0303 (μPC1536C). One very important point to remember with the μPC1534C chip (IC0301) is that pin 28 is not connected – it should be bent underneath prior to insertion or cut off. Nasty effects will occur if pin 28 is connected.

Rapid drifting off tune: This was traced to a number of dry-joints on the tuner – it seemed as if it might have had a fair knock from above. We've traced a number of faults in these machines to dry-joints.

No sound in E-E mode and the channel stuck on 1: The sound was being muted by an output (pin 13) from the μPC1363 channel selection chip IC800 on the front panel. This chip has two inputs, channel up and down, one of which pulses low momentarily to shift the channel. We found that the channel-up input (pin 16) was permanently low, because the 15 kΩ pull-up resistor R0916 on the 'joint PCB' (mounted upright on the inside of the VCR's plastic frame) was open-circuit.

No remote control operation: This was because of a dry-joint at the base of the DTA144 transistor Q0609 on the timer microcomputer PCB.

Samsung VI710

Playback was fine but there was no E-E sound or vision: Inputs from the line or tuner were missing. There was vision at pin 4 of the TA8605N chip IC303, but only hash at pin 10. The amplifier inside the chip had to be all right as it fed both the playback and E-E video through to pin 10. So the cause of the problem had to be in the switching. The chip was stuck in the playback mode as pin 13 remained high. But the PB 5 V was switching correctly – so where was the 5 V coming from? Not via the white-clip network but from within the chip itself. We proved this by disconnecting the pin and finding that the 5 V was still present here. A new chip restored normal operation.

No rewind or fast forward: This was found to be due to absence of the 15 V supply at pin 8 of IC206. R244 (3.3 Ω, 1 W) which feeds this supply to the BA6209 capstan drive chip was open-circuit. After replacing this resistor the machine went into permanent search when play was selected. An open-circuit between the print and pin 3 of CN204 was the cause. This is the 'cap drive' input and the fault could conceivably have occurred during the earlier replacement. A final check showed that normal deck operation had been restored but the E-E signal suffered from a.g.c. overload. The cause of this final fault was a defective tuner.

No erasure: If the problem is that the old sound is left on the tape and there are floating colour blobs on playback of the machine's own recordings look no further than L0504, which is a little oscillator module in a screening can. It's prone to staging a mini bonfire inside. Willow Vale can supply replacements under part number 79710CB.

Carriage out of sync with the rest of the mechanism: There's no separate front-loading motor in these machines, the drive coming from the main mechanism loading motor. Retiming got the machine working but it would snap intermittently as teeth slipped. There was too much lateral play in the front-loading mechanism because of a crack in the right-hand black plastic side piece where the cogs mount.

No display: A few checks in the power supply soon revealed that the 6.8 V zener diode ZD102 was faulty, a replacement putting matters right.

Dead: A check on the 30 V supply at pin 10 of CN101 showed that the voltage here was low. The cause was that R109 (4.7 kΩ) which biases the base of transistor Q102 was open-circuit.

Dead: While plenty of unregulated voltage reached the STK5333 multi-regulator chip no supplies left it. A replacement restored normal operation.

No functions and no display: This was because the STK5333 regulator chip didn't provide any switched voltages. The 25 V input from the bridge rectifier was correct at pin 8, but there was no 30 V input at pin 5. This comes from a separate winding on the mains transformer, via a rectifier diode and 560 Ω resistor (R104). The resistor had gone open-circuit.

No clock display: We found that the 6.8 V zener diode ZD102 in the 30 V regulator circuit on the power supply panel was short-circuit.

Dead deck: When the power switch was pressed the channel indicator in the clock display lit up but not the indicator at the bottom. The deck was dead. The output from the power unit has a PC15 that didn't come on with the power control command, which was present. We found that the 15 V section of the STK5333 power regulator chip was open-circuit.

Failure of E-E or playback picture: Every couple of days this machine would fail to produce an E-E or playback picture. As there was no r.f. output the monitor just displayed noise. When the fault finally showed up in the workshop I noticed that the deck functions worked but there was no test signal and the channels wouldn't search. As a start I checked the power supply outputs at connector CN104. They were all correct. But something seemed to be amiss in the power supply system. The 15 V line feeds the 12 V regulator transistor Q105. A voltage check at Q105's collector produced a reading of 0 V. The cause was a dry-joint at L105. Relief all round: when this coil was prodded the machine came to life. Resoldering L105, also D114 and Q105 to be on the safe side, restored reliable operation.

Chewed tape on eject: The tape was of course not being wound back in. We assumed that the cause of the fault was mechanical and replaced the idler/clutch assembly, then checked the brake and soft-brake assemblies. None of this made any difference. As the subpanel at the back of the deck is prone to dry-joints, causing various symptoms, this was next removed and checked. Once again we drew a blank. Eventually the cause of the trouble turned out to be the BA6209 capstan motor drive chip IC206 – it's on the subpanel we'd just soldered up. A replacement cleared the fault.

No record: Investigation showed that the REC 9 V supply to the preamplifier on the luminance/chrominance panel (at the top) was missing. The reason for this was that diode D0305 at the front left of this panel was open-circuit.

Video head failed?: This machine displayed all the symptoms of a defective video head, but fitting a replacement made no difference. It was beginning to look as though the lower drum may have been the culprit – until we learnt about the machine's history. It had spent a lengthy period at another workshop and had eventually been retrieved by its disgruntled owner who had brought it to us for assessment. We put off changing the lower drum and concentrated on the head amplifier module. When we unplugged the unit to check the rotary transformer connections and continuity we soon realized that the module could easily be reconnected in the wrong position. A sigh of relief was breathed when this proved to be the case. After a clean and service, the machine was returned to its grateful owner.

Machine totally dead, tape stuck inside: We're not familiar with this particular model. Fortunately, we noticed that R109 in the power supply had disintegrated. After fitting a replacement the machine powered up. We then discovered, not surprisingly, that the idler was worn. Changing it didn't look too easy, as the carriage has to be removed. We overcame the problem by turning the machine upside down, removing the idler pulley from its assembly then, after replacing the tyre, refitting it.

Monochrome playback: There was also no vision in the E-E mode. Power supply checks showed that there was no supply to the 3-pin regulator fixed to the lower drum assembly. The cause was the $2\,\Omega$ safety resistor FR02 being open-circuit. No cause for its failure could be found.

Display produced random flashes and there were no other functions: When one of these machines comes in with the no operation symptom you can usually bet that the STK5333 regulator chip is faulty. On this occasion it was OK, the cause of the rather unusual symptom being the $3300\,\mu\text{F}$, 16 V smoothing capacitor C103.

Samsung VI730

Machine stopped in partially laced condition: As the $1.2\,\Omega$, 0.5 W protection resistor (FR101) for the STK5333 voltage regulator was open-circuit the machine was dead. The cause of the failure was a loose fuse clip in the mains plug.

Blank raster with no sound although playback was normal: The supply voltages were correct at the tuner and i.f. cans, but no video emerged from the latter. When pin 1 of IC404 (TA7348) was lifted the video returned on the scope's screen. IC404 had a crack down its centre. The video signal was now being weakly displayed but there was still no sound. This was because the TA7348 audio switching chip IC403 was also faulty. The video coupling capacitor C415 ($10\,\mu\text{F}$, 25 V) was short-circuit and the video input socket's terminating resistor R416 ($75\,\Omega$) was open-circuit. We feel that the customer knows more than he is prepared to reveal about the causes!

Dropouts: The symptom was long, black streaks that ran across the picture. As a start we decided to set up the CCD level control as laid down in the manual. Connect a scope to TP3303 and adjust VR3301 for a video level of 0.6 V peak-to-peak it said. In fact the video signal was missing at TP3303, although it was present at pin 4 of the 1 H delay chip IC3302. The only item in between is a 3 MHz low-pass filter, FL3303, which was open-circuit. We took one from a scrap machine and after setting up the dropout compensation operation was back to normal.

Samsung VI910

Dead: We replaced the blown fuses and switched on. The capstan started to run straight away but the operate switch did nothing. Then the fuse blew again. The cause of the trouble was the mica insulator under Q1 (2SC1983) in the power supply – it was leaky.

Intermittent cutout on playback: No cause could at first be found: the most likely suspect, the reel drive idler, appeared to be fine. After the machine had worked correctly for several days it came to a stop – when we noticed it the tape had unlaced and the machine was in the off mode. We restarted it and it ran for only a few seconds before stopping again. This time we noticed that the power LED went out momentarily just as the machine cut out. We restarted it again, with a meter handy. This time when the fault occurred the machine just stopped dead with the tape still fully laced. We quickly found that the 13 V supply, from which all the other supplies are derived, was missing. The cause was the 3132 V regulator chip IC1. It could be turned on again by applying just a few drops of freezer to its case.

Field roll: The cause was insufficient back tension because of sticky grease on the back-tension post bearing. Cleaning and relubrication provides a cure – the cleaning needs to be thorough.

Weak r.f. output: A damaged r.f. converter was the cause. The VI920 is sold complete with a cloth carry bag and shoulder strap. The owner was in the habit of leaving the r.f. lead connected to the machine when he carried it. When he lowered the bag to the floor the plug would land first, transferring the force to the r.f. socket which as a result had been pushed in. The PCB was damaged, but a repair was possible using a small iron and fine wire.

Samsung VIK310

Would cut out intermittently during record or playback: I found that the take-up was jerky and would very intermittently stop. The cause of this is a stiff 'idler arm' – the felt pad wears.

Mechanism out of alignment: After realigning it I found that the carriage would go back and forth on its own. The BOT sensor, which is mounted on the main PCB, also acts as a tape inserted sensor. It was dry-jointed.

Spilled tape out in play: In addition, fast-forward operation was noisy. On investigation we found that a piece of paper had jammed in the teeth of the take-up reel. Removing it restored correct operation.

Would sometimes jam and go to standby when a tape was inserted: Alternatively, when a cassette was ejected the indicator would sometimes keep flashing and the capstan motor would continue to turn. The mode switch was faulty.

Samsung VIK316

Dead, no functions: Checks in the power supply revealed that the always 5.8 V supply at pin 6 of connector CN02 was missing. The rectifier diode in this supply (D34) was OK, but its 470 μF, 16 V reservoir capacitor C35 had dried up. When a replacement had been fitted the machine sprang to life, but the display was rather dimly lit. The cause of this was another dried-up capacitor, this time C38 (100 μF, 10 V).

Dead: The obvious power supply checks fuses etc were carried out – but everything appeared to be in order. Time to get the circuit and check some voltages. This revealed that the always 5.8 V supply was low at 3.5 V. I decided to replace C35 (400 μF, 16 V) and C36 (330 μF, 16 V), after which the machine fired up and worked perfectly.

Very dim display: The display was invisible in standby, and barely visible when powered up. Otherwise the unit worked fine. The filament supply was at 2.5 V, with 6 V peak-to-peak of hash on it. This pointed to capacitor trouble, and we duly found that C38 (100 μF, 10 V) was open-circuit. The correct filament supply voltage is not quoted in the manual: with the machine working correctly I found that the reading was 5.72 V.

Display disappeared at switch-on: When these VCRs are switched to standby, the display should dim. In this case it disappeared. Capacitors C37 and C38 in the power supply were faulty. We also replaced C35, since this tends to give trouble. For those of you who have not come across failure of C35, the symptom is that the machine starts up then shuts down.

Dull display, otherwise OK: A check on the VF+/− supply produced a reading of 5 V a.c. instead of about 4 V d.c. with 1 V a.c. The fault was caused by C38 (100 μF, 10 V).

Samsung VIK320

Lift would shuttle back and forth with no cassette inserted: On inspection I found that the tape-start sensor was dry-jointed. Resoldering this cured the problem. We've since had the same symptom caused by dry-joints on the LED tower and tape-stop sensor.

Went dead within minutes of installation: The input fuse F101 was black and the DG06M bridge rectifier BD101 was short-circuit.

Does 'weird' things on its own: Watch out for this one. If the machine starts to do weird things by itself, e.g. trying to load with no cassette inserted, or the cassette flap 'flapping' with no tape in, check for dry-joints at the lighthouse.

Dead, no power and no clock: An all too common set of symptoms with these machines. After fitting the power supply repair kit the power came up but the servo was drifting in both playback and record, as though there were no control pulses. A new servo chip (IC201) cured the problem.

Samsung VIK326

Wouldn't tune in any channels: This machine searched for stations but wouldn't find any. When I checked the i.f. output from the tuner during search I found a reasonable signal as the tuner passed through the station, but there was no output from pin 13 of IC401. Replacing this chip cured the fault.

No playback picture: There was just a blank raster. Resoldering all the joints on the head amplifier connector CN302 cured the fault.

Wouldn't come out of standby: The clock display (unset) was present, but when the power-on key was pressed nothing happened. When a cassette was inserted the unit tried to power up then returned to standby. One clue as to the cause of the fault was present – a low level but raucous noise from the power supply can. Voltage checks here showed that the ever 5.9 V supply was low at about 4.8 V. As a result, the 5 V supplies were too low to be of use. A scope check showed that there was a lot of hash on the supply. C35 (470 µF, 16 V) was open-circuit.

Would power up, then return to standby: In fact this machine did it so quickly that we didn't have time to carry out any checks in the power supply. So we placed a short-circuit across the collector and emitter of the power control transistor Q158, thereby keeping the supply lines switched on. In this condition we found that the 5 V supply was low at 3.5 V. The culprit was the 8 V supply's smoothing capacitor C35, which had dried up. It's a 470 µF, 16 V 105°C electrolytic.

Power supply problems: This machine would start up, then trip. If any load on the secondary side of the circuit was disconnected the power supply would run. The cause of the trouble was C35. Thank you Samsung Technical for help with this one.

Blue screen in E-E mode: The tuner and the 33 V stabilizer were both OK. I traced the cause of the problem to bad contacts at the pins of the power supply plug and socket.

Sanyo

Sanyo VHR1100

Reel turntables didn't turn: The machine reverted to stop soon after loading and during the unloading process spilled tape all over the front of the deck. We found that fuse F3001 on board SY1 was open-circuit. An ammeter connected in its place showed that a normal few hundred milliamps passed when the forward functions were selected, but the current rose to over 2 A when the reel motor was asked to go backwards. The output section of the BA6209 reel motor drive chip IC3006 consists of a transistor bridge arrangement. One or more of the transistors here must have gone short-circuit. A new fuse and BA6209 restored the machine to normal operation.

Machine damaged tapes: The two tapes supplied by the customer had creased sections hanging out of the cassette, both very near the beginning of the tape. When a workshop Panasonic tape was inserted, rewound and play was selected everything appeared to be all right. We then tried the customer's Scotch tape. After a few seconds a loop of tape started to grow slowly between the pinch roller and the take-up side of the cassette. The tape was rewound to remove the loop, then taken out. We again tried with the Panasonic tape, in the playback mode. As the tape swung round the swing guide (the one between the pinch roller and the cassette) it could be seen to lift clear then tighten up again cyclically, but it didn't form a loop. The problem was that at the beginning of the tape the take-up drive wasn't going fast enough to keep the tape taught, although the drive could cope further through the tape. The cause was a defective reel drive motor. Interesting, the different effects with the two types of tape.

It threaded, then immediately unthreaded: An intermittent fault on this machine took some time to trace and cure. The machine would sometimes thread up then immediately unthread again, with the head drum turning too slowly. Sometimes during play the drum and capstan would both slow down dramatically to give a screen full of noise and very slurred sound. Any attempt at diagnosis would restore normal operation. We finally found that the cause of the trouble was loss of the subcarrier reference feed at pin 42 of the servo jungle chip IC4001. The feed capacitor C1102 was going open-circuit intermittently.

Won't lace up and/or nasty smell: If you encounter one of these machines that won't lace up and/or produces a nasty smell or puff of smoke when asked to record or play, check R3110 which is associated with IC3005 on the SY-1 board. If it's overheating, the loading motor is drawing excessive current. Depending on how many goes the user has attempted, the motor drive chip IC3005 (BA6238A) may or may not have survived.

Would perform any deck functions: If you come across a dead VCR of this type, one which will not eject a tape, perform any deck functions or respond to the remote control handset, the likelihood is that its on/off key is stuck in. If so it will come up on 'operate' (rather than standby) as soon as the machine is plugged into the mains supply. The cause is usually physical, the on/off button being stuck, jammed or 'gunged up'. Check by removing the front cover.

Machine won't switch on: Although the clock/display works OK. We've had more than one of these VCRs with this fault. Checks have shown that the power supply and the syscon microcomputer chip were doing their stuff, the cause of the problem being that the 12V switch transistor Q5006 is open-circuit. The equivalent books say that a BC328 can be used as a substitute. In practice we've found that this transistor is not man enough and recommend that the correct type is ordered and fitted.

Sanyo VHR1300

Poor pictures, heads badly worn: When the heads were replaced things looked fine. While the machine continued to play the test tape I relocated the PCB that hinges over the mechanism: the drum and capstan servos then began to vary widely. By flexing the board both motors would stop and the machine would unlace. The fault could be provoked by prodding or poking in the servo area of the board to any degree. As the fault was so general I checked the servo d.c. supplies and found that the always 5 V and 15 V voltages disappeared between CN1003 and CN1004. The tracks that link these two connectors run right at the back of the panel, over the r.f. connection sockets, and were all broken in two places each. A point to note if you have a heavy-handed customer.

Intermittent tuning drift: As all channels appeared to be affected we checked the 33 V supply, which was slightly high. We also noticed that the panel (the timer/tuning panel) in the vicinity of the 33 V regulator IC6206 was brown and showed signs of overheating. The supply to this 33 V regulator chip is provided by a constant-current regulator arrangement on the power supply/system control panel. Note that there are two different circuit diagrams in the manual, this was the more complex one. The voltages around the regulator transistor Q5004 (2SA984) didn't agree with those in the manual, but the transistor, along with diode D5003 and zener diode D5004, were all OK. Resistor checks were then carried out. The emitter resistor R5010, which is used to sense the current, was found to be only $270 \, \Omega$ instead of $560 \, \Omega$, the wrong value had been fitted. In addition R5015 was $560 \, \Omega$ instead of $1.8 \, k\Omega$. Both resistors were original parts and had been in the machine for around three–four years. Maybe this was a one-off occurrence, or maybe more machines with these errors will start to show up soon.

Intermittently stopping in play or record: It was then difficult to get the machine going again. We ran the machine in the workshop for four days and it never stopped once. When it was returned the fault immediately occurred – isn't that always the way? – so back it came. This time we were able to see the fault. When play was selected the machine would lace up and would then straight away unlace and stop. Take-up was present, the capstan turned, so did the drum. As the machine started to unlace just after the capstan and take-up started, leaving insufficient time for a cutout operation from either of these functions, I checked the head-switching waveform at pin 28 of IC4002 on the top PCB. It was absent. As it's fed to the system control chip to provide an indication of drum rotation (or lack of it) this was the reason for the machine stopping. The pulses from the PG coil enter IC4002 at pin 25. There were no pulses here either, because the pick-up coil was open-circuit – the slightest pressure would correct the problem. To restore correct operation we had to order and replace the complete stator assembly.

Sanyo VHR135

Would not do more than two or three timed recordings in sequence: The complaint with this machine was that it wouldn't do more than two or three timed recordings in sequence. I called at the house and set it up to do five 2-minute recordings at 2-minute intervals. It went through the sequence without error, so I came to the conclusion that the problem was caused by operator error. I set up a further sequence of half-hour recordings and left to make other calls. When I returned in the late afternoon the machine should have been half way through its last recording period. Instead it was off and the cassette had been ejected. On questioning the customer I discovered that on occasions when a blank cassette was inserted the machine would go into automatic play as though it was a protected cassette. Light dawned. The record protection switch looked OK and produced a good continuity reading when operated, but I fitted a temporary link across it and asked the customer what programme he wanted to record that night. I set up a full six timed recordings and warned the customer that the protection circuit wouldn't operate until I came back. The following day I found that all six items had been recorded without problem. Obviously dodgy switch contacts persuaded the machine, when it had been in operation for some time, that a protected cassette was loaded – hence the eject. Removing the link and cleaning the switch contacts restored normal operation, but to be on the safe side a new switch was ordered and fitted.

Intermittent working: We found that most of the pins at connectors CN301 and CN302 were dry-jointed.

Intermittent failure to take up in play or rewind: The cause of the problem was that the idler was sticking on its shaft. A drop of oil on the pivot shaft was all that was required to restore correct operation.

Large hum bars on E-E picture: The drum was rotating flat out all the time. Scope checks showed that there was 2V of ripple on the 5V supply and 4V of ripple on the 13V supply. The cause was eventually traced to an open-circuit secondary winding on the mains transformer, between pins 78. As a result, only half of this centre-tapped winding was in circuit. When we removed the transformer we saw that one end of the heavy gauge, enamel-coated wire was only wrapped, not soldered, to the terminal. Scraping off the insulating enamel and soldering the pin cured the hum and drum faults.

Sanyo VHR150

Mangles tapes in review mode: This was a simple enough fault, but the cure was very expensive. The machine would sometimes mangle tapes when in the reverse search (review) mode. When we finally got to see what happened, we found that the tape was piling up on the right-hand side of the drum. The cause was excessive friction on the lower drum periphery. We had to replace the lower drum assembly.

Very slow rewind/fast forward: This machine played and searched correctly but there was no or very slow rewind and fast-forward operation. Tests showed that the rewind/FF torque was very low. The clutch seemed to be operating in the wrong mode, supplying play-level torque in the rewind and FF modes. The cause of this was a missing plastic end stop, against which the clutch mode select arm should rest. I inserted a screw in the plastic hole behind the clutch arm to act as an end stop. This restored normal rewind/FF torque.

Continuous crackle on sound: It sounded like an old telephone system. It was present in record and playback, but only in the LP mode. The cure was replacement of the audio chip IC210 – pin 14 had been pulling the LP/SP switching line to 1.2 V (should be 4.9 V).

Sanyo VHR190

Signals would disappear 20 minutes into the recording: After being recorded for about 20 minutes the f.m. signal would disappear. The bad news is that fault-finding in this area is almost impossible, as you can't lift the board out of its can and operate the machine. The good news is that Sanyo can supply a replacement PCB at a reasonable cost. The part number is 613 123 6110. Fitting a replacement cured the fault.

Machine completely dead: When we checked around in the power supply section with an oscilloscope, however, we found that there were needle pulses at the chopper transformer. The cause of the failure was the 14 V rectifier DS101, which was short-circuit.

Intermittent remote control: I stripped the machine down and removed the remote control receiver, but there were no dry-joints here. On investigating further I found a collection of dry-joints at CN712, which connects the two front panels. Resoldering these dry-joints cured the problem.

Winds tape forward in 'rewind' mode: There was an odd fault with this centre-deck machine. When rewind was selected it would wind the tape forwards, although the display said that the machine was in the rewind mode. As we didn't have a service manual, we gave Sanyo a quick call and were told that a forward–reverse switching voltage is applied to pin 1 of the capstan motor drive unit connector. A meter check showed that this was happening. So we were left with the LB1688 motor drive chip as the only possible culprit. Unfortunately it's not available as a separate item. You have to replace the complete motor unit (ouch!). Anyway, this cured the fault.

Sanyo VHR291

Sound slurring and tape creasing: At some previous stage the machine had been stripped to fit a new reel sensor and had not been reassembled correctly. Dismantling it and reassembling it correctly put matters right.

Intermittent loss of colour in record: The fault could be instigated by going to 'pause' and changing channels. When checking around IC101 (LA7395) we found that all its inputs were correct but when changing channels in the record mode the colour-killer would trip. A new LA7395 chip cured the problem.

Failure to eject cassette: The problem could usually be cured by briefly disconnecting the mains power, thus resetting the microcontroller chip. This didn't always work, however, and was causing the user some frustration. This not uncommon symptom pointed to our old friend the mode select switch. But beware! It's buried under the loading motor block. Thus a service manual is almost essential, to be able to reset the timing marks on the cam gears and sprockets. It's worth checking the condition of the loading motor belt while the loading block is out and you have easy access to it.

Intermittent failure to power up, power down and to eject: The cause of the trouble was dry-joints at plug CN712.

Would stop when in play: Although the take-up sensor was working, it wasn't producing a regular series of pulses. The strange thing was that when the machine was turned on its side or upside down it played all right. The increased space between the bottom of the reel and the faulty sensor somehow improved its response. A new sensor cured the fault.

Sanyo VHR291E

Servicing tips: When you've uncased this VCR you have to lift the main PCB CP1 before work can be carried out on the deck mechanism. The board can be mounted vertically at the rear of the chassis, and to facilitate this a long ribbon cable at the right-hand side is connected to the operation/display board TM1 on the front panel, being long enough when unfolded to allow board CP1 to stand upright. The ribbon cable at the left-hand side of board CP1, connected to the audio level meter board AD2, is shorter and must be disconnected. If audio tests are not involved you can operate the deck with this cable disconnected – alternatively if it's necessary to work on panel AD2 use extension cable 'relay jig' VHJ-0088 which is available from Sanyo. It's common, when deck maintenance or cleaning has been carried out to find that, on test after reassembly, there's an audio problem. It shows up as a whine at about 3 kHz, usually in the right-hand channel only, with both recordings and playback. The cause is that board CP1 has been refitted with the long, right-hand side ribbon cable stowed, as seems to be natural, with a single fold in the clear space in the chassis to the right of the deck mechanism. As a result of this, however, the folded end of the cable towards the rear of the case can become sandwiched between Nicam board TM5 at the bottom of the case and the tuner/VIF/r.f. modulator modules on the underside of board CP1. The problem is that signals from the ribbon cable's clock/data lines are then picked up by the other circuit modules, producing the audio whine. Make sure that when board CP1 is refitted the long ribbon cable is rolled up and stowed in the front section of the case, well away from the other circuit modules. There doesn't seem to be any cautionary note about this in the manual, and the first time we came across the problem there was a lot of head scratching before we solved it.

No E-E sound or vision: There was also no sync detection. T6901 on the sync detector subpanel was found to be dry-jointed.

Sanyo VHR3100

Wouldn't accept a tape: The cassette LED (the infrared light source for the tape end sensors) was not emitting because the SWD5 V line was missing. This comes from the 2SA984 switching transistor Q4001 on the SY-1 board. The transistor was open-circuit at its collector.

No results: This is a common complaint with these VCRs and in all cases we've found that R5001, a $2.7\,\Omega$ 0.5 W safety resistor, has been open-circuit. No other fault has been found. With R5001 open-circuit the always −12 V line drops to 5.5 V and the always −5 V line to 1.3 V.

No record sound: The playback sound was OK. A pre-recorded tape produced a good picture and sound but, as the tape-loading and front-loading mechanism operated, the verticals in the E-E picture bent and hum bars appeared on the screen. The EE picture returned to normal as soon as the mechanism drive motors stopped. With the machine in the record mode, once the tape was fully loaded and the hum bars had gone the E-E sound was lost (muted). Hence the no sound on record problem. As we've had power supply faults with these machines we checked all the regulated rails carefully with a digital meter. They were all within 0.2 V of the readings specified in the manual, and no detectable drop occurred in the loading and unloading modes when the hum bars appeared on the screen. As a working machine was available and the boards can all be removed easily we decided to isolate the cause of the fault by panel swapping. The top signals board seemed to be the most likely source of the problem as most of the sound processing is carried out here. But changing this then the system control panel and the front function and tuning boards made no difference. We finally swapped the power supply panel. This cured the no sound on record and the hum bars with the motors running problems. A scope check soon revealed the source of the fault. There was ripple on the 12 V and 13 V rails. We then found that there was a dry-joint on the reservoir capacitor C5001 (2200 µF, 50 V) at the input to the STR7226 regulator chip. As a result the chip could cope under low load conditions, i.e. in the E-E and playback modes, but on record the extra load produced by the bias oscillator drive to the sound and erase heads caused excessive ripple on the switched 12 V line. This upset the sound mute control in IC2001.

Tape loading problems: The cassette lift would load half way then come part of the way back out again. After this the machine would shut down. If the tape was put into the stop position by hand, rewind and fast forward worked normally. But when play was selected the tape would load, the capstan would start to run very fast, then the tape would unlace and the machine would shut down. The voltages all seemed to be OK. We then checked the mode switch by substitution. Next we suspected the system control microcomputer chip, but the fault was still present when a replacement had been fitted. In the end the cause of the fault turned out to be the L4412 digital servo chip IC4001.

No sound in E-E mode and failure to record sound: The picture was fine and playback of pre-recorded tapes was normal. My first thoughts were of a fault in the tuner/i.f. area, but a glance at the rear of the machine revealed all: the external link that couples the tuner audio-out to audio-in was missing. A temporary wire link produced good sound in all modes. When our field engineer returned the machine he found the link under the base of the TV/video stand. It took him a while to live that one down.

Cassette would not go in: In fact it would only go in only about a quarter of an inch then come out again. On investigation we found that the capstan to idler belt had become very soggy: it had glued itself to the capstan pulley which couldn't turn. A good clean-up and some new belts made the machine as good as new.

No rewind: I tackled this fault the hard way. First, I checked the idler assembly and all the parts around it. Then I ordered the service manual. This isn't exactly comprehensive, so it took quite a lot of observation to discover what was wrong. On the underside of the mechanism there's an assembly that consists of a main slide on top of which (with the unit upside down) there's a subslide. This has an elongated hole and is held in place by a pin, a washer and a plastic retaining clip. In the fault condition the subslide rode over the pin because the space between the washer and the subslide was too great. I carried out a temporary repair by fitting a compression spring between the washer and the plate to hold the subslide tighter against the main slide. This worked very well. I then sent a fax to Sanyo to ask what I should have done. A short time later I received a phone call from Sanyo's Technical Department to tell me that the fault was not an unusual one. All that was necessary was to push the retaining pin back against the subslide and fit a circlip on the other side of the deck to hold it in place. Many thanks Sanyo for a prompt, efficient and polite service.

Power and eject buttons didn't work: Everything else worked – there's a power button on the remote control handset. But there was no remote 'eject' function. The two buttons are on a separate PCB from the timer and all the other buttons. They share it with the tracking control, which worked. The cause of the problem was lack of an earth connection to the two switches: the print was intact, but the lead that should have earthed the area to the mechanism was loose inside the unit. It should have been screwed to the top of the cassette carriage.

Dead: Checks in the power supply brought me to the GZB16C zener diode D5009 which was short-circuit. It's connected to the always 13 V line.

Sanyo VHR315

Dead machine kept blowing fuse: The fuse in question was the N38 fuse PR511. By disconnecting the various 5 V rails we discovered that there was an internal short-circuit in the tuner/i.f. unit. A replacement module restored normal operation.

Cassette jammed inside: The usual cause of is a faulty mode switch, which jams the mechanism. After fitting a replacement we put the machine on test and found that it would sometimes leave tape out on eject. A sticky capstan brake was the cause. Replacing the brake pad and cleaning the capstan cured the problem.

Dead, no functions or clock and the power supply was tripping: We found that IC511, a zener regulator on the secondary side of the power supply, was short-circuit.

Sanyo VHR3300

Intermittently chewing tapes and failing to eject the cassette: This was a difficult puzzle to solve. It took some time to discover, while the machine was going through bouts of working and not working, that fuse 5001 on the power supply panel (PW1) was going open-circuit intermittently. When it opened and closed the sequence was that the loading arms retracted but the tape was not wound into the cassette. Eject was then permitted, with some 20 cm of tape at large in the machine. Messy!

Would intermittently go to standby: You could tap the case anywhere and it would turn off, but the fault could be provoked more easily on the servo/syscon board. On removing the bottom cover the cause of the problem was obvious – the board had jumped out of its retaining clips during transit, and was shorting to chassis via the bottom case.

Would intermittently come out of standby: The clocks worked fine but there was sometimes no response from the on button. If you get this problem cheek the fitting of the front PCB and the clearance between it and the plastic front cover. If one of the control switches is held in the others don't work.

No off-air signals: There was a failure to tune, and the orange channel display was both dim and flashed on/off. The rest of the display was also dim, although the machine provided good playback. We found that the −30 V supply to the display/keyboard was low at −12 V because RS008 (47 Ω, safety type) on power panel PW1 was open-circuit. No overload could be found and the replacement resistor has not failed.

Machine had swirling B/W patterns: This machine had fine E-E sound but instead of a picture there were swirling black-and-white patterns that looked as though they were caused by instability. This led us to investigate the vision i.f. section, but we found that a perfectly good composite video signal emerged from the demodulator. The trouble was caused by IC001 (LA7223) in the video circuit. The instruction entering this route-switching chip was correct and the video signal was present at its input pin 7. But all that emerged from the output pin 1 was hash.

Intermittent failure to record: When the fault was present the machine would go straight into play, even when the cassette's safety tab was unbroken. The cause was a broken tab-sensor switch. We often come across this, although the fault is usually more certain. Our man-on-wheels who brought the machine in was puzzled by the 3 seconds each way wind/rewind cycle performed by the machine each time a tape is loaded. This is programmed into the control microcomputer chip.

E-E problem: A common fault with this machine – we've had it four times so far – is failure of IC1001, a route-switching chip, on circuit board VD1. The symptoms are no E-to-E vision signal, with wavy lines and patterns on the monitor's screen, the E-to-E sound being unimpaired. You'll find that a video signal enters this LA7223 chip at pin 7 but the output is not present at pin 1.

Tape snapped in two: We've had three of these machines with the same evil intermittent fault, which snaps a tape completely in two – sometimes! This model apart, it's rare for any VCR to break a tape although it often occurs when efforts are made to remove a tangled tape. With the VHR3300 the disaster occurs at the beginning of tape threading, due to a reel brake problem. Look at the feed spool's hard brake arm. On its left there's a metal pin which is, or should be, pushed by an underdeck lever. The pin can work upwards in the plastic arm to the point where it rides over the lever. Its correct position is where it just clears the topmost surface of the underdeck plastic cam. Fix it in position with superglue or, better, replace the arm assembly.

Head wear: The audio/control head stack used in this and similar models wears at a relatively fast rate – we've replaced many of them. Signs of failure are: sound level fluctuation; tape fussiness with respect to servo lock with a machine's own recordings; and intermittent miscounting by the tape counter. The only cure is a replacement ACE head assembly.

Loading mechanism jammed: Now that these machines have a few years under their belts they are beginning to produce a new fault – the loading mechanism jammed, with the half-load pole failing to get out of the way of the entry guide on its way back towards the cassette during the unthreading process. Replace the lever assembly no. 79 in the mechanical parts (2) diagram. You'll find that its metal pin has become loose or strained.

Machine dead: The 400 mA fuse F5101 was shattered. As there was no obvious short-circuit in the power supply we replaced the fuse and switched on. The fuse remained intact but the machine was still dead. Further investigation showed that R5001 (2.7 Ω) was open-circuit and that all the pins of the STK7226 power supply regulator chip seemed to be dry-jointed. A thorough resoldering, including many other joints elsewhere in the power supply, brought the machine back to normal operation and after a very long soak test we pronounced it fit.

Picture jumping and tape damage: The first thing we noticed was that someone who shouldn't have had been inside it – there were damaged screw heads and the audio/control head was way out of alignment. This was mainly because one of its levelling screws, the one that retains the coil spring, had been sheared off. After rummaging through our box of worn heads to find replacement screws we carried out a rough realignment then tried powering up. Playback was fair, but there was no E-E picture and no recording. At this point we phoned the customer, who told us that a friend had adjusted the machine to improve the sound! The lack of E-E hadn't been noticed as the machine was mainly used to play hired videos. After some discussion about cost we returned to the machine and carried out some scope checks. These confirmed that the tuner and i.f. sections were in good order: composite video was present at the output of tuner block VD1, and was traced around the board until it disappeared into the LA7223 chip IC001 at pin 7. It didn't reappear at pin 1. This is a video/audio switching chip that doesn't seem to be in any wholesaler's list. A replacement was obtained from Sanyo however (part number 409–114–4407). Fitting it restored normal operation and a general clean, lubrication and realignment completed the repair.

Would stop intermittently in play or record: The deck functioned in all modes but would stop intermittently in play or record. The tape counter showed that the tape was moving steadily, so the reel pulses were initially discounted. They were in fact intermittent: cleaning the opto unit cleared the fault. Don't be misled, as I was, by thinking that the reel pulses drive the counter. It's driven by the output from the audio/control head.

Intermittent tape damage: If you pressed stop during fast forward near the start of a tape the tape would spill out into the mechanism. The cure was to replace a spring in the brake trigger area.

Total jam-up of tape loading mechanism: Now that these machines are ageing a total jam-up of the tape loading mechanism is becoming common. It's usually instigated by failure of the half-load arm to move clear of the exit guide assembly during the unlacing process. The cause is the fact that the metal pin on the half-load lever assembly has listed to starboard. Replacement is the only cure: it's item 79 in the exploded deck view in the manual.

Very intermittent failure to accept tape: This is an increasing problem with these middle-aged machines. The tape goes in, half laces, half ejects, goes back down and is then fully ejected! The culprit is the mode switch, which is more accessible in this than in some Sanyo models.

Distorted, whitewash E-E picture: The effect produced by this fault suggested that its cause lay in the i.f. or a.g.c. circuits. Symptoms were a grossly distorted, soot-and-whitewash E-E picture with patterning and loss of sync. In fact the signal that emerged from the vision detector was perfectly good! The cause of the trouble was a faulty vision-switching chip, IC1001 (LA7223).

Sanyo VHR4350

Head drum wouldn't turn: However, voltage was present at the drum motor and it was free to turn. The cause of the problem turned out to be dry-joints at CN823 on the deck-mounted junction PCB. As a result the motor was off earth. You have to remove the deck to gain access for resoldering – do all the joints while you are at it.

Playback drifts to lines and snow: Recordings were similarly affected. The sound didn't vary, and the capstan speed was correct. The cause of the problem was that the head drum wasn't phase locked, because no PG pulses were being fed back into the servo system. The little PG coil inside the drum motor has a printed link to the motor's connection plug and there was a dry-joint in the circuit. It's not difficult to dismantle the motor and repair it.

Eject accompanied by tape looping with consequent tape damage: Remove the bottom cover and take a look at the reel drive system. There's a 'switched clutch' that slides up and down the gear shaft between the spools. On several occasions we've found the clutch to be tight on the shaft.

Loop of tape left at eject: A loop of tape being left at eject is not uncommon with these full-lace machines. The cause in this case was new to me. As the capstan brake was sticking there was excessive braking and lack of reel drive during the unlacing process. Cleaning proved the point, and replacement cured it.

Intermittent drum servo lock: The effect is similar to that of dirty heads, but it drifts in and out of the picture. Check the internal jointing and plug/socket connections to the drum motor's PG coil. You have to dismantle the upper and lower drum assemblies to get at the stator PCB.

Unreliable cassette front loading: With an afflicted machine there's an even chance that a proffered tape will be drawn in then spat out again. The usual cause is simple: loss of tension in the two finger springs that hold the cassette firmly in the front-loading cradle.

At switch-on the machine went into the fast-forward mode automatically: We first checked the mechanism's alignment which was OK. A replacement capstan motor cured the problem.

Wow on sound after after 1 or 2 hours: This fault is more common with Hitachi than Sanyo VCRs! Soon after the fault developed the capstan would start to stop at intervals of a few seconds. Under the fault condition we found that the drive chip, which is mounted on the capstan motor, was too hot to touch. The entire motor assembly had to be replaced.

Tape crunching: A loop of tape left hanging from the cassette flap on eject is an increasing problem with these machines. Capstan brake binding is the cause. Cleaning the brake pad and the periphery of the flywheel solves the problem, but it's perhaps better to clean the flywheel and replace the brake arm.

Sanyo VHR7250

Failure to accept a tape, drum not turning at switch-on: In our experience this has always been because the 13 V supply is low. Check for dry-joints at D5105, D5106 and D5107.

Would play for 5 seconds and then cut out: This machine would play for about 5 seconds then cut out. I noticed that the loading motor continued to run after loading up. A replacement mode switch cured the fault.

Machine wouldn't accept tape: The drum failed to turn. Checks in the power supply showed that the always 13 V output was low. The cause was a dry-joint at D5107.

Sanyo VTC5150

No playback capstan servo lock: Control track pulses were present at pin 23 of IC4001, and the reference pulses were correct at pin 27. There should have been short 4.5 V pulses at pin 24 but these were absent. C4016 was disconnected in case it was leaky, then the i.c. was changed. Still no luck. After a bit more sorting around we found that R4030 (27 kΩ) read about six billion ohms.

Problem with playback: Fast forward and rewind were OK, but when play was selected the tape would lace up then, after a few seconds, unlace. The fact that the loading motor continued to run after the tape was fully laced drew attention to the after-load switch. It closed but the contacts were oxidized. A squirt of Electrolube cleared the trouble.

Would stop after a few moments in any mode: This old Betamax machine led me a dance. Whichever mode was selected it would stop after a few moments. Since the machine would sit there happily in the pause mode the reel sensor system was dearly implicated. There were pulses from the reel sensor optocoupler in the other modes. These pulses were being amplified sufficiently by Q3012 to keep the tape counter working but not sufficiently for the microcontroller chip IC3001. The optocoupler's output pulses were of low amplitude because the LED and photodiode beneath the take-up reel were thick with dust. A good blow-through cured the problem.

Sharp

SHARP VC381
SHARP VC381H
SHARP VC386
SHARP VC681
SHARP VC7300
SHARP VC750
SHARP VC780HM
SHARP VC8300
SHARP VC9300
SHARP VC9700
SHARP VCA100
SHARP VCA105
SHARP VCA105HM
SHARP VCH81H
SHARP VCM20

Sharp VC381

Most functions didn't work: A quick check on the power supply showed that the 13 V and 12 V outputs at pins 1 and 8 were missing. When the power supply was taken apart we were able to check back through the regulators. This revealed that there was no output from the 2SD1308 14 V regulator transistor Q01. I ordered a replacement and fitting this cured the problem. After a few weeks, however, the machine returned with the complaint that it would load but nothing else worked. This was the case and in addition the loading was very slow. A check on the power supply showed that while the 13 V output was correct off load it dropped to 3 V when a cassette was being loaded and the motor was running. It turned out that our new 2SD1308 transistor was faulty, although this was not obvious at the time due to the difficulty of taking measurements with the power supply apart.

E-E signal present in playback mode: Also the video light didn't come on as it should. The reason for this was that the PLBK 12 V line was not present, the cause being Q806 (2SA950). It's on the bottom syscon/ servo board, at front centre.

Misalignment problem: Realignment usually provides a lasting cure, making repairs justifiable. This particular example suffered from intermittent playback chroma. When the colour was present there was patterning on it. The cure was to reset the carrier peak adjustment slightly. In the record mode there was no picture because the dark clip was misadjusted.

Intermittent clock and timer setting: On the odd occasion when this was possible the machine would begin to load under the control of the timer, then unload with the clock resetting to zero. Severe patterning was evident on playback of a tape, to the extent that the picture was almost obliterated. The E-E pictures remained normal. Scope checks showed that some very bad hash was present on the supply rails. The following capacitors were found to be open-circuit: C12 (100 µF, 16 V), C17 (100 µF, 16 V) and C16 (10 µF, 50 V). As a precaution all other capacitors of this type – there aren't many – were removed and tested.

Intermittent loading fault: Sometimes the cassette would be lowered only half way and remain there. If the eject button was then pressed the cassette would be returned. A meter connected across the carriage motor during the loading process showed that the voltage at the earthed terminal would fluctuate then rise to 12 V, thus stopping the motor. The cause of this was soiled contacts in relay RY802 on the main panel. We carefully prized off the cover and gave it a squirt of switch cleaner. This cured the problem.

Sharp VC381H

Intermittent blank raster in play or E-E mode: This may seem to be obvious, check that the test pattern switch on the modulator isn't dirty.

Would only accept cassette half way: The carriage left-hand microswitch was broken. A replacement cured the problem.

Intermittent cut-out: The customer said that the fault often happened, usually preceded in play by a still frame/pause. This suggested that the capstan was stopping. The first fault I found, however, was that the cassette-down switch was defective, causing all deck functions to cut out.

After replacing this the machine was put on soak test. It was ages before the capstan fault showed up. Its cause turned out to be a dry-joint at pin 1 of IC7006. I resoldered all the connections as they were going the same way.

Sharp VC386

Tuning drift: We thought it was caused by either the PTC (PR01) in the power supply regulator circuit, the tuning voltage regulator IC1403 or intermittent tuning potentiometers. After checking on these points, including fitting a replacement tuning potentiometer unit (this has cured previous problems on similar models), we found that the VTL-7C tuner unit was the cause of the trouble. The customer was not impressed with the cost of the tuner.

No signals, varicap voltage awry: This is a nasty fault, which we've had twice so far. It appears only when channel change is attempted, and then only intermittently. The cause is a dry-joint at socket TA on PWB (printed wiring board!) T. This removes the pulses on the control-1 or control-2 line, whereupon IC1401 throws a wobbly.

Intermittent colour in both the record and playback modes: IC501 proved to be defective, but its replacement cleared only the no playback colour fault. Because of chip tolerances the a.f.c. adjustment was wrong. There's no setting-up procedure in the manual but we found that rotating the a.f.c. adjustment control to give 4.7 V at pin 29 of IC501 provided a complete cure.

Sharp VC681

Jammed cassette: The cause was yet another misaligned mode switch. After realigning the switch we tested the machine in all modes. The sound of tape crinkling could be heard in the play mode – the tape was rubbing against the top edge of the reverse guide assembly. This guide is between the pinch roller and the cassette take-up guide. Inspection of the pinch roller revealed that it had distorted to a barrel shape. A new pinch roller cured the crinkling and hopefully prevented several tapes losing their top edges.

Intermittent refusal to eject: The customer had for some time complained about intermittent operation – the machine would occasionally refuse to eject a tape. We'd soak tested it on a couple of occasions but had not been able to find anything wrong. Now the machine was back in the workshop with a note to say that the tape couldn't be ejected. After releasing the tape manually the machine worked perfectly. We suspected problems with the mode switch but decided to check out the mechanics. When we stripped down the master cam we found that there was considerable wear on the inner surface. As a result the mechanism would occasionally stick. We had no further problems after replacing the master cam. Alignment is critical and has to be followed exactly.

Would remain in search mode: When this machine finds a blank portion of tape it switches into the video search mode and quickly skips through until it comes to the next recording. It then reverts to the play mode. We had one of these machines that would remain in the search mode, however: the only way to stop it was to press the play button. The blank detector circuit is centred around the μPC393C chip IC701. It works by detecting the off-tape line sync pulses. All was well at Q701's collector, but although the signal at pin 3 of IC701 changed when a recording was found the signal at pin 1 didn't change. As replacing IC701 made no difference it seemed that R7106 and R7107 at pin 2, the operational amplifier's non-inverting input, were faulty. But their values were spot on. With no faulty component present we could only conclude that the circuit had gone out of tolerance, perhaps due to age. So a modification was called for. After some consideration we decided that the fault could be cured by reducing the voltage at pin 2 by 0.2 V. The modification consisted of adding a 470 kΩ resistor in parallel with R7107. A good soak test showed that this had cured the fault.

Sharp VC7300

Tape slack at the end of rewind: The loop of tape then gets crunched by the cassette flap when the tape is ejected. Much time can be spent on the reel brakes if you don't realize that the cause of the trouble lies elsewhere. At eject the loading motor, under the deck, kicks to take up the slack, via a belt coupling to the spool turntable. If the latter doesn't move, check the loading (short) belt for slippage and the clutch on the loading block assembly for excessive friction.

No picture: Two machines came in with no picture in the E-E or play modes. In both cases I first blamed the modulator then had to start serious fault-finding. I'd no circuit diagrams for the first machine, a Sharp VC7300, but managed to track down the fault to the HA11703 chip. The second machine was a Panasonic NV730 where the chip that carries out the same functions, i.e. head signal amplifier and E-E/video switching, is an AN6337S. It's on the folded luminance-2 panel. Unfortunately this chip seems to be unavailable – and the board is hideously expensive. Worse still, my meter probe slipped whilst I was monitoring the power supply lines. This damaged the tuner and the BN5115 on the demodulator panel, the result being low gain. I was able to replace these items with parts from a scrap NV366, with some modifications. Note that the manual may not correspond with the actual demodulator PCB or the aerial booster/modulator unit.

Playback picture interference as the VCR warmed up: The playback picture showed increasing interference as the machine warmed up – the effect was similar to that produced by an arcing tripler in a colour TV set. The E-E picture was normal and the sound was not affected. We found that the problem was caused by the HA11703 chip 1403 in which the playback f.m. signal is detected.

Sharp VC750

Intermittently stops during play after a few seconds: We had two machines with this fault. Sometimes the capstan didn't turn, other times the brake lever didn't move over so that the idler didn't contact the take-up reel and the tape spilt out. A new mecha-state switch was tried but made no difference. In desperation I rang the nice man at Sharp. He suggested changing the cam gears, and he was right.

Only lower channels could be tuned: There was also tuning drift. A check showed that there was only 10 V at the 33 V regulator chip IC1405. C1411 was leaky.

Machine's mechanism kept jamming: The cause was the loading gears, which were worn. They would jump a tooth, particularly when unlacing. As a result the back-tension arm fouled the movement of the supply guides. Fitting replacement gears cured the trouble.

Tape stuck: The customer's complaint was that this machine would sometimes jam and now had a tape firmly stuck in it. I found that loading gear A had a broken tooth and that the master cam was worn. Replacing these items and the mode switch put matters right.

Mechanical problem: The complaint with this machine was that it 'would sometimes go straight to still in playback or record, then quickly shut down'. On inspection I noticed that the idler didn't always reach the take-up reel. A more detailed look at the operation revealed that a lever, which is directly driven by a cam, didn't go over far enough. The cause was broken plastic in the grease on the cam. The debris hadn't come from the machine.

Sharp VC780HM

Machine would eject any cassette that was inserted: We soon found that the machine could be fooled into accepting a cassette if it was switched off immediately after inserting the cassette then switched on again. Checks at pins 47 and 14 of the IX0234GE system control chip showed that they changed state before the cassette was ejected. The reason for this became clear when we checked at pins 17 and 18: there were no end or start sensor pulses. When attention was turned to the sensors themselves we found that two tiny black squares of tape had been applied. Where had they come from?

Intermittently goes into play from cold then unlaces: The capstan didn't rotate in the fault condition, which occurred about once a day. Because of the fault's unpredictable nature, and the fact that the machine would revert to stop whenever the fault occurred, it was virtually impossible to do any fault-finding. A new cam switch seemed to provide a cure.

Machine would jam while unlacing: A colleague had been looking at this one but couldn't work out the cause of the problem. On dismantling the mechanism and carrying out a closer examination I found that one of the two cams – not the cam-switch one – had a broken wall on its hidden side. Replacement and retiming got the mechanism working properly again.

Sharp VC8300

Capstan and drum servo fault symptoms: The playback picture showed all the symptoms of both a capstan and a drum servo fault and I initially began checking the circuits that are common to these two loops, i.e. the supply lines, IC701, IC702, IC703 etc. The scope showed that all the relevant ramps, sample pulses and sample/hold d.c. outputs were present, although they were varying wildly as the loops were unlocked. This is usually an indication that the i.c.s are in fact working, but as time went by I was driven to replace IC701 and IC702, only to find that the fault was still present. At this point I decided to try a different line of approach. If I could prove that a drum fault was causing the capstan to unlock, or vice versa, I would have narrowed down the possibilities by 50 per cent. This was my biggest mistake, a further 2 hours were wasted. How did I go about it? By disconnecting the servo loops one at a time and using a variable d.c. supply instead. All this did was to prove that there was indeed a common cause, but what? Then I saw it. The f.e.t. Q703 in the drum sample/hold circuit is biased from the same point as Q707 in the capstan sample/hold circuit. A check on the d.c. conditions revealed that the gate potentials were both low. Further checks led me to the $10\,\mu F$ tantalum capacitor C731 which read $10\,k\Omega$ when measured out of circuit. Needless to say the celebrations went on for some time. Perhaps I'd have found this one sooner if the circuit had been drawn larger, as the common supply via R769 is not at all clear. That's my excuse, anyway!

Intermittent loss of sound in the E-E and playback modes: A lot of heating, probing and flexing were necessary before we discovered the dry-joint to be on the mother board where the audio module plugs in.

Distorted own recordings: This machine worked fine in the playback mode but its own recordings produced a very distorted, monochrome picture. Most of the i.c.s on the Y/C board had new solder on them so, expecting a difficult fault, out came the coffee, manual and scope. The video waveform at TP204 was squashed, but readjusting the white and dark clip level controls cured that – yes, the phantom twiddler had struck again! I now had a monochrome picture on record and the frequency counter was required to get the VXO and a.f.c. controls right. But there was still no colour. A chroma signal was present at pin 3 of plug CB, so attention was directed to the head amplifier PCB where I found that the record chroma current control had been turned right down.

No display: This was fairly quickly traced to an open-circuit digitron heater. A simple fault, but one that in practice is not as common as you might expect.

No carriage operation: The T2.5 A fuse on the power panel was open-circuit. When we replaced the fuse and loaded a cassette the machine very slowly laced up before blowing the fuse again. The loading motor proved to be the cause. Amongst other things it had also damaged the STA401 loading motor drive chip I805.

No record or playback colour: After a number of initial checks I got round to scoping the playback chroma signal and noticed that the channel 2 chroma was excessively noisy. This made the colour-killer operate. If you looked carefully you could pick out the noise on the channel 2 f.m. signal, so I carried out checks in the f.m. and head areas, all to no avail. I then recalled a similar fault with a VC7700, where a d.c.–d.c. converter in the power supply was defective, radiating noise along the chassis lines. But in that instance the noise had affected both f.m. channels. Nevertheless I followed up the hunch and disconnected the chassis earth lead near the full erase head. Colour was immediately restored. When the now floating earth lead was scoped a 45 kHz spike waveform at about 2 V peak-to-peak was seen. Comparing this signal with the d.c.–d.c. chopper converter (Q905) signal confirmed that the chopper was indeed the source of the noise. When I scoped the converter's outputs I found that the noise was on the 15 V line. Replacing C929 (330 µF) provided a complete cure. The lesson is that for no colour with these older Sharp machines, i.e. Models VC7700, VC381 and VC8300, the chroma signal should be checked to see whether it's noisy. If it is the chopper circuit is suspect.

Sharp VC9300

Loss of capstan servo lock: This is becoming a common problem with these machines. The symptom is often noise bars that move slowly through the picture although the sound doesn't wow. If you check the capstan lock voltage at TP3 when the flywheel is slowed by hand you'll find that it doesn't alter. Usually waveform TP1 is missing, due either to D707 being leaky or R475 (82 kΩ) open-circuit. If the drum and capstan servos are both 'wowing', check that the PB 50 Hz signal is steady. If the countdown chip is faulty the signal's mark – space ratio can vary.

No output from modulator: We took it apart carefully and found that L2 in the 12 V supply was open-circuit. This fault can also be caused by an open-circuit (base-to-emitter) transistor Q803 on the mechacon board, which does the supply switching.

No E-E or playback vision: This machine came in from another dealer with the suggestion that the 12 V supply was missing. Sound was OK but there was no E-E or playback vision. Scope checks showed that the video signal was OK up to the HA11703 chip IC402 but didn't emerge at pin 16 of this chip. An incorrect d.c. voltage reading (12 V) at this pin led us to check Q406 (2SC945) which was short-circuit collector-to-base.

Hence the 12 V at pin 16 of IC402. A new transistor put matters right – fortunately the chip hadn't been damaged.

Dew indicator blinking: Also there were no deck functions. Pin 19 (dew sensor input) of the microcontroller chip IC801 was high because the sensor was open-circuit. It's mounted on the cassette flap opener bracket, near the pinch wheel.

No switch-on: It took us rather a long time to locate the cause of this fault. The syscon 'on' command turned on Q9008 in the power supply but Q9005 failed to come on. Thus the regulator wasn't being latched on. The culprit was the electrolytic capacitor C9008, which is a decoupler in Q9005's emitter circuit.

Capstan motor slows intermittently: After fitting a new switch on the cassette housing and servicing the deck we found that the capstan motor would slow down intermittently. The cause was eventually traced to a dry-joint on the f.e.t. Q705.

Capstan motor would stop after a couple of seconds: In the past we've had problems with the STK5725 power supply regulator 19002, so this is where we started. Its 13 V and 18 V outputs were OK, but the 11 V output at pin 1 was missing although the 12.7 V input was present at pin 2. We next found that the power control input at pin 3 was permanently high: it should be low for power-on. Shorting the base of Q9003 to chassis via R9021 brought some life back to the VCR. The power-on low signal comes from pin 24 of the main microcontroller chip I5002. Checks here showed that the 5 V supply at pin 64 was missing. This comes from regulator transistor Q5002, whose 13 V collector supply was missing. Yet the 13 V rail was OK at the power supply end. The syscon PCB in this model uses double-sided print. The track to Q5002's collector is on the component side, where the fault lay. Q5002 and the surrounding components were covered in brown glue that holds a passing bunch of wires. When we'd removed the glue we found that the track to Q5002's collector had been eaten away. A fine-wire link restored normal operation. Now how's this for coincidence? A week later the same dealer brought in another of these machines, this time with a clock fault. The time could be set but the colons didn't flash and the clock didn't count. A check at pins 2–3 of 15002 showed that the 32.768 kHz clock oscillator had stopped. We confidently fitted a new crystal but the clock still didn't work. Further checks in this area showed that R5047 (56 kΩ) was open-circuit. Guess what: it's underneath the same blob of glue around Q5002! When we removed the glue we found that

R5047 was without its lead-out wires. A replacement put matters right. It seems that this moisture-absorbing, corrosive glue is like a time bomb ticking away in these old machines.

Tips on removing the head: I thought everyone knew this dodge, but when a friend rang up and said that he'd undone the screws but couldn't remove the head I thought I'd better pass it on. In this machine there's a little cover over the wires that connect the heads and the rotating transformer. To change the drum you have to remove this cover, which is held by quarter-inch long screws. You can then get to and undo the half-inch long head securing screws. If the head won't budge, simply put the longer screws down the screening cover holes. The thread is the same but as the head screws are longer they bottom on the housing and thus force the head off. I suggest tightening the screws to the bottom by hand, then a further quarter of a turn alternately with a screwdriver until the head breaks free.

Tape would not lace: Rewind and fast forward were OK. When either play or record was selected, however, there was motor noise but the tape remained unlaced, the machine entering the forward mode at a slightly faster speed than normal playback. Fortunately the cause of the fault was nothing more than a stretched lace-up belt. It's under the deck assembly, at the front left-hand side.

Refused to accept a tape: This machine refused to accept a tape because the 'tape-in detect' leaf switch, which is mounted on top of the carriage, was buckled. A new switch restored normal operation. It's available from Willow Vale under part number 27354BT.

Sharp VC9700

Machine intermittently dead with no clock display: Tests around the STK772B chopper regulator chip I9001 in the fault condition showed that there was 31 V at pin 7 but no 13 V output at pin 6. Pins 3 and 4 feed the internal pulse-width modulator and the voltages here were both negative – clearly wrong! These pins are also associated with the overcurrent protection circuit (Q9001). As it was difficult to determine whether the chip was responsible or Q9001 was telling it to shut down we disconnected pins 3 and 4 – this seemed to be the best way of tackling things. The negative voltages remained and fitting a replacement chip put the machine back on the road. It was the second machine of this type in as many weeks that needed replacement of the STK772B chip.

Failure to accept a cassette: So long as operating power etc. is available, this fault is usually caused by a faulty cassette-in switch. Not this time, however! The main microcomputer chip 1801 was receiving the cassette-in message, and indeed was telling the associated output expansion chip 1803 that the message was present. Output pin 12 (CASS-M-CTL) of 1803 was going high, but not pin 17 (CASS-M-DRIVE) of 1805. The message was being lost in R892 because Q810, which prevents conflicting loading and cassette motor actions, had gone short-circuit from collector to emitter, thus holding down the CASS-M-ON line.

Clock went off after 30 seconds: A large red thing on the print side of the audio panel was getting very hot: a shot of freezer here would bring back the clock for a few seconds after which it would fade away again. A check on the resistance of this positive-temperature coefficient thermistor, which is marked 4R7, produced a reading of $50\,\Omega$ when cold and a few hundred ohms when hot. A replacement obtained from a scrap machine restored the correct display brightness.

Sharp VCA100

Capstan motor refused to work: We found that the 15 V and 5 V supplies were present and correct and that the 2.5 V reference voltage was correct, but the capstan drive voltage was low at 1.5 V instead of 2.4 V. The capstan stop signal at pin 57 of the system control chip IC827 was then found to be permanently low. When this monster chip was replaced the capstan motor worked correctly.

Intermittent refusal to play/record: The capstan and drum would both be motionless although the machine had loaded and the display showed the play symbol. The cause of the fault turned out to be poor riveting on the print that connects pin 10 of IC801 to AE4 en route to the cam switch, which modulates the voltage supplied to IC801 and thus confirms the mode position. At least it wasn't the mode switch again! The problem was cured by fitting a bridging wire over the defective print.

Fast forward in stop mode would start fast-forward search: If you pressed fast forward when this machine was in the stop mode it would automatically go into fast-forward search. The loading belt was in very poor condition, but the cause of the problem turned out to be the mode switch. After stripping, cleaning and retiming, normal results were obtained.

Sharp VCA105

Tape jammed, faulty mechanics: We get quite a few of these machines in the workshop with the complaint that either the mechanical functions are faulty or, as with this machine, a tape is jammed inside. Dirty mode switch contacts are the usual cause. The best cure is to remove the switch assembly, dismantle the switch, then clean the contacts using a fibre pen or a cotton bud moistened with methylated spirit. Simply squirting cleaner into the switch does not always provide a permanent remedy. The markers on the rotary centrepiece must be aligned when the assembly is refitted – otherwise the assembly will not go back into place. When we'd done all this we connected the machine to the mains supply and inserted a cassette. Everything seemed to be fine until play was selected. We then found that the capstan ran far too fast. The control chip had failed. This meant a new capstan motor, more than doubling the cost of the repair.

Would shut down: This machine would power on and initialize, then shut down with no functions and no display. We couldn't find any incorrect voltages. Replacing the timer chip IC5001 cured the fault. A useful tip with a fault like this is to remove all plugs from the front timer PCB and insert a pre-recorded tape with the safety tab removed. The machine will accept, load and play this tape, proving that the deck and syscon are OK.

Mode switch/master cam problems: The mode switch and master cam assembly cause various problems with this model. The suspect cam is black, replacements being white. Unlike earlier models, replacement is not too difficult. To confuse matters, a modified assembly was introduced after serial number 659812 and in later models that have a mode switch with a yellow centre. These require a different cam and switch. They are not interchangeable.

Sharp VCA105HM

Back-tension arm jamming the mechanism when the machine unlaces: The cause of the fault is not the tape missing the arm during loading but wear and strain on the operating lever, which is under the sliding cam assembly that runs across the front of the deck mechanism. We change the tension arm and the operating lever, part numbers MLEVP0134GEZZ and MLEVP0133GEZZ, or Willow Vale 27349TR and 27349TA.

Machine badly damaged tapes: This was because the loading arms/ guide poles didn't return to their correct rest position when unlacing was complete – they sat directly below where the tape emerged from the cassette's flap. The timing between them and the sector gear was one tooth out. Retiming provided a cure but the gear had to be replaced as it was damaged. Following advice in a previous fault report (above) I also replaced the back-tension post lever and gear – it's conceivable that this was the root cause of the problem.

Tape intermittently gets stuck during eject: Two of these machines have been regular visitors to the workshop over the past couple of months, both with this same intermittent fault. If the tape was wound out manually everything would be all right until some weeks later when the customer would return the machine with another tape jammed inside. We approached Sharp Technical on several occasions, as a result of which various parts were replaced, including some kind of modification PCB on the cassette housing, but the fault persisted. As we'd never seen the fault occur it was difficult to know what to do. Finally both machines arrived at the shop together. A careful examination with the machines side by side, one with the tape ejected and the other with the tape stuck, showed that both main decks were in the same position mechanically. This could mean only one thing, that the cause of the fault was the cassette carriage. Very careful investigation of the housing showed that just before the cassette lift arm, item 308, came to the vertical position the clutch could jump out, disengaging the drive to the lift assembly. It seemed certain that the clutch latching bar, item 319, was the cause of the trouble but to be on the safe side and avoid further comebacks we ordered two new cassette housings. Examination of the new housings showed that the latching bar and clutch operating lever (item 321) have been redesigned. We have subsequently fitted just these two items, with no further problems. Part numbers are latching bar MLEVP-0140GEZZ (item 319); operating lever MLEVP0139GEZZ (item 321). The new latching bar has a rectangular cutout, the cutout in the old latching bar being angled at one side.

Jammed tape: Check for foreign bodies in the machine then check whether the tension arm assembly is bent, preventing the loading arms moving to the unloaded positions. If the latter is the cause of the problem, replace the tension arm assembly, band brake and spring.

Just snow in E-E mode: Playback was OK. When the up/down channel search button was pressed there was a normal pulse-width modulated square wave at the base of Q1451, but there was no voltage at its collector (or the tuner's VT pin) because the 33 V regulator chip IC951 was short-circuit.

Sharp VCH81H

Tries to load a cassette when none is inserted: The IR emitter was short-circuit.

Cassette jammed in mechnism: After extracting it manually and confirming that there was no mechanical damage I inserted another cassette, but it wouldn't play and there was no capstan rotation. The supply voltages and control signals on the motor board were correct. A new motor assembly got the machine working again. Fortunately the assembly is not expensive and is easy to change.

Fails to make timed recordings: This machine had been seen by two of my colleagues during the previous couple of months for the same reason. The power supply had been checked thoroughly and the machine had been given a service, yet here it was again on the bench with the same problem. Bearing in mind all that had been done, I wondered where to start. In the end I decided to replace the mode switch, even though the other modes all seemed to be OK. The machine then tested OK – but it always did! We returned it and asked the customer to report back if any more problems were encountered. A phone call to him some months later proved that all was well.

Sharp VCM20

No clock display and no functions: When you encounter these symptoms in this model you will probably find that Q901 is short-circuit and R904 open-circuit. To prevent a recurrence of the fault Sharp has introduced a heatsink for Q901. The part number for this is PRDAF1065UMFW.

Q901 blew intermittently: Even after fitting a heatsink Q901 blew intermittently. In addition the chopper transformer would buzz. The cause of the problem was dry-joints on the optocoupler IC901.

Loss of response to jog shuttle: If there is intermittent loss of response to jog shuttle, or functions are erratic, check for dry-joints at plug and socket AO and OA.

Sony

```
SONY SLC6
SONY SLC9
SONY SLF1
SONY SLV270
SONY SLV353
SONY SLV373
SONY SLV415
SONY SLV425
SONY SLV615
SONY SLV625
```

Sony SLC6

Same tape speed in search and play: We found that the voltage across the reel motor was the same in both modes, but the service manual told us that this was correct. The cause of the fault was traced to the fact that the reel motor was being loaded by the relay pulley and spindle, which was binding on the bearing (drive transfer from the reel motor to the top of the deck). Cleaning and lubricating the spindle and bearing cured the fault.

Intermittent noisy rewind: It made a sound like tape chewing. We found that the large fast-forward pulley (item 424) had moved sideways, twisting the fast-forward belt, because of wear in the fast-forward arm assembly. Replacing the arm assembly (item 421) cured the trouble.

Lack of capstan servo lock: The problem was worse in play than in record. The cause was C7 (0.22 μF). Sony say that this is a common problem and in addition recommend changing C8 which is also 0.22 μF.

Rolling noise bars and sound varies: When play was selected, every 2 seconds a noise bar filled the screen and the sound slowed down. Obviously a capstan servo fault. The pulses from the control head are amplified to 7 V peak-to-peak and should be present at test point 7 on board SS9. A scope check showed that they were OK at this point. Voltage checks around the servo chip then suggested that everything was in order here. A cure was provided by slight adjustment of the capstan free-run preset RV001.

Sony SLC9

Accepts a cassette but fails to load: Instead, the eject light flashes. If you thread the tape manually everything works fine until you press the eject button, whereupon the tape unthreads but the cassette fails to eject, once more accompanied by the eject LED flashing. The cause of these symptoms lies with the unload-end switch, just in front of the cassette. It will need adjustment or replacement.

Cassette lift problems: The customer had tried to remove a jammed cassette and in doing so had put everything out of sequence. I got the lift working after replacing the broken cogs, then the fun started. When I switched the machine on the rewind motor started to run then the eject light flashed. If a cassette was loaded manually it would thread and play, but when eject was pressed it would unthread but not unload. A cold check on the operation of all the switches failed to reveal anything amiss, but after a lot of searching the cause of the problem turned out to be the unthreading switch which had gone high resistance. It's mounted alongside the loading rings.

No clock display due to a faulty d.c.–d.c. converter: The converter was on the rear-mounted power supply PCB. Unfortunately the cost of a replacement unit is over £20 trade. By the time that labour had been added the charge would have been outside the customer's budget. So we decided to open up the old unit to see if it could be repaired. The only difficult job was unsoldering the tin can that surrounds the PCB inside. We did this by applying heat from a miniature blow-torch powered by lighter fuel. Once we'd got inside we found that a 2SD789 transistor had an open-circuit emitter terminal. A 2SD774 made an excellent substitute as we didn't have a 2SD789 in stock. For good measure we replaced the four electrolytics (10 μF, 16 V; 10 μF, 50 V; 10 μF, 50 V; 330 μF, 16 V) as tests showed that they were well down and possibly the cause of the transistor failure. After reassembling the case we refitted the converter and gave the VCR a two-day soak test. The results were excellent and our charge was within the customer's £50 limit.

Intermittent loss of heads: My latest restoration – it took so long that I could hardly call it a repair! – was to the local school's last Betamax VCR. The school has a lot of programmes on Betamax tape: now that the last working machine had lost its heads these couldn't even be copied over to VHS. As you often find with an old machine, this one had another, intermittent fault. But the school had been living with it. Nevertheless before buying new heads I thought it best to find the cause of the fault, as the part responsible might no longer be available (the heads are available to order from CHS). When the machine was switched on from cold, no deck function would continue for more than 3 seconds: play, fast forward and rewind all terminated after a few seconds. Once the machine had been on for a few minutes the fault would clear and not reappear until next day. A look through back issues of *Television* magazine brought the suggestion that the deck state switches can give trouble. But a week spent monitoring them proved that they were without blame. I then checked the reel tacho pulses. Sure enough the take-up reel tacho pulses (the TFG signal at test point TP001) were missing in the fault condition. The voltages at the take-up pulse opamp were found to be lower than those at the supply pulse opamp. C004 (10 μF, 16 V) turned out to be leaky. The heads are now on order!

Sony SLF1

Drum spinning at high speed: This machine was brought in by a merchant seaman who used it on board ship. This explained the rusty aerial socket! The symptoms were as follows: the drum and take-up reel were spinning at high speed in the forward direction while the supply reel rotated in the reverse direction. The syscon in these machines is fairly complex, but when you get to know it and apply a logical fault-finding procedure you find that it rarely causes difficulty. Not long after delving into the syscon I found myself heading towards the reel servo where the switching transistor Q204 was short-circuit.

Wouldn't play: We found that the pinch press lever had become disengaged from the pinch solenoid lever. When a new press lever had been fitted – the original one had a worn plastic arm – the machine played for about 2/3 seconds then cut out. We then found that the take-up torque was low. Since fast-forward operation was perfect it seemed that the cause of the fault was servo rather than motor trouble. While checking the waveforms in the reel servo I found the rather unlikely cause of the fault – a speck of solder was bridging two contacts on adjacent print lands. At first sight it looked like a single length of track, but the short effectively joined pin 1 of IC201 (supply FG) to pin 29 of IC601 (syscon-2). Fortunately no lasting damage had been done. The short must have been present from new and it's remarkable that it had only now showed up. Our customer accepted the estimate but refused a second one for the drum surfaces causing the usual rewind trouble – apparently he rewinds his tapes in another machine!

Cassette lid wouldn't stay latched down: On examination I found that the latching assembly, on the left-hand side of the mechanism viewed from the front, was being held in the 'carriage down' state. The lacing mechanism appeared to be displaced so I wound the loading motor to where I thought it should be and the lid then closed. When the machine was switched on again the mechanism returned to its previous state – unlacing continued for too long. Maybe the unlace end sensing switch? It was OK but following this path I soon came to Q611 and Q613 which had both exploded. These are the unlacing end sense switching transistors on the syscon PCB. A lot of heat had presumably been involved as there were numerous dry-joints.

Sony SLV270

No display, deck solenoid clicked: Checks showed that all the outputs from the power supply were varying. C1326 (47 µF, 25 V) was the cause of the trouble.

The machine was dead, no display: This Grundig clone was dead. Checks in the power supply showed that C1325 (1 µF, 100 V) measured only 0.7 µF while C1326 (47 µF, 63 V) measured 19 µF and was very leaky.

Sony SLV353

Damages or snags the tape during the eject operation, often intermittently: Feel the half-loading arm: it will probably be very stiff on its shaft, as a result of which it will be slow to retract. If it doesn't get there by the time the cassette moves up, it's bad news for the tape! The cure is to clean and lubricate the arm's shaft and bearing.

Only partial lace-up: The tape went in but would only partially lace up because the post limiter had seized on its pivot. Dismantling it and relubrication cured the fault.

Tape creasing: The first item to check is the pinch roller. Replace it if worn. The other thing to check if necessary is that arm assembly RVS is well lubricated, clean and moving freely so that the tape is removed from the cassette smoothly.

Sony SLV373

Tape chewing: Sony machines are not generally given to tape chewing. This one would damage the tape at the end of high-speed rewind, however, especially if you went straight to eject. The cause of the trouble was that the take-up reel was being inadequately braked. This was because the felt pad on the brake shoe was skew-whiff. As a result it failed to contact the turntable rim. We restuck it with superglue.

Diagonal dark bands on recordings: This fault caused an immense amount of head scratching until we made a call to Sony Technical in Cumbernauld. Even though I sometimes struggle to understand their broad Scottish accents, they certainly came up trumps with this one. The symptom was diagonal dark bands on recordings only: the picture was perfect with E-E operation and playback of a pre-recorded tape. Curiously, the fault clears when the top, right-hand PCB is hinged up. The offending component is L252. Apparently an incorrect type that radiates to the erase circuitry was fitted. The part number for the replacement is 1–412–092–11.

No colour on playback of pre-recorded tapes: This new machine had apparently worked for a week before the fault revealed itself. I tried making a recording and sure enough the playback produced good colour. But there was no colour when I played back the recording in a known good machine – our old faithful Ferguson 3V29. So the Sony machine was working to its own standard. There was only very slight colour with playback of pre-recorded tapes. I ordered a reference manual and hoped that the fault would go away by itself. Well, the manual came but the fault didn't go away. I dived in at the HA118016NT chroma processing chip IC801 on the YC board, checking all the waveforms and d.c. voltage while playing a pre-recorded tape. The conditions at pin 19 were very wrong: the d.c. voltage was low and the waveform was completely different from that shown in the manual. Sony calls this waveform C ROT. It should be a square wave at about 4.5 V peak-to-peak. But it didn't look like a square wave at all. So I traced it back to the head amplifier board where I found that the print at pin 6 of plug CN004 was broken. It was obvious that someone had been at it before, as the soldering around this plug was in an appalling condition for a new machine. After repairing the print and generally tidying up the plug all tapes played back correctly.

Would not playback own recordings: This machine would play back pre-recorded tapes perfectly in either the SP or LP mode but with one of its own recordings only the blue mute raster was present – there was no vision signal on the tape. Scope checks showed that the f.m. luminance plus chroma signal was present at pin 18 of the head amplifier/switching chip IC001, but there was no output to any of the heads. The d.c. switching conditions around the chip were correct but at whichever outputs were selected the d.c. voltage was high – 3.5 V instead of 2.2 V. There were also incorrect (low) voltages at pins 16 (head select) and 26 (25 Hz pulse input). The chip was faulty.

Displayed 'P6' and rewind didn't work: The cure was to replace the two reel turntable photosensors PH001 and PH002 (part number 8–759–144–33) on board MD40 beneath the deck mechanism. It's best to replace the two as the cost is minimal and any further trouble in this area is prevented. They are easier than it may at first seem to replace.

Intermittent tape chewing: The customer provided three sample tapes that showed they were looping on eject. The cause was the capstan motor, whose baseplate had warped. A new motor solved the problem. I understand that Sony is aware of this problem and that replacement baseparts are available.

Sony SLV415

Would go into fast-forward video search when a tape was loaded: The symptoms with this machine were odd. A tape could be loaded but the machine then went into fast-forward video search with no other functions available – although twisting the jog-shuttle control backwards and forwards would sometimes give the pause mode. The tape couldn't be ejected: it had to be wound out manually after disconnection from the mains supply. Another symptom was that none of the buttons on the remote control unit had any effect, though an IR test showed that the device was working. With very little experience of this model and no manual, we decided to go straight to the deck mechanical position (mode) switch. But after giving this a good clean the faults remained the same. Then a thought came to mind: the jog-shuttle control at the front of the machine is more or less just a big mode switch and could be faulty. Fortunately it plugs in. Disconnecting it brought the machine back to life, the remote control unit now worked, and all functions were available. When we opened the jog-shuttle control up we found that it was very tarnished inside. Cleaning it failed to cure the problem, a replacement from Sony (part number 1–572–662–11) being required. It costs less than £10. After a good clean the machine performed excellently.

Dead or would display a very bright MON: This was a nasty one! When the machine was connected to the mains supply one of two things would happen: there would be either no sign of life or a very bright MON – and nothing else at all – would appear in the fluorescent display panel. In neither case would the front-panel keys or the remote control system do anything at all. We found that the power supply worked but didn't get an 'on' command from the system control department. It transpired that the 4.19 MHz crystal X1, which is mounted alongside the micro-controller chip on the front panel, was responsible: its oscillation was intermittent, in fact 'ragged' when viewed with an oscilloscope.

Drum servo unstable: It hunted, especially when the machine was cold. When I applied a puff of freezer to the AN3814K drum motor drive chip the drum virtually stopped – it was left just twitching backwards and forwards, even with no tape inserted in the machine. A replacement drive chip, part number 8–759–420–83, cured the problem.

No picture and a smell of burning: When we took the top off we saw that the drum motor wasn't rotating. It was pulsing instead, but if spun would build up speed. After finding that the motor drive chip was OK (by replacing it) we suspected the Hall chip within the lower drum and, taking the easy course, replaced the lot. The fault was still present, however. The cause of the trouble was actually C014 (0.1 μF). Failure of C013 or C015 would have had the same effect. It's best to check all three capacitors by replacement.

Sony SLV425

Failed to record sound: The cause was the bias oscillator, which in these machines is separate from the erase oscillator. CY1255 (47 μF, 63 V) was open-circuit.

Poor playback, would revert to blue screen: The reported fault was that the playback picture was poor and reverted intermittently to a blue screen. There's a Sony technical tip for a similar fault, to replace the EPROM ICY270 (part number G79801093), but doing this made no difference. Good job the machine was under guarantee – the chip costs around £30 trade. The usual reason for reversion to a blue screen is a noisy signal. So we checked the video playback output at pin 7 of CIC2160. It was a good, clean signal. From here the signal is passed to CIC2520, which provides signal delay and is part of the dropout compensation circuit. The input at pin 6 was OK, but there was a severely distorted output at pin 4: it should have been delayed video but looked more like high-amplitude (4 V peak-to-peak) digital noise of some kind. A check showed that the clock pulse input at pin 7 was fine. What next? Fortunately we had another of these machines in for a different fault. So we hoisted it over in order to compare the voltages and waveforms around CIC2520. With this second machine the signal that emerged at pin 4 was indeed normal, delayed video. Before condemning the chip we decided to check the conditions at the other three pins (excluding the supply and chassis pins) – pins 1, 5 and 8. In the faulty machine there was a 5 V clock-frequency sine-wave at pin 8, the other machine having just a d.c. voltage here. The only component connected to this pin is a 10 nF surface-mounted capacitor, CC2527. To my relief when a 10 nF capacitor was bridged across it the problem cleared. Fitting a replacement restored the machine to good health.

Intermittent failure to power up and play correctly: If the machine's power switch was pressed when the fault was present the fluorescent display would light, giving the impression that the machine was powered up, but the power LED would remain red rather than glowing green. The green power LED is controlled by the +5VF line, which comes up only when the +12VF line is active. This line is controlled by T108, CT110, CIC130 and the syscon chip CIC200. Basically, when the +12VF line goes high this feeds to the power supply and brings up the +5VF line which powers the rest of the machine. T108 and CT110 are controlled by pin 7 of CIC130: the conditions at this pin didn't alter when the power switch was pressed. A check was then made on the inputs to this chip from CIC200. These did seem to alter when the power switch was pressed. As CIC130's supply and chassis lines were intact, we found and fitted a replacement. It's a surface-mounted chip, and care has to be taken as there are quite a few surrounding components – hot-air soldering guns are becoming essential for successfully and safely removing such devices. With the new chip fitted the machine worked perfectly. After spending the afternoon on test, it was pronounced fit.

Tapes only half load: I've had a few of these Grundig clones that use the Panasonic G mechanism all with the same fault. When a tape is inserted it starts to lace but fails to reach the half-load position: it then unlaces and is ejected. In each case the cause has been an open-circuit solenoid.

Sony SLV615

No reverse functions: This machine wouldn't carry out any reverse functions (rewind, review, reverse slo-mo): within a few seconds of one of these being selected the deck would shut down. The obvious suspect was the right-hand end sensor, but both this and the nearby operational-amplifier buffer chip proved to be innocent. The 100-pin flatpack syscon/servo chip was the cause of the trouble.

No rewind or reverse picture search: We've had this fault several times with these machines. When rewind was selected the machine rewound the tape for a second then stopped. The same thing happened in the review mode, but every other function was fine. We suspected something opto-ish, such as an end sensor, reel sensor etc. So we scoped the outputs from the supply and take-up spool sensors. The low-frequency take-up reel sensor pulses appeared to be perfect, with correct amplitude. Things were different with the output from the supply reel sensor PH001, however. Pulses were present, but at only half the correct amplitude. Switching the scope to d.c. input showed that the pulses were still well above 0 V at their lowest level. After fitting a new optosensor (part number 8–759–144- 33) the machine was back in good working order.

Loss of playback picture: This was only a blue mute screen but the hi-fi sound OK! When pause was selected a very poor-quality still-frame picture appeared. This suggested that the video heads were badly worn. In view of the age of the machine, its infrequent use and the good-quality hi-fi sound, however, we resisted the temptation to replace the very expensive upper drum. An interesting display was obtained when the scope was connected to the head amplifier chip's output. The waveform didn't look quite right and appeared to have something superimposed on it. All was revealed when the head rotation was momentarily interrupted and the waveform remained. It transpired that one of the four head amplifiers was producing a 12 MHz signal. A new HA118019NT head amplifier chip put things right.

Sony SLV625

No output from one head: This fault almost caused me to order a new drum – after the normal cleaning and checks had been carried out. But I decided to take off the drum and use a magnifier to see whether I could find anything amiss. A non-soldered connection was found on the lower drum, and when this had been resoldered the whole machine worked perfectly. This goes to show how important it is always to check around, even when the fault appears to be such an obvious one.

Machine would play or record for about half an hour then shut down displaying 'L' in the hours part of the clock: The spool-rotation sensors are a weak spot in this and similar Sony VCRs. After the fault occurred the machine would shut down as soon as any tape forward mode was selected, although it was quite happy to run in the reverse modes. The cure was to replace both the optocouplers, HP001 and HP002, under the deck. Type number is PS6002.

Machine chewed tapes on eject: The cause was loss of the 9 V supply because Q203 had burnt out.

Cassettes not loading properly: Some cassettes, but only a few, would go in but wouldn't go all the way down. After a few seconds they would come back up. After inserting one of these cassettes many times and watching what happened I noticed that the left-hand carriage release lever was bent back towards the cassette. Straightening it cured the fault.

Tries to load, then unlaces: Even if the machine did manage to go into the record or playback mode it would intermittently revert to stop. The cause of the trouble was bad connections at both ends of the drum motor plugs.

Tatung

TATUNG TVR6111
TATUNG TVR6122

Tatung TVR6111

Intermittent ejection of cassettes: This model is similar to those in the Amstrad range of several years ago. A symptom that's becoming common is intermittent ejection of cassettes, either when one is inserted or at random during play. The cause is dry-joints at the tags of the three microswitches (start, in, out) on the horizontal PCB fixed to the right of the front-loading gantry.

Reel drive intermittently fails to engage when fast forward or rewind is selected: If you get this symptom, check that lever trigger 260 is free to slide along brake plate 261. If it's stiff, the metal stop for the brake plate (formed from the deck plate) needs to be bent very slightly to the right as you view the underside of the deck from the front. The numbers quoted above are taken from the exploded deck diagram in the service manual. This machine also appears under the Amstrad banner.

Intermittent failure to rewind: There are doubtless other Amstrad VCRs and more makes and models that use this mechanism. A very common fault with it now is intermittent failure to rewind or wind fast forwards. The motor whirrs but the reel-drive gears don't move into engagement. The cause of the problem is failure of the trigger lever (item 260 in the exploded view of the deck) to flip back into position in the stop mode. Cure it by filing a fraction of a millimetre off the end of its skirt, as shown in Figure 11.

In Mauritius we have the Casio VX4000/Funai VIP2000 etc. that use the same deck. In the stop mode, plastic lever 1 (see Figure 12) comes to rest by the metal post, where the rubber damper is fitted. If this damper

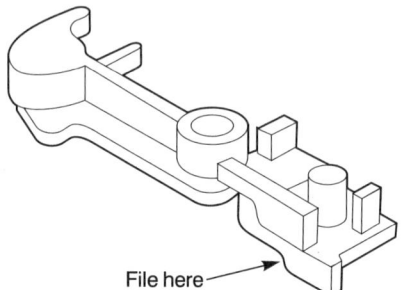

File here

Figure 11 *Where to file the trigger lever on a Tatung TVR6111/Amstrad VCR9410/etc.*

is worn out, plastic lever 1 rests just a little bit behind its correct location (more to the left, closer to the metal post), thus making it impossible for the trigger lever to flip back into position. As a result you get the fault condition mentioned above. The correct cure is to replace the rubber damper. If it's not available a plastic sleeve (e.g. e.h.t. cable sleeve) cut to length will do nicely.

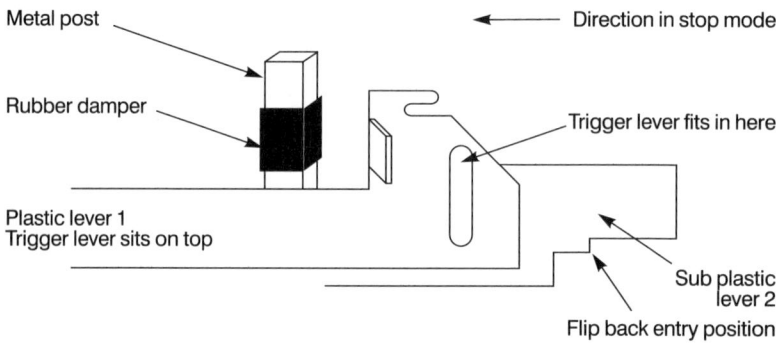

Metal post

Direction in stop mode

Rubber damper

Trigger lever fits in here

Plastic lever 1
Trigger lever sits on top

Sub plastic
lever 2

Flip back entry position

Figure 12 *Position of the rubber damper that causes the intermittent rewind/fast-forward fault.*

Intermittent or permanent no-go: If as well as this the display panel is black and there are no mechanical functions, it may well be that the power supply is working all right but crystal X801 is dry-jointed at one or both legs. It's mounted to the right of the FDP on the front panel.

Tatung TVR6122

Sound flutter: This is a common problem – it's a longitudinal sound track machine. The trouble is usually worst during playback of its own recordings, and sometimes gets worse as the machine warms up. The way to cure it is as follows: replace the pinch roller; get the take-up tension down to about 80 g/cm by fair means or foul; lubricate both spool spindles and the back-tension pole pivot; fit a new back-tension band and reset the back tension; and clean the entire tape path. Replacement of the reel drive clutch, part number 250814, is also advised.

Tuning drift: In addition we found that when a new channel was selected it took several seconds to arrive. The stabilized tuning voltage supply was low because the 2SA1038 transistor Q1001 in the 50 V supply line was faulty. It's mounted at the top edge of the PSU-stabilizer board.

Toshiba

```
TOSHIBA V109
TOSHIBA V110
TOSHIBA V110B
TOSHIBA V212
TOSHIBA V309
TOSHIBA V404
TOSHIBA V55
TOSHIBA V57
TOSHIBA V703
TOSHIBA V83
TOSHIBA V93
```

Toshiba V109

Damaging tapes: This full-lace machine was accused of damaging tapes, as a result of which the heads were now clogged. After cleaning these we found that the cause of the problem was very slightly weak reel drive due to a worn drive clutch. It's a delight to change this item from underneath.

Laced-up tape jammed in machine: Checks showed that the on/off 9 V supply was missing. The repair consisted of replacing IC811 in the power supply and, as a precaution, the mode switch.

Couldn't eject a tape: When I tried the machine out I found that there were no mechanical functions at all. Replacing the 2SA1297Y transistor Q082 restored the machine to life.

Toshiba V110

Tuning is off: Check the 2.7 V zener diode DT53 in the video 5 V supply regulator circuit. It tends to become leaky. Other possible symptoms when this diode is leaky are no playback colour or distorted playback and E-E pictures.

Machine dead, no 12 V standby supply: We found that resistor RP14 in the power supply was hot: well it would be with a dead short across the 12 V rail. The cause was the 15 V zener diode DP011, which is connected across the 12 V line to provide protection in the event of TP03 going short-circuit. A check showed that TP03 was all right, and a long soak test brought no other possible cause of DP011's failure to light.

'Clock went first now dead': A check on the power supply showed that all voltages were present and correct. When the main board was lifted we saw that there's a large heatsink on the left-hand side. Two transistors here, TT52 and TT53, and IC IT14 were dry-jointed.

No fast forward or rewind: We found that the pin had broken off the white lever in the loading block. The complete loading block had to be replaced as the part is not available separately.

No play, fast forward, rewind etc.: The machine would try to go into a mode then, after a few seconds it would revert to standby. When hand-winding the loading block I found that it would jam. On stripping it down I discovered that the main cam was damaged. To put this right you have to replace the full loading block assembly.

Pinch roller would not engage: After this the machine would then go to standby. When the tape was ejected we noticed that the drum was rotating too fast. A check at BT33 (drum PG/FG in) showed that the waveform was missing. This is the drum tacho – a square-wave should be present here. The optocoupler on the drum motor PCB had failed.

Toshiba V110B

No E-to-E output, clock display wrong: There was no E-to-E output, just a blank raster in play and the clock display showed wrong characters. An initial check around the power supplies revealed that the U8 (5 V) line was high at 8.8 V. We found that the ZPD2.7 V zener diode DT53 was open-circuit. When a replacement was fitted the U8 supply was back at 5 V (check it at the collector of TT53) and the faults had cleared.

Will not take in a cassette: If there's no capstan rotation check whether transistor TT68, type BC557, is open-circuit.

No E-E tuning: If there's just snow on the screen, check for 30 V at power supply test point BP08. If it's missing the items to check are DP04 (ZTK33B) and RP03 (2.2 kΩ).

No results and no display: The cause of the fault was obviously in the power supply, and it didn't take us long to find that the ZPD6V8 zener diode DP09 was open-circuit. In addition DP011 (ZPY15) was short-circuit.

Would go into stop mode when mains supply switched on: After a lot of tests we discovered that replacing the end and supply sensors restored normal operation.

Toshiba V212

Would fail to accept or eject a tape: It appeared to load and unload rather slowly. A new loading motor put matters right.

Intermittently poor playback: The fault could be instigated by tapping the head preamplifier. We found that CQ05 was faulty but have also had CQ06 go open-circuit.

Would eject after a few seconds: The usual cause of this is the loading motor. On this occasion, however, the cause was tinfoil that had been jammed in the loading area. There was also a squeal on rewind. We found that the take-up reel was dry – a little oil silenced the squeal. Next, the playback picture pulled from side to side, as though there was a drum fault. The cause was dust on the plastic cap beneath the drum – it provides information on the drum speed and position, via an optical pick-up. Finally the audio head and pinch roller had to be replaced.

Toshiba V309

Tape chewing: Checks with a dummy cassette showed that everything went round correctly, but when a tape was inserted and play was selected the tape was damaged because of excessive take-up torque – in fact the play torque was nearer what you would expect for fast forward. We found that the voltage at the output pin of the reel-motor drive chip was 15 V instead of 3.4 V. The cause was traced to Q625 being short-circuit collector-to-emitter.

Would neither play nor eject tape: There was one in the machine and it would try to rewind. The timing of the loading gear was a mile out. After setting it up the machine worked correctly. A week later it came back with the same fault, however. This happened again after a further week. A new loading motor assembly finally cured the problem.

Drum runs too fast intermittently: Check for dry-joints at P509 on the main video PCB, also IC501 for bad connections.

No functions and dew symbol showed: A check on the dew line input to microcontroller chip IC601 showed the voltage to be correct at 4.6 V. Replacing the chip put matters right.

Would stop in playback or record: This would occur after anything from 20 minutes to 2 hours. The cause was high reel motor drain current, although the motor provided very fast rewind and wasn't particularly noisy. A new motor cleared the fault, which can also occur with the V109 and V209.

Toshiba V404

No front display: If the machine plays all right but there's no front display, check whether RP051 (39 Ω, safety) in the power supply is open-circuit. A possible cause of its failure is that the ribbon cable to connector PK02 on the front panel is incorrectly routed, with the ribbon chafing on the cabinet top. The resistor's part number is 70041116.

Machine is dead: If the power supply outputs are low and pulsing, replace CP008 (100 μF, 25 V) and CP007 (10 μF, 50 V).

Dead but for ticking power supply: The cause was CP008 (100 μF, 25 V).

Toshiba V55

Intermittently dead machine: For an intermittently dead machine – even the clock goes off – check for dry-joints at the a.c. input to the power supply module.

Wouldn't load cassette: E-E operation was OK. My first check was on the power supply outputs, which were all OK. A scope check was then made on the inputs to the microcomputer chip. It was receiving instructions but was it carrying them out? I scoped the outputs but these didn't make much sense. Try a different approach: maybe no power is going to the motors? Three changeover switch chips feed 13 V to the motors. Check and find that there's nothing at IC204/5 because circuit protector CP1 is open-circuit. As there were no shorts I replaced CP1. At switch-on 13 V appeared at the switches. A cassette could now be inserted and the drum and capstan motors rotated, but there was no lace-up. Feel each switch chip and find that one is too hot. Is it faulty? Switch off and remove the chip to check it. Checks not conclusive. Decide to try feeding each switch output with 10 V from an external power supply – with the VCR switched off of course. Find that all the motors except the mode one run. Remove leads from mode motor and connect to power supply. Motor still doesn't run even with the drive belt removed. Replace motor and i.c. switches and find that all is now well.

Wouldn't switch-on: However the display segments all lit. On investigation we found that there was a dry-joint at pin 1 of connector CN1 on the regulator PCB. Resoldering the plug connection cured the fault.

Intermittent rewind/review: The cause turned out to be the capacitor across the start sensor. It was short-circuit.

Toshiba V57

Intermittent operation: 'Went haywire, then OK except that it wouldn't eject, now no functions' read the report. This is becoming a common fault. CP1 (ICP-F15) goes open-circuit due to failure of the M54544L cassette motor driver chip. Another reason for this can be a defective motor. The same two possibilities apply with the loading motor and its drive chip, which are also fed via the same ICP. When CP1 goes open-circuit the 13 V supply to both circuits is removed.

Capstan would stop rotating: It rotated for a few seconds when this machine was switched on, then it shut down. We found that the circuit protector CP1 for the switched 12 V supply was open-circuit, a replacement restoring normal operation. The current through CP1 peaked at only 110 mA during lace-up, so there didn't seem to be any overload. After soak testing it for a day we pronounced the machine fit.

Machine needed new heads: We fitted a new drum and carried out a full mechanical service. After a couple of weeks, however, the customer reported that there was a problem with the sound – any sustained note suffered from wow. As a new pinch roller made no difference, I took the machine back to the workshop. A scope check showed that the capstan servo would periodically hunt. The control pulses and signals from the capstan motor were present, and servo alignment failed to improve matters. So a new motor was tried. This again made no difference. The culprit turned out to be the capstan belt. The one I'd fitted as part of the service obviously had a tight spot, although it felt all right and there were no obvious signs of poor manufacture.

Toshiba V703

Dead or intermittent operation: Replace all the small electrolytics in the power supply with 105°C types. Now here's a real puzzle. The last one of these machines we had in for repair with the above symptoms had another fault in addition to the power supply problem. The system would crash when a new E180 cassette was loaded with the record protection tab intact. If the tab had been removed, the tape would go into auto-play and all would be well until a wind function (fast forward, rewind, cue or review) was selected. The system would then crash. If the tape was wound in another machine until only 30 minutes or so of tape was left on the supply reel, and this cassette was then inserted, the machine would behave normally with all functions selectable. Once it was accepted, the tape could be rewound to the start and played normally. C60 tapes would not play regardless of how little tape remained. The rotation sensors were OK, and the tape-remaining readout was always correct. Here's the solution. The customer had decided to help the flagging mechanics along by removing the top and bottom covers – 'to give it a bit of a push'. In doing so he had broken off the master cam's first opto blind. Fitting a new cam cleared all the symptoms. Is there anyone out there who can explain why the machine reacts in this way?

No E-E picture or baseband video from scart sockets: The on-screen graphics and the E-E sound were OK. Scope checks along the video path showed that the signal was present at pin 8 of ICF01 on the terminal board and TP201 but not at TP203. A visual check revealed a crack in the print to pin 1 of plug 201 on the motherboard.

No power: There was no display, no functions could be selected and a slight whistle from the power supply could be heard. A check on the supply lines showed that the ever 14 V supply was missing. The simple cause was that protector Z821 was open-circuit.

Toshiba V83

Intermittent clock display: This would fail about an hour after switching the machine on. The fault was found to be the 4 MHz clock crystal XX01.

Intermittent shutdown in record or playback: We've recently had three of these machines with this problem. Observation of the deck at the moment of failure showed that the take-up reel had stopped, allowing slack tape to spill on to the deck. The problem was in each case cured by replacing the reel motor, part number 70326539.

Tape keeps being ejected: If the machine then goes to standby with the cassette indicator flashing, suspect ICX01 (TMP47C410AN6775) of being faulty. Check it by replacement.

No video record: The sound was being recorded but not the video. It wasn't just a matter of head cleaning as we initially thought. Checks showed that the REC 7 V supply at pin 6 of P202 on the head amplifier pack was missing. The regulated 12 V supply was correct but there was no output from the record buffer transistor Q131 because its feed resistor R165 (22 Ω, fusible) was open-circuit. Replacing this item restored normal operation.

Display dimly lit: All functions were showing. We checked at pin 42 of the clock chip ICX01 – this is called the back-up +B line. It should be at 5 V permanently, but there was no voltage. We traced the line back to the Selector Timer 2 board (the tuning control board to you and me) and checked the ever 12 V line here. This line feeds a 5.6 V zener diode via a 390 Ω resistor and was correct. Aha, I thought, a short-circuit zener diode. Nine times out of ten it probably would be, but I always seem to get the odd time. The zener diode turned out to be OK. A PST520C chip (ICL10), the reset generator, is also connected to this line. The resistance between its input and deck read about 20 Ω. Fitting a replacement cured the problem.

Capstan motor running too fast: A check on the drive voltage showed that it was high at about 10–11 V instead of 6.7 V. Checks around the servo chip IC501 showed that although the voltage at pin 14 (capstan a.p.c.) was correct at 3.3 V the voltage at pin 15 was only 0.9 V instead of 3.3 V. Scope checks at pins 19 and 20 (CTL in and out) showed that the control pulses were of correct amplitude although the frequency was of course high because of the excessive tape speed. The tracking input at pin 28 varied the length of the waveform, so all seemed to be correct here. The next check was on the FG pulses at TP518. The waveform here had gaps in it and varied a little in amplitude. Unfortunately I ignored this, putting the irregularity down to the motor's increased and wowing speed. Wrong decision! So after replacing TC501 and finding that the fault remained as before I had a closer look at the FG pulses. When I dismantled the capstan assembly I found that the coil which forms the stator of the pulse generator was dry-jointed at the point where the enamelled copper wire is connected to the terminal.

Toshiba V93

No clock or other display: I'm sorry to be vague about this one, but we don't have the manual. The basic fault was no clock or other display, although the deck functions were OK ZL62, a Wickman fuse on the bottom panel, was found to be open-circuit, replacement bringing the machine back to life. There's a small can, beneath which an oscillator resides, on this panel (timer-2/i.f. and prescaler). The coil has a little metal top hat as screening, glued into place. This cap falls off. I'm not sure whether a change of inductance occurs to open-circuit the fuse or whether it's simply a matter of a short-circuit due to the metal contacting something in the circuit. I'm led to believe that the problem is a common one.

Chewed tapes: This machine's owner, or rather her two small children of two and four, were very upset. It had chewed a couple of their tapes. The reason for this was obvious when the fault developed: the take-up reel stopped but the capstan continued to run, thus making a right mess of Noddy and Big Ears. As we've had problems with it on earlier models suspicion fell on the reel motor. Sure enough if it was given a sharp knock with the handle of a screwdriver it stopped. A replacement put matters right, but unfortunately Rod, Jane and Freddie were no more.

Would play but no clock display: This means that the tuning doesn't work either. The cause of the fault is the d.c.–d.c. converter unit, which develops an internal short. As a result its circuit protector goes open-circuit.

No display: There was a normal tape playback picture but no front display and only snow in the E-E mode. The cause was traced to circuit protector 2L62 (ICP-N10) on the timer-2 PCB being open-circuit. A long soak test after fitting a replacement proved that all was well, with no other cause apparently being present to make the CP go open-circuit.

Television index/directory and faults discs plus hard copy indexes and reprints service

Index disc

Version 8 of the computerised Index to *Televison* magazine covers Volumes 38 to 49 (1988–1999). It has thousands of references to TV, VCR, CD, satellite and monitor fault reports and articles, with synopses. It also contains a TV/VCR spares guide, an advertiser's list, a compendium of useful web resources and a directory of trade and professional organisations. The software is quick and easy to use, and runs on any PC with Microsoft Windows. Price is £36 (supplied on a 3.5" HD disc). Those with previous versions can obtain an upgraded version for £16. Please quote the serial number of the original disc. See the CD-ROM package and upgrade offer below.

Fault Report discs

Each disc contains the full text for television, VCR, monitor, camcorder, satellite TV and CD fault reports published in individual volumes of *Television*, giving you easy access to this vital information. Note that the discs cannot be used on their own, only in conjunction with the Index disc: you load the contents of the Fault Report disc on to your computer's hard disc, then access it via the Index disc. Fault Report discs are now available for:

Vol 38 (Nov 1987–Oct 1988); Vol 39 (Nov 1988–Oct 1989);
Vol 40 (Nov 1989–Oct 1990); Vol 41 (Nov 1990–Oct 1991);
Vol 42 (Nov 1991–Oct 1992); Vol 43 (Nov 1992–Oct 1993);
Vol 44 (Nov 1993–Oct 1994); Vol 45 (Nov 1994–Oct 1995);
Vol 46 (Nov 1995–Oct 1996); Vol 47 (Nov 1996–Oct 1997);
Vol 48 (Nov 1997–Oct 1998); Vol 49 (Nov 1998–Oct 1999).

Price £15 each (supplied on 3.5" HD discs).

Fault-finding guide discs

These discs are packed with the text of vital fault-finding information from *Television* – fault-finding articles on particular TV chassis, VCRs and camcorders, Test Cases, What a Life! and Service Briefs. There are now three volumes 1, 2 and 3. They are accessed via the Index disc. Price £15 each (supplied on 3.5" HD discs).

Complete package on CD-ROM

The Index and all the Fault Report and Fault Finding Guide discs are available on one CD-ROM at a price of £196 (this represents a huge saving). Customers who have the previous CD-ROM can upgrade on CD-ROM for £46 (other customers call for a quotation). Please quote the serial number of your disc when you order.

Reprints and hard copy indexes

Reprints of articles from *Television* back to 1986 are also available: ordering information is provided with the Index, or can be obtained from the address below. Hard copy indexes of *Television* are available for Volumes 38 to 49 at £3.50 each.

All the above prices include UK postage and VAT where applicable. Add an extra £1 postage for non-UK EC orders, or £5 for non-EC overseas orders. Cheques should be made payable to SoftCopy Ltd. Access, Visa or MasterCard Credit Cards are accepted. Allow up to 28 days for delivery (UK).

SoftCopy Limited
1 Vineries Close, Cheltenham, GL53 0NU, UK
Telephone: 01242 241 455
Fax: 01242 241 468
e-mail: sales@softcopy.co.uk
Web site: http://www.softcopy.co.uk